REPRODUCTIVE BEHAVIOR

ADVANCES IN BEHAVIORAL BIOLOGY

Volume 1 • BRAIN CHEMISTRY AND MENTAL DISEASE
Edited by Beng T. Ho and William M. McIsaac • 1971

Volume 2 • NEUROBIOLOGY OF THE AMYGDALA
Edited by Basil E. Eleftheriou • 1972

Volume 3 • AGING AND THE BRAIN
Edited by Charles M. Gaitz • 1972

Volume 4 • THE CHEMISTRY OF MOOD, MOTIVATION, AND MEMORY
Edited by James L. McGaugh • 1972

Volume 5 • INTERDISCIPLINARY INVESTIGATION OF THE BRAIN
Edited by J. P. Nicholson • 1972

Volume 6 • PSYCHOPHARMACOLOGY AND AGING
Edited by Carl Eisdorfer and William E. Fann • 1973

Volume 7 • CONTROL OF POSTURE AND LOCOMOTION
Edited by R. B. Stein, K. G. Pearson, R. S. Smith,
and J. B. Redford • 1973

Volume 8 • DRUGS AND THE DEVELOPING BRAIN
Edited by Antonia Vernadakis and Norman Weiner • 1974

Volume 9 • PERSPECTIVES IN PRIMATE BIOLOGY
Edited by A. B. Chiarelli • 1974

Volume 10 • NEUROHUMORAL CODING OF BRAIN FUNCTION
Edited by R. D. Myers and René Raúl Drucker-Colín • 1974

Volume 11 • REPRODUCTIVE BEHAVIOR
Edited by William Montagna and William A. Sadler • 1974

A Continuation Order Plan is available for this series. A continuation order will bring
delivery of each new volume immediately upon publication. Volumes are billed only upon
actual shipment. For further information please contact the publisher.

REPRODUCTIVE BEHAVIOR

Edited by

William Montagna

Oregon Regional Primate Research Center
Beaverton, Oregon

and

William A. Sadler

Center for Population Research
National Institute of Child Health
and Human Development
National Institutes of Health
Bethesda, Maryland

PLENUM PRESS • NEW YORK AND LONDON

Library of Congress Cataloging in Publication Data

Conference on Reproductive Behavior, Oregon Regional
  Primate Research Center, 1973.
  Reproductive behavior.

  (Advances in behavioral biology, v. 11)
  Papers from the conference planned by the Center for Population Research
of the National Institute of Child Health and Development, and the Oregon
Regional Primate Research Center.
  Includes bibliographies.
  1. Sexual behavior in animals—Congresses. 2. Reproduction—Congresses.
I. Montagna, William, ed. II. Sadler, William A., ed. III. United States. National
Institute of Child Health and Human Development. Center for Population
Research. IV. Oregon Regional Primate Research Center, Beaverton. V. Title.
[DNLM: 1. Reproduction—Congresses. 2. Sex behavior—Congresses. 3. Sex
behavior, Animal—Congresses. W3 AD215 v. 11 1973 / WQ205 C753r 1973]
QL761.C645   1973   599'.05'6   74-14728
ISBN 0-306-37911-2

Proceedings of a Conference on Reproductive Behavior held at
the Oregon Regional Primate Research Center in Beaverton, Oregon,
July 16-17, 1973

© 1974 Plenum Press, New York
A Division of Plenum Publishing Corporation
227 West 17th Street, New York, N.Y. 10011

United Kingdom edition published by Plenum Press, London
A Division of Plenum Publishing Company, Ltd.
4a Lower John Street, London W1R 3PD, England

Printed in the United States of America

Foreword

Sexual compatibility between male and female partners is in-
dispensable to normal and successful fertilization in mammals.
Thus, the genes from males and females whose sexual behavior is
characterized by awkwardness, ineptness, and miscues are eliminated from the gene pool of the species. In human societies,
this compatibility is not always evident; and the behavior that
precedes and accompanies copulation and fertilization is exceedingly complex and affected by many variables. As in most other
species of animals, the entire repertoire of reproductive behavior
of man is not well understood by man. When viewed, discussed, or
reported, the topic is too often and most unfortunately regarded
as an amalgam of emotion, mysticism, and biology.

In the past, such emotion-charged approaches to the biological fact of reproduction did much to obfuscate the subject; and
as a result, much of the array of hormonal, neural, psychological,
and social variables that control and insure the successful reproduction of the human species remains even now in Victorian ignor-
ance. But with the recent rash of books and scientific treatises
on the subject, some progress has been made in elucidating human
reproduction and associated sexual behavior. However, so entrench-
ed are some of our social taboos that the danger still lurks of
equating social acceptance of the words with an understanding--
all too lacking--of the process to which they refer.

It is precisely the purpose of this book and of the studies
that preceded it to elucidate the various aspects of sexual behavior in man, nonhuman primates, and other mammals and to explain,
if possible, the mechanisms that control them. Herein are the proceedings of a Conference on Reproductive Behavior held on July 16
and 17, 1973, at the Oregon Regional Primate Research Center
(ORPRC) in Beaverton, Oregon. The Conference was supported by the
Center for Population Research (CPR) of the National Institute of
Child Health and Development (NICHD), and was planned jointly by
CPR and ORPRC to be held in association with the third annual

meeting of the Directors of the CPR-supported Population Research Centers and Program Projects. The authors of these articles are leading scientists in a field that covers the entire range of disciplines from neuroendocrinology to human reproductive behavior. They were chosen for their scientific acumen and achievement, the timeliness of their approach, and the thoroughness of their research. The format of the Conference was designed to stimulate discussion among scientists whose collaboration in population research has for too long been curtailed by the disparity of their disciplines and the discreteness of their individual approaches. The articles in this volume are grouped according to the Conference session in which each was presented. The three sessions were: The Central Nervous System and Sexual Behavior, chaired by Julian M. Davidson of the Department of Physiology of Stanford University; Pheromones and Hormones in Reproductive Behavior, chaired by Charles H. Phoenix of the ORPRC; and Behavioral and Social Determinants of Reproductive Behavior, chaired by Norman T. Adler of the University of Pennsylvania Department of Psychology. To conclude the Conference, a special lecture, Human Sexuality and Evolution, was given by the distinguished psychologist Franklin A. Beach, of the University of California in Berkeley; we are much indebted to him for his illuminating presentation.

The primary goal of the Conference was to disseminate new and significant research on reproductive behavior. Another goal was to synthesize a basis for predicting the acceptance, success, and impact of future attempts to regulate fertility and alleviate infertility. The first goal was achieved almost from the outset by the selection of participants who are specialists in population research. Whether and when the second goal will be attained is contingent upon a continuation of the dialogue begun at the Conference between biomedical, behavioral, and social scientists engaged in one common pursuit--population research. The success of this Conference, therefore, like that of any other, can be assured if the initial dialogue thus engendered leads to increased collaboration and coordination of efforts by these scientists and their colleagues.

We have been assisted in the preparation of these proceedings for publication by so many that to try to name them all would be to risk missing a few. Because of our singular indebtedness to them, however, we must mention Sonja Young of the CPR and Rosemary Low, Margaret Barss, Margaret Shininger, Hildegarde Peterson, and Linda Cooper of ORPRC.

William Montagna, Ph.D.

March 1974   William A. Sadler, Ph.D.

Contents

NEURAL CONTROL OF GONADOTROPIN SECRETION

S.M. McCann

Department of Physiology, Southwestern Medical School

University of Texas Health Science Center, Dallas

INTRODUCTION

The system that mediates gonadotropin release is intimately related to the neural system which mediates sexual behavior. In all species, sexual behavior ultimately results from the release of sex hormones, which is, in turn, induced by the previous release of gonadotropins. Furthermore, there is a close temporal relationship. For example, in the rat, the preovulatory discharge of gonadotropins occurs approximately two hours before the onset of sexual receptivity (Moss and Cooper, 1973). Thus, sexual receptivity is timed to occur just before ovulation to ensure fertilization of the freshly shed ova. In addition, luteinizing hormone (LH) releasing factor (LRF) can induce mating behavior (Moss and McCann, 1973). When LRF is released into the hypophysial portal vessels just before the pre-ovulatory release of LH, it may also be released from axon collater-als directly into the preoptic sexual receptivity center. Lastly, in species which ovulate in response to mating, such as the rabbit, cat, and ferret, the act of mating can induce a discharge of gonado-tropins via the release of LH-releasing factor (Everett, 1964). Coitus-induced release of gonadotropins appears to occur also in the rat (Moss and Cooper, 1973; Spies and Niswender, 1971) and may also occur in other spontaneously ovulating species.

The current status of our knowledge of the neural control of gonadotropin release will be briefly reviewed. The major emphasis

[1] Research in the author's laboratory was supported by NIH grants AM-10073 and HD-05151, the Ford Foundation, the Texas Population Institute, and the Robert Welch Foundation.

1

will be on the hypothalamic regulatory centers but we will also consider briefly recent evidence for the importance of brain stem and limbic system pathways and will allude to the possible importance of the pineal gland. The dramatic progress in this field over the past few years has been due in large part to the development of specific, sensitive, and precise radioimmunoassays for the gonadotropins and prolactin. Nearly all the hormonal measurements to be discussed here were performed by radioimmunoassay.

ELECTRICAL AND ELECTROCHEMICAL STIMULATION
OF THE PREOPTIC-HYPOTHALAMIC REGION

Electrical stimulation of the preoptic region has long been known to induce ovulation (Haterius, 1937; Markee et al., 1946), and the development of radioimmunoassay has made it possible to quantify the response of LH and to measure the release of follicle-stimulating hormone (FSH). Stimulation at sites in a band of tissue extending from the medial preoptic area through the anterior hypothalamus caudally to the median eminence-arcuate region evoked ovulation and produced a dramatic elevation in plasma LH concentrations (S.P. Kalra et al., 1971; Cramer and Barraclough, 1971). When a slightly more caudally located region from the anterior hypothalamus to the median eminence-arcuate region was stimulated, both FSH and LH were released (S.P. Kalra et al., 1971). Selective release of LH without FSH and in two animals of FSH followed stimulation in the medial preoptic and the anterior hypothalamic areas, respectively. These results suggest that the areas controlling the release of the gonadotropins overlap, but that the region controlling LH release extends farther rostrally than that controlling FSH release.

These results have also demonstrated that in the normal animal, the preovulatory discharge of FSH and LH is in excess of that required to evoke ovulation (S.P. Kalra et al., 1971; Cramer and Barraclough, 1971). Ovulation has followed stimulations which produced considerably less gonadotropin than that released at proestrus. Furthermore, ovulation can be induced with the release of only LH since it occurred after preoptic stimulations in which no FSH was released.

THE EFFECTS OF HYPOTHALAMIC LESIONS
ON GONADOTROPIN AND PROLACTIN RELEASE

Thus, it is no surprise that when lesions are made in regions which, when stimulated, augment gonadotropin release, the release of gonadotropins is impaired. The impairment is most severe when lesions are made in the median eminence-arcuate region because this would remove the pituitary from neural control. In animals thus treated, there is a cessation of cyclic gonadal activity, failure of

follicular development, and the persistence of large, functional
corpora lutea (Taleisnik and McCann, 1961). As a result of removing the negative feedback action of gonadal steroids in ovariectomized animals with already elevated levels of gonadotropins, these
lesions led to a dramatic decline of about 95% from preoperative
concentrations of plasma LH (Bishop, Fawcett et al., 1972). The 75%
decline from preoperative levels of FSH was less striking. Thus,
eliminating hypothalamic control causes an almost complete cessation
of LH release and a less inhibited FSH release. The relative inability of the lesions to inhibit FSH release may mean either that
the pituitary has an intrinsic capacity to release FSH or that the
lesions were incomplete, even though no normal neural tissue was
seen.

On the basis of the presence of functional corpora lutea and
the ability of median eminence lesions to induce lactational changes
in the mammary glands (McCann and Friedman, 1960), some have postulated that plasma prolactin concentrations are elevated in these
animals (Nikitovitch-Winer, 1960; DeVoe et al., 1965); and this
suggestion has been borne out by measurements of plasma prolactin
in animals with median eminence lesions (Bishop, Fawcett et al.,
1972; Chen et al., 1970). The values are nearly as high as those
in lactating females. Since the suckling stimulus produces maximal
release of prolactin (Krulich et al., 1970), the removal of neural
control of the gland apparently leads to almost maximal prolactin
release. The net hypothalamic influence on gonadotropins, therefore,
is stimulatory; on prolactin, it is inhibitory.

On the other hand, if lesions are placed in the suprachiasmatic
region, a different syndrome, called constant vaginal estrus, results
(Taleisnik and McCann, 1961; Hillarp, 1949). The animals fail
to ovulate; but the ovaries are filled with large follicles which,
as constant vaginal cornification indicates, secrete estrogen.
Apparently, these lesions have interrupted the pathways involved
in ovulation. Measurement of plasma gonadotropins in these animals
has shown that the levels of FSH and LH are slightly elevated
(Bishop, Fawcett et al., 1972). The cause of this elevation is
unknown, but it may be the estrogen secreted by the follicles. Even
though the pathways involved in the preovulatory discharge of gonadotropins are interrupted by these lesions, the negative feedback of
gonadal steroids is at least partially intact. Ovariectomy is
followed by a rise in the concentrations of plasma FSH and LH but
not to as high titers as in animals with intact brains. Conversely,
estrogen lowers the elevated plasma levels in castrate animals with
suprachiasmatic lesions and in animals with intact brains (Bishop,
Jakra et al., 1972). The elevation in prolactin levels in animals
with suprachiasmatic lesions is almost certainly caused by the
secretion of estrogen from the follicles since it returned to
normal after ovariectomy and since small doses of estrogen elevated
prolactin in normal animals and those with suprachiasmatic lesions
(Bishop, Fawcett et al., 1972; Bishop, Kalra, et al., 1972). As a

result of the secretion of estrogen and prolactin, some mammary
development occurred in these animals (Taleisnik and McCann, 1961)

The preovulatory discharge of gonadotropins in the rat is
undoubtedly brought about by the stimulatory effect of gonadal
steroids, estrogen, and possibly progesterone, acting together with
a cyclic timing mechanism on the afternoon of proestrus (Everett,
1964). Barraclough and Gorski (1966) postulated that the preoptic
region constitutes a cyclic center which is involved in the preovulatory release of gonadotropins and that the basal tuberal region constitutes a tonic center which is involved in the basal release of
gonadotropins and is the site of the negative feedback action of
gonadal steroids. The lesions and stimulation studies just described
are certainly in accord with this theory.

THE EFFECT OF CUTS WITH THE HALASZ KNIFE

In another approach to the study of hypothalamic control of
gonadotropin release, cuts with a specially designed knife developed
by Halasz and Pupp (1965) interrupted the connections between various
parts of the hypothalamus. If so-called "frontal cuts" are made,
which separate the preoptic region from the anterior hypothalamus,
a constant vaginal estrus syndrome resembling that which develops
after suprachiasmatic lesions is produced. This suggests that fibers
from the preoptic region project to the median eminence-arcuate
region to mediate the preovulatory discharge of gonadotropins.
Halasz and his co-workers have produced evidence which suggests that
the stimulus for ovulation arises from the preoptic region since cuts
which eliminated lateral and posterior connections to the basal hypothalamus were ineffective in blocking ovulation, which continued to
occur after the preoptic region had been separated from more rostral
structures (Halasz and Gorski, 1967; Koves and Halasz, 1970).

EFFECTS OF IMPLANTS OF GONADAL STEROIDS IN THE
PREOPTIC OR HYPOTHALAMIC-PITUITARY REGION

The results of implanting or microinjecting minute amounts of
gonadal steroids into these regions have corroborated the feedback
actions of gonadal steroids on all of these sites. The fact that
implants of estrogen into the preoptic region have been followed by
the induction of precocious puberty (Smith and Davidson, 1968) suggests that in this site estrogen acts to stimulate gonadotropin
release.

There is also evidence that implantation of the steroid in the
pituitary gland itself augments gonadotropin release (Davidson et al.,
1970). On the other hand, long-term implants in the basal tuberal
region (Ramirez et al., 1964) or into the pituitary (Bogdanove, 1963)

clearly inhibit gonadotropin release. Several workers have suggested
that steroid implanted in the basal tuberal region is taken up into
the hypophysial portal vessels and transported down to the pituitary
to exert its effects there (Ramirez et al., 1964; Bogdanove, 1963).
This doubtlessly occurs to a limited extent; however, recent studies
indicate that only small quantities of substances placed in the third
ventricle actually reach the anterior lobe (Ondo et al., 1972).
Furthermore, releasing factors injected into the anterior lobe are
clearly more effective than when injected into the median eminence
(Dhariwal et al., 1969).

Recent studies, in which 5 ng of estradiol benzoate were microinjected into the third ventricle, have shown that this minute dose
of estradiol inhibits LH release in the ovariectomized rat and that
this inhibition is associated with the normal responsiveness of the
pituitary gland to LRF (Orias et al., 1974). This clearly
establishes a site for the negative feedback in the CNS since the
same dose of estradiol administered into the systemic circulation
was ineffective. Implants of estradiol into the hypothalamus are
associated with a decrease in the LRF stored in this tissue (Chowers
and McCann, 1965); and the stored LRF diminished after systemic
injection of large doses (Piacsek and Meites, 1966) or small physiological doses of estrogen (Ajika et al., 1972). This constitutes
additional evidence that the feedback occurs at the hypothalamic
level. Lastly, the elevated levels of LRF in the portal blood of
long-term castrate animals (Ben-Jonathan et al., 1973) indicate
that under normal conditions steroids act at the hypothalamic level
to inhibit the release of the releasing factor.

Negative feedback is also exerted on the anterior lobe itself.
Implants of estrogen (Ramirez et al., 1964; Chowers and McCann, 1967)
and testosterone into the gland inhibit LH and FSH release, respectively; in addition, testosterone implants induce elevated levels of
FSH in the pituitary gland (Kamberi and McCann, 1972; Kingsley and
Bogdanove, 1973). These implants also inhibit castration cell
development; the inhibition is inversely proportional to the distance
of the cell from the implant (Bogdanove, 1963).

THE CIRCADIAN CLOCK INVOLVED IN RELEASING GONADOTROPINS

Gonadal steroids interact with a circadian clock in the rat to
produce a discharge of gonadotropins on the afternoon of proestrus.
For example, if an ovariectomized animal is primed with estrogen,the
the initial response is a lowering of plasma LH as a result of negative feedback (P.S. Kalra et al., 1973). If, on the second or third
day after the administration of estradiol, either progesterone or a
second injection of estradiol is given, a discharge of gonadotropins
takes place on the afternoon of that day mimicking the preovulatory
discharge in quantity and timing (Caligaris et al., 1971; P.S. Kalra

et al., 1972). If the time of day of the second injection of
steroids is changed, however, the discharge does not take place
since it can take place only between 3:00 and 5:00 p.m. when the
animals are on a conventional light-dark cycle (lights on 5:00 a.m.
to 7:00 p.m.). If a sufficient dose of estradiol alone is injected,
a repetitive discharge occurs every afternoon for several days.
Consequently, the cyclic timing mechanism can discharge every afternoon; and this discharge results in a release of gonadotropins only
in the presence of the proper steroid milieu, which can consist of
high levels of estrogen alone or intermediate levels of estrogen
plus progesterone. This response can be blocked by pretreatment
with large doses of progesterone (Caligaris et al., 1968).

UPTAKE AND LOCALIZATION OF ESTROGEN IN THE
PREOPTIC, HYPOTHALAMIC-PITUITARY REGION

Estrogen is avidly concentrated by the anterior pituitary,
preoptic region, and hypothalamus (Pfaff et al., in press; Stumpf,
1968). As in other tissues, the steroid is taken up, bound to a
cytosol receptor, and then transported to the nucleus where it
probably alters RNA synthesis. Radioautography has demonstrated an
accumulation of the steroid in the medial preoptic region and in the
arcuate nucleus (Pfaff et al., 1974; Stumpf, 1968), areas probably
involved in the positive and negative feedbacks of estrogen,
respectively. Estrogen is also taken up by the medial amygdaloid
nucleus, which, on the basis of stimulation and lesion experiments
(see below), is also thought to be involved in gonadotropin control.

HYPOTHALAMIC GONADOTROPIN-RELEASING FACTORS

Crude hypothalamic extracts from various species stimulate the
release of LH and to a lesser extent FSH from the pituitary (McCann
et al., 1960; Igarashi and McCann, 1964). The LH-releasing factor
(LRF) has been purified (Guillemin et al., 1963; Dhariwal et al.,
1965; Schally and Bowers, 1964), separated from other hypothalamic
releasing and inhibiting factors, and finally isolated and synthesized (Matsuo et al., 1971; Burgus et al., 1972). The decapeptide
LRF can release LH in all species so far examined, including man
(Kastin et al., 1972).

Early work, in which the release of gonadotropins was estimated
by bioassay, indicated a separate FSH-releasing factor in hypothalamic extracts (Dhariwal et al., 1967; Schally et al., 1966;
Jutisz, 1970). With the development of radioimmunoassay, it has been
difficult to obtain evidence for a separate FSH-releasing factor;
however, several groups have reported hypothalamic fractions which
produce a more or less selective release of FSH (Fawcett, 1973;
Johansson et al., 1973). In addition, crude hypothalamic extracts

obtained at certain stages of the estrous cycle and after treatment
with estrogen produce a higher ratio of FSH to LH release when
incubated with pituitaries in vitro (Fawcett, 1973; Libertun et al.,
1974). These findings, coupled with evidence from the stimulation
and lesion experiments cited above, which appear to indicate discrete
LH- and FSH-controlling centers in the hypothalamus, lead to the
conclusion that a distinct FSH-releasing factor will eventually
be isolated.

Natural LRF or the synthetic decapeptide produces a dramatic
release of LH within a few minutes of its intravenous injection in
a variety of species including man. Except at very high doses, an
acute pulse of LRF provokes an almost exclusive LH release in the
rat; however, FSH release can be promoted if infusions of the hormone are given (Arimura et al., 1972; Libertun, Drias, and McCann,
1974) or if it is administered subcutaneously to slow down absorption
(Zeballos and McCann, 1974). Even in these situations the relative
amounts of LH released are much greater.

LOCALIZATION OF GONADOTROPIN-RELEASING FACTORS

In early bioassay studies, FSH-releasing activity was localized
to the median eminence-arcuate region when frozen sections of the
hypothalamus were extracted and assayed in vitro for releasing
activity (Watanabe and McCann, 1968). On the other hand, LRF was
extracted from a region extending from the suprachiasmatic area,
caudally and ventrally to the median eminence-arcuate region
(Crighton et al., 1970). Both activities were found in the pituitary
stalk but not in the neural lobe itself. More recent assays in
which gonadotropins released by the extracts were measured by radioimmunoassay have indicated that both activities can be extracted
from a co-extensive region extending from the preoptic region
rostrally to the median eminence-arcuate area caudally (Krulich et
al., 1971). These observations obviously do not agree with the
results of hypothalamic stimulation in the preoptic region which
selectively released LH. The discrepancy may be explained by the
fact that assay of gonadotropin-releasing factors in vitro potentiates
the FSH-releasing action of the extracts by allowing a prolonged
exposure to releasing factors.

Lesions in the suprachiasmatic region were followed by a decline
in LRF content in the median eminence-arcuate region to values
roughly 50% of those in controls (Schneider et al., 1969). We postulate that these lesions destroy the cell bodies of LRF neurons with
relatively long axons which extend caudally to the median eminence
and terminate in juxtaposition to the hypophysial portal capillaries.
Since activity in the median eminence did not disappear after these
lesions, we postulate that the cell bodies of some of the LRF neurons
are located more caudally, perhaps in the arcuate nucleus, with

relatively short axons which project to the median eminence. Mess and co-workers have reported decreases in FSH-releasing factor content in the median eminence after lesions in the paraventricular region and have suggested that neurons which contain FSH-releasing factor have cell bodies in the region of the paraventricular nucleus and axons which project to the median eminence (Mess, 1969).

EFFECTS OF GONADAL STEROIDS ON THE RESPONSIVENESS OF THE PITUITARY TO LRF

If the pituitary is a site for feedback actions of gonadal steroids, altered levels of these steroids would be expected to influence the responsiveness of the gland to LRF; in a number of species, this is the case. For example, after the intravenous administration of minute doses of estradiol, sufficient to lower plasma LH in ovariectomized rats, there was a diminished response to the intravenous injection of LRF (Negro-Vilar et al., 1973). Similar findings have been reported in man (S.S.C. Yen, personal communication). This establishes the pituitary as a physiological site for the negative feedback of gonadal steroids. We have earlier reviewed evidence that there is also a hypothalamic site.

The enhanced sensitivity to LRF which is observed within several hours after acute castration in the female rat reflects this inhibitory action (K.J. Cooper and S.M. McCann, unpublished data). Similarly, castration is followed by an enhanced response to LRF in the male which indicates that testicular steroids suppress the pituitary responsiveness to the hormone (Watson et al., 1971).

When the responsiveness of the rat pituitary to a single intravenous injection of 100 ng of estradiol benzoate is followed over a period of time, the early inhibition at one to two hours is replaced at eight hours by such a sensitization that much higher levels of LH result after infusions of LRF (Libertun, Orias, and McCann, 1974). Thus, estrogen has a biphasic effect on the sensitivity of the pituitary to LRF, an early diminished sensitivity followed eight hours later by increased sensitivity.

This increased sensitization is probably responsible for the increased sensitivity to the hormone just before the preovulatory discharge of LH observed in a number of species including man (Cooper et al., 1973; Reeves et al., 1971; Yen et al., 1972). The time-course of sensitization in the rat fits with that of plasma estrogen levels which reach maximum values about eight hours before the preovulatory discharge (S.P. Kalra, P.S. Kalra, and Abrams, 1973). These observations establish the pituitary as a physiological site for both the negative and the positive feedback of estrogen.

At proestrus there is an enhanced LH discharge in response to

electrochemical stimulation of the preoptic region (S.P. Kalra and
McCann, 1973b), partly because of the sensitization of the pituitary
by ovarian estrogens. It may also reflect an increased synthesis
of LRF on the day of proestrus since the levels of LRF stored in the
basal hypothalamus appeared to rise on proestrus before the preovulatory discharge of gonadotropins (S.P. Kalra, Krulich, and McCann,
1973).

Some clear-cut effects of estrogen and testosterone on the
responsiveness of the pituitary to LRF have been observed; but under
most circumstances at least, progesterone appears to have little or
no effect on the responsiveness of the gland (Libertun, Orias, and
McCann 1974). However, we have recently observed increased sensitivity to LRF at the time of an LH surge provoked by a subcutaneous
injection of progesterone (C. Libertun and S.M. McCann, unpublished
data).

POSSIBLE SYNAPTIC TRANSMITTERS WHICH ALTER GONADOTROPIN RELEASE

Synaptic contacts with other neurons must mediate the influence
of the CNS on the discharge of the gonadotropin-releasing factors.
Anatomically, evidence for abundant noradrenergic terminals in the
preoptic region has been obtained by the histochemical fluorescence
technique (Fuxe and Hökfelt, 1969). In the arcuate region, this
technique has demonstrated a dopaminergic pathway. The cell bodies
of the dopaminergic neurons lie in the arcuate nucleus and their
axons project to the external layer of the median eminence. Since
norepinephrine can also be extracted from this region, noradrenergic
terminals are probably also present here. Recent studies with knife
cuts indicate that the norepinephrine in the basal tuberal region is
contained in terminals of neurons whose cell bodies lie outside the
hypothalamus since the norepinephrine disappears when the basal
tuberal region has been isolated (Weiner et al., 1972).

For many years there has been evidence that adrenergic blocking
drugs could interfere with gonadotropin release (Everett, 1964;
Sawyer et al., 1947). Recent studies have shown that dopamine acts
either in vitro or in vivo to provoke the release of FSH-releasing
factor, LRF, and prolactin-inhibiting factor (PIF) (McCann et al.,
1972; Porter et al., 1972); norepinephrine was less effective, and
epinephrine the least effective in these studies.

Further evidence that dopamine acts as a transmitter to stimulate the release of PIF was obtained with drugs which alter the
synthesis of catecholamines and with adrenergic blocking agents
(Donoso et al., 1971). On the other hand, similar studies suggested
that a noradrenergic synapse is involved in the stimulatory effect
of both estrogen and progesterone on gonadotropin release and in the
preovulatory discharge of gonadotropins (S.P. Kalra and McCann, 1974;
P.S. Kalra and McCann, 1973). This synapse may lie in the preoptic region since stimulations of this area failed to induce LH

release when norepinephrine synthesis was inhibited (S.P. Kalra and
McCann, 1973a). Similar studies in castrate animals provide further
evidence that norepinephrine is involved but also suggested a role
for dopamine, at least in the postcastration rise in FSH (Ojeda and
McCann, 1972).

There is no evidence that catecholamines act directly on the
anterior pituitary to alter the release of FSH or LH (McCann et al.,
1972); however, minute doses of catecholamines can inhibit the
release of prolactin from pituitaries incubated in vitro (Birge et
al., 1970; MacLeod, 1969; Quijada et al., 1974). Dopamine (DA)
infused into cannulated portal vessels had no effect on gonadotropin
and prolactin release (Porter et al., 1972). Thus, it is not clear
whether DA has a direct inhibitory effect on pituitary prolactin
release. In animals with elevated plasma prolactin concentrations
as a result of median eminence lesions, L-DOPA lowered prolactin
(Donaso et al., 1973). Apparently, L-DOPA was taken up by the
pituitary and converted by 1-aromatic acid decarboxylase into dopamine, which then inhibited prolactin release. If dopamine actually
is released into portal vessels, which so far has not been demonstrated (Ruf et al., 1971), it may play a role in inhibiting prolactin by direct action on the pituitary.

There is evidence of cholinergic synapses in the gonadotropin-
controlling pathways. In the pioneering work of Markee et al.
(1952), atropine was shown to block ovulation. Recent studies have
demonstrated that atropine injected either sc or into the third
ventricle in much smaller doses can block the proestrus discharge of
gonadotropins and prolactin (Libertun and McCann, 1973); however,
intraventricular injections of carbachol have not uniformly produced
immediate increases in plasma gonadotropins (C. Libertun, unpublished). Therefore, additional work is required before a cholinergic link in gonadotropin control can be established.

Terminals of serotonergic neurons are found in the anterior
hypothalamus by the histochemical fluorescence technique; and, when
administered into the third ventricle, serotonin inhibits the release
of gonadotropins and stimulates the release of prolactin (McCann et
al., 1972; Porter et al., 1972). In our studies, blocking serotonin
biosynthesis with parachlorophenylalanine has had little effect on
gonadotropin or prolactin release (Donoso et al., 1971); but Kordon,
Blake, and Sawyer (1972) have reported that parachlorophenylalanine
interferes with the suckling-induced rise in prolactin. Further
studies are necessary to determine whether serotonin acts as a synaptic transmitter to inhibit gonadotropin and augment prolactin release.

The pineal principle, melatonin, can, in large doses at least,
inhibit gonadotropin (Reiter, 1973) and augment prolactin release
(Porter et al., 1972). Removal of the pineal has resulted in eliminating an early-morning rise in prolactin in male rats (Rønnekleiv et

al., 1973) and on occasion has resulted in slightly elevated gonadotropin levels (Rønnekleiv and McCann, 1973). Therefore, the pineal may exert an inhibitory effect on gonadotropin release.

EFFECTS OF PROSTAGLANDINS ON GONADOTROPIN AND PROLACTIN RELEASE

Prostaglandins appear to be involved at a number of loci in the reproductive tract. Recent studies demonstrated that prostaglandin E_2 elevated plasma LH within 15 minutes of its injection into the third ventricle (Harms et al., 1973). Prostaglandin E_1 had a lesser effect, and $F_{1\alpha}$ and F_2 were without effect. By contrast, prolactin was elevated by intraventricular prostaglandin E_2 but not by the other prostaglandins (Harms et al., 1973). Conversely, when prostaglandin synthesis was inhibited by indomethacin, ovulation was blocked (Grinwich et al., 1972); and preliminary studies show that indomethacin injected into the third ventricle lowered plasma LH (P.G. Harms and S.R. Ojeda, unpublished). When cyclic adenosine monophosphate (cAMP) and its derivative (DBcAMP) were microinjected into the third ventricle, they depressed prolactin without modifying LH (Ojeda and McCann, 1973), an indication that cyclic nucleotides are involved in the release of PIF or the inhibition of PRF. Prostaglandins and cyclic nucleotides may even mediate the release of hypothalamic neurohormones in response to adrenergic input.

EFFECT OF THE LIMBIC SYSTEM ON GONADOTROPIN RELEASE

Although it is still too early to be sure of the ultimate role of the limbic system in the control of gonadotropin release, it is clear that stimulations and lesions in limbic structures do alter the release of gonadotropins. For example, stimulations in the medial amygdala evoke ovulation (Velasco and Taleisnik, 1969a), but lesions in the structure have been reported either to enhance or inhibit gonadotropin release (Elwers and Critchlow, 1960; Lawton and Sawyer, 1970). This ambiguity may be resolved by a careful study of the localization of function in the amygdala; alternatively, different means of producing lesions may explain the discrepancies. Implants of estradiol into medial amygdala evoked lactational changes in rabbits, a finding which suggests that they evoked prolactin release (Tindal et al., 1967); and recently, they have been shown to evoke LH release (P.S. Kalra and McCann, 1972).

Stimulations in the hippocampus appear to inhibit LH release (Velasco and Taleisnik, 1969b), but again conflicting results have been obtained with lesions. The hippocampus may be an inhibitory area, but additional work is required to determine its importance.

CONCLUSIONS

Gonadotropin release is ultimately responsible for the release of gonadal steroids which in turn induce sexual receptivity at the appropriate stage of the estrus cycle. The preovulatory release of gonadotropins is so timed that ovulation coincides with the period of sexual receptivity that will ensure fertilization of the ova. Recent studies suggest that the release of LRF into the preoptic region at this time is involved in the induction of sexual receptivity.

Hypothalamic regulation of gonadotropin release is brought about by dual centers: a tonic release center located in the arcuate region which regulates resting gonadotropin release and is tonically inhibited by gonadal steroids; and a phasic center of the preoptic region which evokes the preovulatory discharge of gonadotropins in response to an interaction between the stimulatory effects of gonadal steroids, estrogen and possibly progesterone, and a cyclic-timing mechanism, at least in certain rodents such as the rat.

The discharge of gonadotropin-releasing factors brings about the alterations in gonadotropin release, but the responses of the pituitary to LRF are modulated by gonadal steroids. Estrogen has a biphasic effect, producing an initial inhibition which is followed by sensitization at the time of the preovulatory release.

The release of the gonadotropin-releasing factors is under adrenergic control. The preovulatory discharge may be brought about by increased impulse traffic across nonadrenergic synapses in the preoptic and/or anterior hypothalamic area at the time of the firing of the cyclic clock mechanism. Adrenergic transmission is probably also involved in the negative feedback of gonadal steroids. Both norepinephrine and dopamine may be involved.

On the other hand, strong evidence suggests that dopamine is a transmitter to release a prolactin-inhibiting factor which brings about the tonic inhibitory hypothalamic control over this pituitary hormone. Both pineal principles and serotonin can inhibit gonadotropin release and may play a physiological role. Evidence is beginning to accumulate for a role of prostaglandins as intermediaries in the release of gonadotropin-releasing factors. Influences from the limbic system can also modify gonadotropin release, but their role in the normal regulatory process remains to be determined.

REFERENCES

Ajika, K., Krulich, L., Fawcett, C.P., and McCann, S.M., 1972,
      Effects of estrogen on plasma and pituitary gonadotropins and
      prolactin, and on hypothalamic releasing and inhibiting factors,

Neuroendocrinology 9:304-315.

Arimura, A., Debeljuk, L., and Schally, A.V., 1972, Stimulation of
    FSH release in vivo by prolonged infusion of synthetic LH-RH,
    Endocrinology 91:529-532.

Barraclough, C.A., 1966, Modifications in the CNS regulation of
    reproduction after exposure of prepubertal rats to steroid
    hormones, Rec. Prog. Horm. Res. 22:503-539.

Ben-Jonathan, N., Mical, R.S., and Porter, J.C., 1973, Superfusion
    of hemipituitaries with portal blood. I. LRF secretion in
    castrated and diestrous rats, Endocrinology 93:497-503.

Birge, C.A., Jacobs, L.S., Hammeo, C.T., and Daughaday, W.H., 1970,
    Catecholamine inhibition of prolactin secretion by isolated
    rat adenohypophyses, Endocrinology 86:120-130.

Bishop, W., Fawcett, C.P., Krulich, L., and McCann, S.M., 1972, Acute
    and chronic effects of hypothalamic lesions on the release of
    FSH, LH and prolactin in intact and castrated rats, Endocrin-
    ology 91:643-656.

Bishop, W., Kalra, P.S., Fawcett, C.P., Krulich, L., and McCann, S.
    M., 1972, The effects of hypothalamic lesions on the release of
    gonadotropins and prolactin in response to estrogen and pro-
    gesterone treatment in female rats, Endocrinology 91:1404-1410.

Bogdanove, E.M., 1963, Direct gonad-pituitary feedback: an analysis
    of effects of cranial estrogenic depots on gonadotrophin secre-
    tion, Endocrinology 73:696-712.

Burgus, R., Butcher, M., Amoss, M., Ling, N., Monahan, M., Rivier,
    J., Fellows, R., Blackwell, R., Vale, W., and Guillemin, R.,
    1972, Primary structure of the ovine hypothalamic luteinizing
    hormone-releasing factor (LRF), Proc. Nat. Acad. Sci. 69:278-
    282.

Caligaris, L., Astrada, J.J., and Taleisnik, S., 1968, Stimulating
    and inhibiting effects of progesterone on the release of LH,
    Acta Endocrinol. (Copenhagen) 59:177-185.

Caligaris, L., Astrada, J.J., and Taleisnik, S., 1971, Release of
    LH induced by estrogen injection into ovariectomized rats,
    Endocrinology 88:810-815.

Chen, C.L., Amenomori, Y., Lu, K.L., Voogt, J.L., and Meites, J.,
    1970, Serum prolactin levels in rats with pituitary transplants
    or hypothalamic lesions, Neuroendocrinology 6:220-227.

Chowers, I., and McCann, S.M., 1965, Content of LH-releasing factor
    and LH during the estrous cycle and after changes in gonadal
    steroid titers, Endocrinology 76:700-708.

Chowers, I., and McCann, S.M., 1967, Comparison of effect of hypo-
    thalamic and pituitary implants of estrogen and testosterone
    on the reproductive system and adrenal of female rats, Proc.
    Soc. Exptl. Biol. Med. 124:260-266.

Cooper, K.J., Fawcett, C.P., and McCann, S.M., 1973, Variations in
    pituitary responsiveness to luteinizing hormone releasing
    factor during the rat oestrous cycle, J. Endocrin. 57:187-188.

Cramer, O.M., and Barraclough, C.A., 1971, Effect of electrical
    stimulation of the preoptic area on plasma LH concentrations
    in proestrous rats, Endocrinology 88:1175-1183.

Crighton, D.B., Schneider, H.P.G., and McCann, S.M., 1970, Locali-
    zation of LH-releasing factor in the hypothalamus and neuro-
    hypophysis as determined by *in vitro* assay, Endocrinology, 87:
    323-329.

Davidson, J.M., Weick, R.F., Smith, E.R., and Dominguez, R., 1970,
    Feedback mechanisms in relation to ovulation, Fed. Proc. 29:
    1900.

DeVoe, W.F., Ramirez, V.D., and McCann, S.M., 1965, Induction of
    mammary secretion by hypothalamic lesions in male rats, Endo-
    crinology 78:158-164.

Dhariwal, A.P.S., Antunes-Rodrigues, J., and McCann, S.M., 1965,
    Purification of ovine luteinizing hormone-releasing factor by
    gel filtration and ion-exchange chromatography, Proc. Soc.
    Exptl. Biol. Med. 118:999-1003.

Dhariwal, A.P.S., Russell, S., McCann, S.M., and Yates, F.E., 1969,
    Assay of corticotrophin releasing factors by injection into the
    anterior pituitary of intact rats, Endocrinology 84:544-546.

Dhariwal, A.P.S., Watanabe, S., Antunes-Rodrigues, J., and McCann,
    S.M., 1967, Chromatographic behavior of follicle stimulating
    hormone-releasing factor of Sephadex and carboxyl methyl cel-
    lulose, Neuroendocrinology 2:294-303.

Donoso, A.O., Bishop, W., Fawcett, C.P., Krulich, L., and McCann, S.
    M., 1971, Effects of drugs that modify brain monoamine concen-
    trations on plasma gonadotropin and prolactin levels in the rat,
    Endocrinology 89:774-784.

Donoso, A.O., Bishop, W., and McCann, S.M., 1973, The effects of drugs which modify catecholamine synthesis on serum prolactin in rats with median eminence lesions, Proc. Soc. Exptl. Biol. Med. 143:360-363.

Elwers, M., and Critchlow, B.V., 1960, Precocious ovarian stimulation following hypothalamic and amygdaloid lesions in the rat. Amer. J. Physiol. 198:381-385.

Everett, J.W., 1964, Central neural control of reproductive functions of the adenohypophysis, Physiol. Rev. 44:373-431.

Fawcett, C.P., 1973, Discussion of paper #24, in *Proceedings of the Conference on Hypothalamic Hypophysiotropic Hormones* (C. Gual and E. Rosemberg, eds.), p. 111, Excerpta Medica, Amsterdam.

Fuxe, K., and Hökfelt, T., 1969, Catecholamines in the hypothalamus and the pituitary gland, in *Frontiers in Neuroendocrinology* (L. Martini and W.F. Ganong, eds.), pp. 47-98, Oxford University Press, New York.

Grinwich, D.L., Kennedy, T.G., and Armstrong, D.T., 1972, Dissociation of ovulatory and steroidogenic actions of luteinizing hormone in rabbits with indomethacin, an inhibitor of prostaglandin biosynthesis, Prostaglandins 1:89-96.

Guillemin, R., Jutisz, M., and Sakiz, E., 1963, Purification partielle d'un facteur hypothalamique (LRF) stimulant la sécrétion de l'hormone hypophysaire de lutéinisation (LH), C.R. Acad. des Sciences 256(D)(Paris):504-507.

Halasz, B., and Gorski, R., 1967, Gonadotrophic hormone secretion in female rats after partial or total interruption of neural afferents to the medial basal hypothalamus, Endocrinology 80: 608-622.

Halasz, B., and Pupp, L., 1965, Hormone secretion of the anterior pituitary gland after physical interruption of all nervous pathways to the hypophysiotropic area, Endocrinology 77:553-562.

Harms, P.G., Ojeda, S.R., and McCann, S.M., 1973, Prostaglandin involvement in hypothalamic control of gonadotropin and prolactin release, Science 181:760-761.

Haterius, H.O., 1937, Studies on a neurohypophyseal mechanism influencing gonadotrophic activity, Cold Spring Harbor Symp. Quant. Biol. 5:280-288.

Hillarp, N.A., 1949, Studies on the localization of hypothalamic centres controlling the gonadotrophic function of the hypophysis,

Acta Endocrin. 2:11-23.

Igarashi, M., and McCann, S.M., 1964, A hypothalamic follicle stim-
    ulating hormone releasing factor, Endocrinology 74:446-452.

Johansson, K.N.G., Currie, B.L., Folkers, K., and Bowers, C.Y., 1973,
    Biological evidence that separate hypothalamic hormones release
    the follicle stimulating and luteinizing hormones, Biochem.
    Biophys. Res. Commun. 50:20-26.

Jutisz, M., 1970, Purification and chemistry of gonadotropin-releas-
    ing factors, in The Human Testis (E. Rosemberg and C.A.
    Paulsen, eds.), pp. 207-228, Plenum Press, New York.

Kalra, P.S., and McCann, S.M., 1972, Effects of CNS implants of
    ovarian steroids on gonadotropin release, in Abstracts, IVth
    International Congress of Endocrinology (Washington, D.C.,
    June, 1972), Int. Cong. Ser. No. 256, abstr. #292, p. 118,
    Excerpta Medica, Amsterdam.

Kalra, P.S., and McCann, S.M., 1973, Involvement of catecholamines
    in feedback mechanisms, Prog. in Brain Research 9:185-198.

Kalra, P.S., Fawcett, C.P., Krulich, L., and McCann, S.M., 1973, The
    effects of gonadal steroids on plasma gonadotropins and pro-
    lactin in the rat, Endocrinology 92:1256-1268.

Kalra, P.S., Kalra, S.P., Krulich, L., Fawcett, C.P., and McCann, S.
    M., 1972, Involvement of norepinephrine in transmission of the
    stimulatory influence of progesterone on gonadotropin release,
    Endocrinology 90:1168-1176.

Kalra, S.P., and McCann, S.M., 1973a, Effect of drugs modifying
    catecholamine synthesis on LH release from preoptic stimulation
    in the rat, Endocrinology 93:356-362.

Kalra, S.P., and McCann, S.M., 1973b, Variations in the release of
    LH in response to electrochemical stimulation of preoptic area
    and of medial basal hypothalamus during the estrous cycle of
    the rat, Endocrinology 93:665-669.

Kalra, S.P., and McCann, S.M., 1974, Effects of drugs modifying
    catecholamine synthesis on plasma LH and ovulation in the rat,
    Neuroendocrinology (in press).

Kalra, S.P., Ajika, K., Krulich, L., Fawcett, C.P., Quijada, M., and
    McCann, S.M., 1971, Effects of hypothalamic and preoptic elec-
    trochemical stimulation on gonadotropin and prolactin release
    in proestrous rats, Endocrinology 88:1150-1158.

Kalra, S.P., Kalra, P.S., and Abrams, R.M., 1973, Peripheral blood
    LH, estrogen (E$_2$), and progesterone (P) during the rat estrous
    cycle: correlation with pituitary and hypothalamic function,
    Fed. Proc. 32(3):282Abs. (abstr. #380).

Kalra, S.P., Krulich, L., and McCann, S.M., 1973, Changes in gonado-
    tropin releasing factor content in the rat hypothalamus follow-
    ing electrochemical stimulation of anterior hypothalamic area
    and during the estrous cycle, Neuroendocrinology 12:321-333.

Kamberi, I.A., and McCann, S.M., 1972, Effects of implants of test-
    osterone in the median eminence and pituitary on FSH secretion.
    Neuroendocrinology 9:20-29.

Kastin, A.J., Gual, C., and Schally, A.V., 1972, Clinical experiments
    with hypothalamic releasing hormones. Part 2. Luteinizing
    hormone-releasing hormone and other hypophysiotropic releasing
    hormones, Rec. Prog. Horm. Res. 28:201-227.

Kingsley, T.R., and Bogdanove, E.M., 1973, Direct feedback of andro-
    gens: Localized effects of intrapituitary implants of androgens
    on gonadotropic cell and hormone stores, Endocrinology 93:1398-
    1410.

Kordon, C.A., Blake, C.A., and Sawyer, C.H., 1972, Participation of
    serotonin-containing neurons in the suckling-induced rise in
    plasma prolactin levels in lactating rats, in Abstracts, IVth
    International Congress of Endocrinology (Washington, D.C.,
    June, 1972), Int. Cong. Ser. No. 256, abstr. #126, p. 51,
    Excerpta Medica, Amsterdam.

Koves, K., and Halasz, B., 1970, Location of the neural structures
    triggering ovulation in the rat, Neuroendocrinology 6:180-193.

Krulich, L., Kuhn, E., Illner, P., and McCann, S.M., 1970, Blood
    prolactin levels in lactating rats, Fed. Proc. 29:579.

Krulich, L., Quijada, M., Illner, P., and McCann, S.M., 1971, The
    distribution of hypothalamic hypophysiotropic factors in the
    hypothalamus of the rat, Proc. XXV Intl. Cong. Physiol. Sci.
    9:326.

Lawton, I.E., and Sawyer, C.H., 1970, Role of amygdala in regulating
    LH secretion in the adult female rat, Amer. J. Physiol. 218:
    622-625.

Libertun, C., and McCann, S.M., 1973, Blockade of the release of
    gonadotropins and prolactin by subcutaneous or intraventricular
    injection of atropine in male and female rats, Endocrinology
    92:1714-1724.

Libertun, C., Cooper, K.J., Fawcett, C.P., and McCann, S.M., 1974, Effects of ovariectomy and steroid treatment on hypophysial sensitivity to purified LH-releasing factor (LRF), Endocrinology (in press).

Libertun, C., Orias, R., and McCann, S.M., 1974, Biphasic effect of estrogen on the sensitivity of the pituitary to luteinizing hormone releasing factor (LRF), Endocrinology (in press).

MacLeod, R.M., 1969, Influence of norepinephrine and catecholamine-depleting agents on the synthesis and release of prolactin and growth hormone, Endocrinology 85:916-923.

Markee, J.E., Everett, J.W., and Sawyer, C.H., 1952, The relationship of the nervous system to the release of gonadotrophin and the regulation of the sex cycle, Rec. Prog. Horm. Res. 7: 139-163.

Markee, J.E., Sawyer, C.H., and Hollinshead, W.H., 1946, Activation of the anterior hypophysis by electrical stimulation in the rabbit, Endocrinology 38:345-357.

Matsuo, H., Baba, Y., Nair, R.M.G., Arimura, A., and Schally, A.V., 1971, Structure of the porcine LH- and FSH-releasing hormone. I. The proposed amino acid sequence, Biochem. Biophys. Res. Commun. 43:1334-1339.

McCann, S.M., and Friedman, H.M., 1960, The effect of hypothalamic lesions on the secretion of luteotrophin, Endocrinology, 67:597-608.

McCann, S.M., Kalra, S.P., Kalra, P.S., Bishop, W., Donoso, A.O., Schneider, H.P.G., Fawcett, C.P., and Krulich, L., 1972, The role of monoamines in the control of gonadotropin and prolactin secretion, in *Brain-Endocrine Interaction. Median Eminence: Structure and Function* (K.M. Knigge, D.E. Scott, and S. Weindl, eds.), pp. 224-235, S. Karger, Basel.

McCann, S.M., Taleisnik, S., and Friedman, H.M., 1960, LH-releasing activity in hypothalamic extracts, Proc. Soc. Exp. Biol. Med. 104:432-434.

Mess, B., 1969, Site and onset of production of releasing factors, in *Progress in Endocrinology* (C. Gual, ed.), pp. 564-570, Excerpta Medica, Amsterdam.

Moss, R.L., and Cooper, K.J., 1973, Temporal relationship of spontaneous and coitus-induced release of luteinizing hormone in the normal cyclic rat, Endocrinology 92:1748-1753.

Moss, R.L., and McCann, S.M., 1973, Induction of mating behavior in rats by luteinizing hormone releasing factor, Science 181:177-179.

Negro-Vilar, A., Orias, R., and McCann, S.M., 1973, Evidence for a pituitary site of action for the acute inhibition of LH release by estrogen in the rat, Endocrinology 92:1680-1684.

Nikitovitch-Winer, M., 1960, The influence of the hypothalamus on leuteotrophin secretion in the rat, Mem. Soc. Endocrin. 9:70-72.

Ojeda, S.R., and McCann, S.M., 1972, Evidence for participation of a catecholaminergic mechanism in the post-castration rise in plasma gonadotropins, Neuroendocrinology 12:295-315.

Ojeda, S.R., and McCann, S.M., 1973, Participation of cyclic nucleotides in the hypothalamic control of LH and prolactin release, in *Program, 55th Annual Meeting, The Endocrine Society* (Chicago, Ill.), abstr. #176, p. A-136.

Ondo, J.G., Mical, R.S., and Porter, J.C., 1972, Passage of radioactive substances from CSF to hypophysial portal blood, Endocrinology 91:1239-1246.

Orias, R., Negro-Vilar, A., Libertun, C., and McCann, S.M., 1974, Inhibitory effect on LH release of estradiol injected into the third ventricle, Endocrinology (in press).

Pfaff, D.W., Lewis, C., Diakow, C.D., and Keiner, M., 1972, Neurophysiological analysis of mating behavior responses as hormone-sensitive reflexes, in *Progress in Physiological Psychology*, vol. 5 (E. Stellar and J.M. Sprague, eds.), pp. 253-297, Academic Press, New York.

Piacsek, B.E., and Meites, J., 1966, Effects of castration and gonadal hormones on hypothalamic content of luteinizing hormone releasing factor (LRF), Endocrinology 79:432-439.

Porter, J.C., Kamberi, I.A., and Ondo, J.G., 1972, Role of biogenic amines and cerebrospinal fluid in the neurovascular transmittal of hypophysiotropic substances, in *Brain-Endocrine Interaction. Median Eminence: Structure and Function* (K.M. Knigge, D.E. Scott, and A. Weindl, eds.), pp. 245-253, S. Karger, Basel.

Quijada, M., Illner, P., Krulich, L., and McCann, S.M., 1974, The effect of catecholamines on hormone release from anterior pituitaries and ventral hypothalami incubated *in vitro*, Neuroendocrinology 13:151–163.

Ramirez, V.D., Abrams, R., and McCann, S.M., 1964, Effect of estradiol implants in the hypothalamo–hypophysial region of the rat
on secretion of LH, Endocrinology 75:243–248.

Reeves, J.J., Arimura, A., and Schally, A.V., 1971, Pituitary respon-
siveness to purified luteinizing hormone–releasing hormone
(LH–RH) at various stages of the estrous cycle in sheep, J.
Animal Sci. 32:123–126.

Reiter, R.J., 1974, Pineal–anterior pituitary gland relationships,
in Endocrine Physiology (S.M. McCann, ed.), MTP Publishing Co.,
London (in press).

Rønnekleiv, O.K., and McCann, S.M., 1973, The effect of pinealectomy
on plasma gonadotropins and prolactin in the rat. The Physiol-
ogist 16:436.

Rønnekleiv, O.K., Krulich, L., and McCann, S.M., 1973, An early
morning surge of prolactin in the male rat and its abolition
by pinealectomy, Endocrinology, 92:1339–1342.

Ruf, K.B., Dreifuss, J.J., and Carr, P.J., 1971, Absence of measurable
amounts of epinephrine, norepinephrine and dopamine in rat
hypophysial portal blood during various phases of the oestrous
cycle, J. Neuro–Visceral Relations 10:65–73.

Sawyer, C.H., Markee, J.E., and Hollinshead, W.H., 1947, Inhibition
of ovulation in rabbit by adrenergic–blocking agent dibenamine,
Endocrinology 41:395–402.

Schally, A.V., and Bowers, C.Y., 1964, Purification of luteinizing
hormone–releasing factor from bovine hypothalamus, Endocrinology
75:608–614.

Schally, A.V., Saito, T., Arimura, A., Muller, E.E., Bowers, C.Y.,
and White, W.P., 1966, Purification of follicle stimulating
hormone–releasing factor (FSH–RF) from bovine hypothalamus,
Endocrinology 79:1087–1094, 1966.

Schneider, H.P.G., Crighton, D.B., and McCann, S.M., 1969, Suprachiasmatic LH–releasing factor, Neuroendocrinology 5:271–280.

Smith, E.R., and Davidson, J.M., 1968, Role of estrogen in the cerebral control of puberty in female rats, Endocrinology 82:100–
108.

Spies, H.G., and Niswender, G.D., 1971, Levels of prolactin, LH and
FSH in the serum of intact and pelvic–neurectomized rats, Endo-

crinology 88:937-943.

Stumpf, W., 1968, Estradiol-concentrating neurons. Topography in
    the hypothalamus by dry-mount autoradiography, Science 162:
    1001-1003.

Taleisnik, S., and McCann, S.M., 1961, Effects of hypothalamic
    lesions on the secretion and storage of hypophysial luteinizing
    hormone, Endocrinology 68:263-272.

Tindal, J.S., Knaggs, G.S., and Turvey, A., 1967, Central nervous
    control of prolactin secretion in the rabbit: Effect of local
    oestrogen implants in the amygdaloid complex, J. Endocrin.
    37:279-287.

Velasco, M.E., and Taleisnik, S., 1969a, Release of gonadotropins
    induced by amygdaloid stimulation in the rat, Endocrinology
    84:143-139.

Velasco, M.E., and Taleisnik, S., 1969b, Effect of hippocampal stim-
    ulation on the release of gonadotropin, Endocrinology 85:1153-
    1159.

Watanabe, S., and McCann, S.M., 1968, Localization of follicle stim-
    ulating hormone-releasing factor in the hypothalamus and neuro-
    hypophysis as determined by *in vitro* assay, Endocrinology 82:
    664-673.

Watson, J.T., Krulich, L., and McCann, S.M., 1971, Effect of crude
    rat hypothalamic extract on serum gonadotropin and prolactin
    levels in normal and orchidectomized male rats, Endocrinology
    89:1412-1417.

Weiner, R.I., Gorski, R.A., and Sawyer, C.H., 1972, Hypothalamic
    catecholamines and pituitary gonadotropic function, in *Brain-
    Endocrine Interaction. Median Eminence: Structure and
    Function* (K.M. Knigge, D.E. Scott, and A. Weindl, eds.), pp.
    236-244, S. Karger, Basel.

Yen, S.S.C., Van den Berg, G., Rebar, R., and Ehara, Y., 1972, Vari-
    ation of pituitary responsiveness to synthetic LRF during dif-
    ferent phases of the menstrual cycle, J. Clin. Endo. Metab.
    35:931-934.

Zeballos, G., and McCann, S.M., 1974, Rapid, long-lasting and sensi-
    tive responses to synthetic LRF after its subcutaneous adminis-
    tration in the rat, Proc. Soc. Exptl. Biol. Med. (in press).

NEUROHORMONAL CONTROL OF MALE SEXUAL BEHAVIOR

Lynwood G. Clemens

Department of Zoology

Michigan State University, East Lansing

Testosterone not only sustains male sexual behavior in castrated male rats and induces male sexual responses in the female but also increases female responses (Whalen and Hardy, 1970; Pfaff, 1970). Similarly, estrogen facilitates sexual receptivity in both males and females and increases male sexual behavior in the gonadectomized rat (Pfaff, 1970).

Several hypotheses have been formulated to explain these paradoxical effects of testosterone and estradiol on male and female sexual behavior. According to one, the hormone receptor sites related to male and female sexual behavior are sufficiently nonspecific as to be activated by either estradiol or testosterone. According to another, both male and female sexual responses are ultimately made possible by estrogenic steroids (McDonald et al., 1970). Thus the intensification of male or female behavior after testosterone treatment is due to the conversion of testosterone to estrogen, presumably estradiol. The purpose of this paper is to report some experiments which were designed to extend and challenge these hypotheses in two areas of sexual behavior: 1) steroid control of sexual differentiation of male mating behavior; and 2) hormonal activation of copulatory behavior in the male adult.

SEX DIFFERENTIATION OF MALE SEXUAL BEHAVIOR

A central feature of male copulatory behavior in the rodent is the "mount," which consists of clasping the copulatory partner with the forepaws and thrusting with the pelvis. When the mount is oriented from the rear of the copulatory partner, vaginal penetration is achieved, provided, of course, that the mounter has sufficient phallic endowment and the "mountee" has complementary morphology and posture. In our analysis of behavioral sex differentiation, we have been primarily concerned with mounting behavior as such, rather than with immediate genital consequences which will be dealt with in detail in a later section.

In the life of a rodent, gonadal hormones are necessary for mounting behavior during two distinct periods. During adulthood, they play a major role in activating and maintaining optimal levels of masculine coital behavior. Again, gonadal steroids are also essential during a limited time in early development if the organism is to develop the potential for male sexual behavior. In rodents, this period occurs shortly after the onset of sexual differentiation (Clemens, 1973) which is defined here as the age when, in the male fetus, the gonad begins to differentiate into a testis. For the guinea pig, sexual differentiation begins around Day 25 of fetal life; and the onset of behavioral sexual differentiation is about Day 30 (Goy et al., 1964). For the hamster, sexual differentiation begins on Day 12 of fetal life (Price and Ortiz, 1965); and the period of behavioral sex differentiation begins on Day 16, which is also the day of its birth. Consequently, the hamster behavioral sex differentiation occurs postnatally; and the dependence of male coital responses on exposure to gonadal hormone during early development can be easily demonstrated. For example, if the male hamster is castrated on the day of birth, he will not show male sexual behavior as an adult even if given androgen replacement therapy. However, if the Day 1 castrated male (or female) is treated with testosterone on Day 2, 3, or 4 of postnatal life, the probability of his showing male sexual behavior when he is treated with testosterone as an adult is heightened. Mounting behavior is seldom if ever shown by the normal female hamster even if she is treated as an adult with massive doses of testosterone. The only treatment known so far to induce mounting in the female hamster is neonatal administration of gonadal hormones.

Before analyzing steroid control of behavioral sex differentiation in the hamster, we ought first to establish that the golden hamster is not unique in its dependence on gonadal hormones for the development of male sexual behavior.

Unlike the female hamster, the female rat (Whalen and Edwards, 1967), guinea pig (Young, 1961), and laboratory mouse (Edwards and Burge, 1971) all show high levels of mounting behavior when treated with gonadal hormones as adults and placed with a sexually receptive female. Such occurrences of mounting behavior in these normal females suggest that in some species mounting behavior develops independently of exposure to androgen during early development (Beach, 1971). On the other hand, mounting in normal females may indicate that in many species the female is normally exposed to androgens during development.

In collaboration with Linda Coniglio, we examined whether the female rat is exposed to androgens during uterine development and whether those androgens are produced by the male fetus. Rats are normally born in litters of six to twelve pups per litter after a 23-day gestation. Since the sex ratio of males to females is approximately 50/50, most female rats develop in a uterus containing male fetuses. Moreover, since the fetal rat testis begins secreting androgen four to five days before parturition, the female fetus is developing in an environment where androgens are being actively produced.

The first step in this analysis was to determine whether having males and therefore fetal androgens in the uterus was related to mounting in the adult female rat. Females delivered by caesarean section 24 hours before they were normally due were selected from litters of known sex composition and tested as adults for mounting behavior. (Caesarean delivery allowed precise determination of sex ratio before the mother could cannibalize some of her pups.)

Removing one uterine horn from the mother 10 to 13 days after impregnation increased the probability of obtaining some litters containing all females; consequently, all pups in any one litter came from the same uterine horn. After delivery, females were generally fostered to mothers that had normally delivered a litter within the previous one to three days. This procedure allowed all female offspring to be reared in a bisexual litter situation. Females were ovariectomized at 50 to 55 days of age and tested for mounting behavior two weeks later. Each female was placed with a sexually receptive female, and the number of mounts achieved by the experimental animal in a 10-minute period was scored. The first test (pretest) was given before adult hormone treatment; three subsequent tests were given at weekly intervals after daily injections of 250 µg of testosterone propionate (TP). The relation between the number of males in a uterus and the percent of females that displayed mounting behavior is shown in Table 1 (Clemens and Coniglio, 1971).

Females that had developed in a uterus containing three or more males had a higher probability of mounting than females that had developed in a uterus with fewer males and often began mounting after less TP treatment. Three females from litters containing no males failed to mount in any of the tests.

There was no signficant correlation between litter size or sex ratio and the probability of mounting by female nor between their mounting *frequency* and the number of males per litter. This absence of a relationship between the number of males in the litter and the mounting frequency of their adult female siblings suggests that the presence of a male fetus in the uterus is by itself not sufficient to insure mounting in the adult female. This idea, in

TABLE 1.   *Relation of the Number of Males per Litter to Frequency of*
Females Showing Mounting as Adults

Males/ Litter	N	Pretest	Percent of Females Mounting after Daily TP Treatment for: 7 days	14 days	21 days
0	3	0	0	0	0
1	2	0	0	0	0
2	5	0	0	0	20
3	8	0	12	62	75
4	10	0	40	40	60
5	9	0	22	55	88

N: no. of females per group

turn, suggests that it may be the proximity of the female fetus to the male fetus during uterine development that influences her adult behavior potential more than just the number of males in the uterus.

To test this hypothesis, litters were delivered by caesarean section and the position in the uterus of each female fetus in relation to each male fetus was determined. As an independent check of the possibility that normal females are exposed to prenatal androgenic stimulation, the anogenital distance of each pup was measured at the time of delivery. This measure is a sensitive index of androgenic stimulation. Some of the mothers were also treated with the potent antiandrogen, Sch 13521 (4' nitro-3'-trifluoromethylisobutyranilide), from Day 10 to 22 of gestation. The results of the morphological measures are shown in Table 2.

When antiandrogens was not administered to the mother, the anogenital distance of the females varied inversely with the proximity of the female to a male fetus. The closer to the male fetus, the greater the anogenital distance; therefore females developing in proximity to males are probably exposed to higher levels of androgen stimulation than those developing at a distance. Analysis of variance across the anogenital distance scores for the normal mothers in Table 2 was significant at the 0.001 level. Females born in all-female litters had a larger anogenital distance than females that had developed at some distance from a male (e.g., group NC2 in Table 2). However, the body weight of females from all-female litters was also greater than in any of the other

*TABLE 2. Relation of Anogenital Distance (AgD) in Female Rats to
          the Location of a Male Fetus in Utero**

Group	Uterine Position	N	AgD ± S.D. (cm)	Weight ± S.D. (g)
NC1	F	5	2.9 ± 0.1	6.5 ± 0.1
NC2	FFFM	6	2.6 ± 0.2	6.2 ± 0.1
NC3	FFM	14	3.0 ± 0.1	6.0 ± 0.1
NC4	FM	36	3.1 ± 0.1	6.2 ± 0.1
NC5	MFM	14	3.3 ± 0.1	6.1 ± 0.2
Sch1	F	7	3.0 ± 0.0	
Sch2	FFFM	4	2.9 ± 0.0	4.9 ± 0.1
Sch3	FFM	6	3.0 ± 0.0	4.8 ± 0.2
Sch4	FM	15	3.0 ± 0.0	4.5 ± 0.4
Sch5	MFM	5	3.0 ± 0.0	4.5 ± 0.1

*For each row, the means refer to the females designated F in the
Uterine Position column. N: no. of experimental females per group;
NC: pups from normal control mothers; Sch: pups from mothers
treated with antiandrogen; F: female fetus; M: male fetus

groups; this increased body size may have accounted for the larger
anogenital distance. Presumably the body weight of these females
was greater because they came from very small litters. Body weight
did not vary significantly across the litters containing males.
Treatment of the mother with antiandrogen eliminated the relationship between anogenital distance and proximity to a male fetus
(Sch groups, Table 2).

The mean mount frequencies of females from known uterine
positions are given in Table 3 for offspring of mothers not
treated with antiandrogen. Behavior testing was not conducted
on female offspring of females given the antiandrogen. Because
of the high mortality rates in our caesarean-delivered females,
we lack enough behavioral data to warrant any far-reaching
conclusions. However, what data we have tantalizingly suggest that
the closer a female was to a male during uterine development, the
higher the level of mounting achieved in the four-week test period

TABLE 3. *Relation of Mean Mount Frequency in the Female Rat Relative*
         to the Uterine Location of Male Sibs*

NC Group	Uterine Position	N	Mean Mount Frequency
1	F	3	1.9
2	FFFM	3	2.0
3	FFM	6	5.8
4	FM	12	5.1
5	MFM	3	10.4

*Abbreviations as in Table 2

(Table 3). Females that had developed between two males in utero
achieved the highest mean frequencies of mounting; whereas females
that had developed in a uterus containing no males or were at least
two females distant from a male mounted the least. The mounting
scores of females that had developed next to one male or only one
female removed were intermediate. The differences reported in
Table 3 are not statistically signficant because of the small number
of animals involved; however, they are consistent with the data
on anogenital distance and with the theory that adult mounting
behavior is facilitated by exposure to androgens during the period
of sexual differentiation.

The fetal male hormones probably reach the female fetus in
these single uterine horn preparations by way of diffusion across
the amniotic membranes of neighboring fetuses. This process would
explain the graded effect in relation to increased distance from
the male. Studies using intra-amniotic injections of labelled
steroid have also demonstrated the diffusion of steroids across the
amniotic membranes (Fels and Bosch, 1971).

The data from this study of "wombmates" suggest that adult
male sexual behavior in the female rat depends upon exposure to
androgen during early development. This probability is also supported by other studies which show that treatment of the pregnant rat
with antiandrogens reduces the level of male sexual behavior in
male and female offspring when they reach adulthood (Nadler, 1969;
Ward and Ranz, 1972). The difference between rats and hamsters in
this regard may be a consequence of species differences in development. The onset of sexual differentiation, indicated by the time
of testicular differentiation, is 11.5 days of postcoital age for
the hamster and 13.0 days for the rat (Price and Ortiz, 1965);
thus the female rat spends nine days in a uterus in which her male
sibs are secreting androgen. The female hamster, however, whose
birth occurs 4.5 days after the onset of testicular activity, may

escape any significant exposure to the testicular secretions of her sibs. Additional evidence indicates that even if the female hamster is exposed to testosterone prenatally, she may not be competent to respond to it (Nucci and Beach, 1971).

STEROID SPECIFICITY IN THE DIFFERENTIATION OF MALE SEXUAL BEHAVIOR

The fact that testosterone is readily metabolized to other steroid hormones including estradiol (Dorfman and Ungar, 1965; Ryan et al., 1972; Naftolin et al., 1971a, b) poses a problem in determining how testosterone achieves its effect. Are the effects of testosterone treatment due to testosterone itself or to some metabolite of testosterone? Recent work by McDonald and his co-workers (1970) indicates that androgens which are not aromatized to estrogen do not induce male sexual behavior in long-term castrated rats. Consequently, the conversion of testosterone to estrogen may play an important role in the action of testosterone on male sexual behavior. Consistent with this suggestion are the findings of Whalen and Luttge (1971) that androgens which are not aromatized to estrogen fail to maintain copulatory behavior in recently castrated male rats. Further support for this observation comes from the findings of both Davidson (1969) and Pfaff (1971) that estradiol benzoate facilitates male sexual behavior in the rat.

To determine whether the conversion of testosterone to estrogen is important in the sexual differentiation of male sexual behavior, we treated newborn hamsters with various androgens and estrogens and then tested them for male sexual behavior as adults. Figure 1 shows the level of mounting behavior achieved by male hamsters castrated on the day of birth and treated on Days 2 through 4 with various androgens. As adults, these animals were tested for mounting by being placed with a sexually receptive female for ten minutes. Each animal was tested five times: once before the onset of daily testosterone propionate replacement therapy and then 7, 14, 21, and 28 days after the onset of daily injections of 300 µg TP.

Those animals treated neonatally with dihydrotestosterone (DHT) or androsterone (neither of which is aromatized to estrogen), the control vehicle oil, or sodium propionate failed to show appreciable levels of male sexual behavior as adults. Animals treated with TP or the alcohol form of testosterone (T) all showed mounting behavior.

Since these data were consistent with the idea that conversion of androgens to estrogen is important for masculinization, we also tested animals that had been castrated and treated neonatally with estrogens (Fig. 2). Treatment with 25 µg estradiol, diethylstilbestrol, or estradiol benzoate for three days neonatally induced the potential for high levels of mounting behavior when the adults were treated with TP. The 2 µg dosage of estradiol in the neonate failed

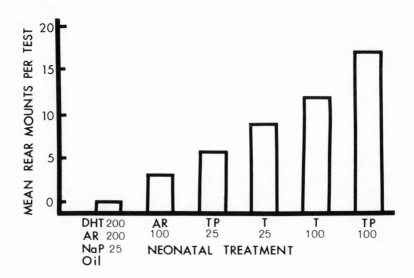

Fig. 1. Mean mounting frequency of Day 1 castrated male hamsters as a function of neonatal treatment with various androgens. Bars represent mean mount frequencies after 28 days of TP treatment as adults. Subscript numbers refer to daily dosage in µg on postnatal Days 2-4. DHT: dihydrotestosterone; AR: androsterone; NaP: sodium propionate; TP: testosterone propionate; T: testosterone (alcohol form). (after Coniglio, Paup, and Clemens, 1973)

to induce mounting, but 2 µg of diethylstilbestrol or estradiol benzoate were effective.

Thus testosterone and estrogens induced behavioral masculinization, but DHT and androsterone did not. Since the latter forms of androgen are not aromatized to estrogen, estrogen metabolites of testosterone may be the intermediates of testosterone action.

Results of tests with female hamsters treated neonatally with the various androgens or estrogens parallel those with the male: estrogens and testosterone, as well as androstenedione, induced the potential for male sexual behavior; whereas DHT and androsterone did not. However, the latter steroids were not entirely without effect; morphologically they had potent masculinizing effects. Figure 3 shows the average length of the clitoral bone and cartilage for female hamsters treated neonatally with androgens or estrogens. All of the androgens significantly increased the length of the clitoral bone; whereas the measure for the estrogen groups did not differ significantly from those of the oil controls.

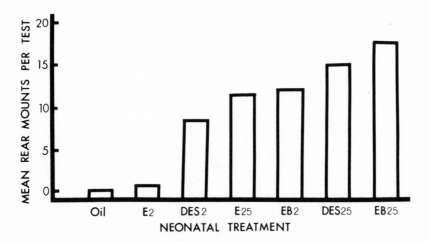

Fig. 2. Mean mounting frequency of Day 1 castrated male hamsters
as a function of neonatal treatment with various estrogens. Bars
represent mean frequencies after 28 days of TP treatment. Subscript
numbers refer to daily dosage in μg on postnatal Days 2-4. E: estradiol; DES: diethylstilbestrol; EB: estradiol benzoate. (after
Coniglio, Paup, and Clemens, 1973)

     Figure 3 also indicates that genital morphology is dissociated
from behavior. The shaded columns represent groups which failed
to show mounting behavior; yet these groups had significantly greater
genital virilization than the estrogen-treated animals which mounted
at high frequencies. A similar dissociation was found between
measures of anogenital distance (AgD) and male sexual behavior.
All the androgens increased the AgD, but only the females treated
with testosterone or androstenedione mounted. This dissociation
between behavioral and morphological measures suggests that the
behavioral effects of our neonatal treatment are a result of changes
in the central nervous system rather than in the genitalia.

INHIBITION OF TESTOSTERONE ACTION

     Thus far we have demonstrated that estrogens and androgens
which can be aromatized to estrogen (testosterone and androsten-
dione) are equally effective neonatally in inducing the potential
for masculine behavior. The data do not necessarily prove that
testosterone must be converted to estrogen in order to be active.
They may merely indicate that the hormone receptors for sexual
behavior are sufficiently nonspecific as to be activated by estrogens,
testosterone, and androstenedione but not by DHT and androsterone.
To resolve this problem, we treated neonatal female hamsters with

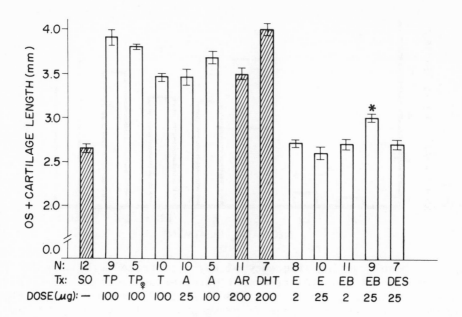

Fig. 3. Relation of neonatal treatment with various steroids to clitoral bone (OS) and cartilage length in female hamsters. Shaded columns denote groups that failed to show male sexual behavior. SO: sesame oil; TP: testosterone propionate; T: testosterone; A: androstenedione; AR: androsterone; DHT: dihydrotestosterone; E: estrogen; EB: estradiol benzoate; DES: diethylstilbestrol; *: group received 21 days of TP treatment just before sacrifice. (from Paup, Coniglio, and Clemens, in press)

testosterone and at the same time with compounds whose pharmacological actions inhibit testosterone conversion to estrogen. If testosterone must be converted to estrogen, then these substances should prevent masculinization of behavior by testosterone.

Barbiturates and steroids share similar if not the same metabolic pathways (Conney and Burns, 1962). Some investigators have even suggested that barbiturates serve as alternate substrates for the enzymes which normally metabolize steroids. For example, pretreatment with barbiturates increases testosterone hydroxylation (Conney and Klutch, 1963), presumably as a result of enzyme induction by the barbiturate (Kuntzman, 1964). Similarly, pretreatment with testosterone enhances barbiturate metabolism as measured by sleep time in response to a standard dose of barbiturate (Conney and Burns, 1962).

This enhanced metabolism of steroid as a result of pretreatment with barbiturate is only one aspect of a two-phase process. The effect of barbiturates on drug and steroid metabolism is apparently biphasic. For the first 12 to 24 hours after treatment, barbiturates inhibit hydroxylation possibly by sequestering the hydroxylase for itself. Once the barbiturate has cleared, however, the enzymes are apparently available at increased levels. Whatever the mechanism involved, 24 to 48 hours after barbiturate treatment metabolism proceeds at an increased rate (Anders, 1971).

Another compound noted for its ability to block hydroxylation is SKF 525A, which apparently uses the same enzyme systems as barbiturates and steroids. It also has a biphasic effect: an initial inhibition of metabolism followed 18 to 24 hours later by an enhancement of hydroxylation (Anders, 1971). Since SKF 525A is without the anesthetic properties of pentobarbital, any blocking effect is not easily attributable to neural depressant action.

To test whether testosterone must be converted to estrogen to achieve behavioral masculinization, we treated neonatal female hamsters with testosterone and either pentobarbital or SKF 525A. We anticipated that these compounds would delay hydroxylation of testosterone and consequent aromatization for at least 24 hours. Since aromatization could occur after that time, we had to administer the testosterone at the end of the developmental period for masculinization. In a previous experiment (Coniglio and Clemens, unpublished observation), we determined that postnatal Days 3 and 4 are the last days in which behavioral masculinization can be achieved with daily injections of 100 μg of testosterone. Thus our experimental females were treated on Days 3 and 4 of postnatal development with 100 μg of testosterone given subcutaneously, in combination with 0.5 mg pentobarbital, 70 μg SKF 525A, or the saline vehicle given intraperitoneally. Whereas some conversion of any remaining testosterone might have occurred beginning 24 hours after the Day 4 injection, this point would be Day 5 of postnatal development and therfore after the time that masculinization could be achieved.

The neonatally treated females were ovariectomized as adults and tested for mounting behavior after daily administration of TP. Both pentobarbital and SKF 525A reduced the levels of masculinization achieved by testosterone treatment (Fig. 4). In both of these groups, the mount frequency during TP treatment was below that of the control group, which had been treated with testosterone and saline. The low mounting scores of the pentobarbital and the SKF 525A groups suggest that prevention of testosterone metabolism reduces its masculinizing influence.

When combined with our findings of the effects of diverse androgens on mounting behavior, these results on inhibition of

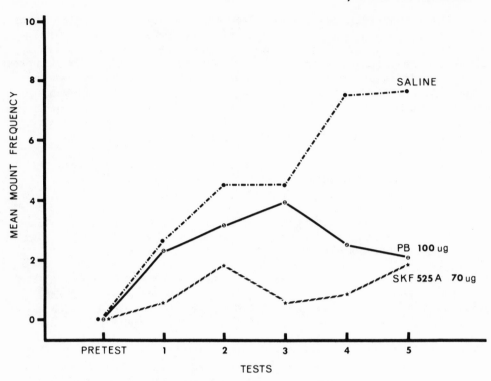

Fig. 4. Influence of pentobarbital (PB) and SKF 525A on neonatal treatment of female hamsters with testosterone. Each group was treated with 100 μg testosterone on Days 3 and 4 of postnatal development, accompanied by treatment with saline, PB, or SKF 525A.

testosterone metabolism argue strongly that differentiation of the potential for male sexual behavior is achieved by some estrogen metabolite of testosterone. Barbiturates have also been shown to block the effect of testosterone but not of estrogen on hypophyseal-gonadal function when the testosterone is administered within 12 hours of the barbiturate treatment, i.e. during the metabolic inhibitory phase (Sutherland and Gorski, 1972). In addition, McDonald and Doughty (1972) have shown that when treatment with MER 25, a potent antiestrogen, preceded neonatal TP treatment in the rat, female rats failed to develop the anovulatory syndrome characteristic of "androgenized" females (Barraclough, 1961).

STEROID SPECIFICITY AND MALE SEXUAL BEHAVIOR IN THE ADULT

We extended our analysis of diverse steroids and male sexual behavior by comparing the effectiveness of four androgens in maintaining male sexual behavior of hamsters castrated as adults (Christensen et al., 1973). These sexually experienced males received 500 µg daily of testosterone, androstenedione, DHT, or An and were tested every two weeks for sexual behavior. The effectiveness of the experimental treatments on mounting behavior is shown in Figure 5.

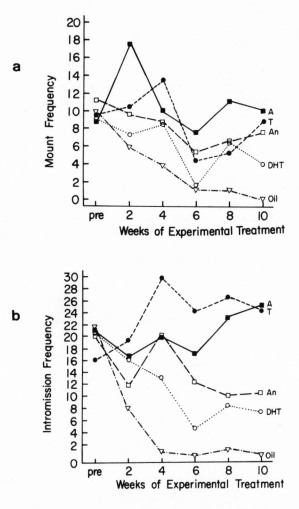

Fig. 5. a) Influence of daily androgen treatment on mounting behavior of castrated male hamsters. b) Influence of daily androgen treatment on intromission frequency in castrated male hamsters. Androgen therapy was initiated immediately after castration of the sexually experienced males. A: androstenedione; T: testosterone; An: androsterone; DHT: dihydrotestosterone. (from Christensen et al., 1973)

Only testosterone and androstenedione maintained intromission frequency at a level significantly above the oil control. Analysis of repeated measures indicated that over the entire treatment period testosterone and androstenedione induced significantly more mounting than androsterone and DHT. At the 10-week test, DHT animals were not significantly different from oil controls; however, all other groups were mounting at significantly higher frequencies than the controls. Thus, testosterone and androstenedione were highly effective in maintaining both intromission and mounting behavior; whereas androsterone and DHT were far less potent. However, androsterone-induced mounting levels were higher than those of controls; and the rate of decrease for DHT animals was slower than for oil controls, an indication of some degree of behavioral maintenance.

Gross histological examination of the penis indicated that the length of the penile papillae did not differ for the four androgen-treated groups. This finding suggests that maintenance of genital morphology by androsterone and DHT may have facilitated sexual behavior. On the other hand, DHT and androsterone may have had weak central actions that facilitated male sexual behavior.

The experimental paradigm in which copulating animals are castrated and maintained on a particular steroid optimizes the possibility of facilitatory effects from genital feedback since animals continue to copulate after castration and thereby achieve genital contact. An alternative to this experimental paradigm is one in which males are castrated and experimental steroid treatment is not initiated until the male has ceased copulating. In this instance, we determine whether the experimental treatment can restore or induce copulatory behavior. This restoration paradigm prevents facilitation of copulation by genital stimulation derived from the female until after the male begins to mount. Using this procedure, we compared the effect of androsterone and testosterone on male sexual behavior in males that had been castrated at 30 days of age. Behavior testing was initiated at 60 days of age, and hormone treatment began after four tests in which baseline mounting frequency was established. Treatment with androsterone induced some mounting behavior after four weeks of daily administration of 300 µg (Fig. 6). However, testosterone was significantly more effective than androsterone in inducing male sexual response and thus more effective than androsterone in inducing and maintaining mounting behavior. The observed effect of androsterone, however, would not be expected if mounting depended entirely upon estrogen metabolites because there is no known pathway by which androsterone is converted to estrogen.

Additional evidence suggests that estrogen is not the active metabolite that supports male sex behavior (Whalen and Battie, 1972). The antiestrogen CI-628 (CN 55, 945-27; Parke, Davis) blocked the testosterone induction of lordosis in the adult rat but failed consistently to decrease male sexual behavior in four males treated

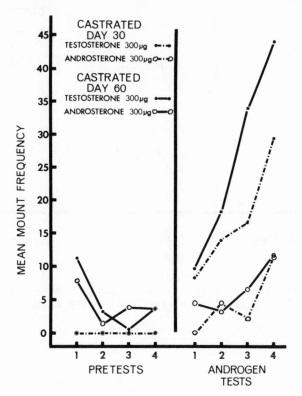

Fig. 6. Influence of daily androgen treatment on mounting behavior in long-term castrated male hamsters.

daily for eight weeks with 300 µg TP plus 1 mg CI-628.

    With these considerations in mind, we next investigated the influence of estrogen itself on male sexual behavior. Male hamsters were castrated at postnatal Days 1, 10, 30, or 50. To insure that males castrated on Day 1 would have a potential for mounting, they were given 300 µg TP on Day 4 of postnatal life; this treatment is known to induce the potential for mounting (Carter et al., 1971). As adults, the males in all groups were then given four pretests to determine baseline mounting. After this, each male was treated with 6 µg/day of estradiol benzoate (EB) for four weeks and tested for mounting 7, 14, 21, and 28 days after the initiation of estradiol treatment. After the 28-day test, two groups (Days 1 and 10) were tested for an additional three weeks without hormone to be certain that the earlier increases in mounting were the result of EB and not of repeated testing. The results of these tests are shown in Figure 7. Daily injections of 6 µg of estradiol benzoate resulted

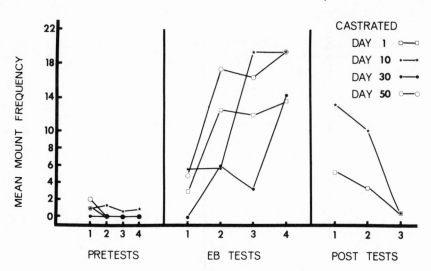

Fig. 7. Influence of daily estradiol benzoate (EB) treatment
(6 μg/day) on mounting behavior of 4 groups of castrated male
hamsters. Day 1 castrates were given 300 μg TP on Day 4 of postnatal
life.

in high levels of mounting for all groups. Upon cessation of EB
treatment, mounting decreased over the next three weeks. Thus 6 μg
EB/day were highly effective in inducing male sexual behavior in
the hamster. However, comparison of the mounting scores obtained
with EB with those obtained with TP treatment shows that the levels
of mounting are higher for the 300 μg TP groups than for the 6 μg
EB groups.

HORMONE SPECIFICITY IN THE ANDROGENIZED FEMALE HAMSTER

If the conversion of testosterone to estrogen in an important
step in testosterone action, estrogen should be more potent than
testosterone, and androsterone, which is not converted to estrogen,
should be ineffective. But as we have seen, 6 μg doses of EB induced mounting, but not as effectively as 300 μg of testosterone.
Moreover, the fact that androsterone also promoted male sexual
behavior is at variance with a conversion hypothesis.

Several explanations are available. (1) Although in the system
tested thus far androsterone is not known to have estrogen metabolites, such conversion might occur in the hamster after 28 days of

300 µg/day. (2) Androsterone itself may have a central action upon sexual behavior. (3) Androsterone treatment may induce secretion of active agents from the adrenal gland. (4) the dose of estrogen used in the present study may have been too small to achieve a meaningful comparison with 300 µg TP. (5) The virilizing effects of androsterone, as well as of testosterone, upon genital morphology may have facilitated male sexual behavior.

If this last explanation is correct, then androsterone and testosterone should be less effective in the androgenized female since she is far less virilized morphologically than the male preparations we have thus far tested. Therefore, we compared the effects of androsterone with testosterone and estrogen on the androgenized female hamster.

After being injected with 25, 75, or 300 µg TP on Day 4 of postnatal life, females were tested for mounting behavior as adults. The tests for mounting behavior were preceded by ovariectomy and tests for female sexual behavior after treatment with estrogen and progesterone. Results of the lordosis tests will be reported elsewhere.

The tests for mounting were divided into four phases. In Phase 1, the females in each of the three neonatal TP dosage groups were tested for mounting behavior without any hormone therapy. In Phase 2, the females in each of the three TP groups were subdivided into two groups and treated daily with 6 µg of EB or 6 µg of diethylstilbestrol (DES) for 30 days. Periodic tests were made to determine the effects of these estrogens upon mounting. After the last test at the end of the 30-day treatment, the females were treated daily with sesame oil (Phase 3) for 20 days and tested for mounting. In Phase 4, the females from each neonatal TP group were again randomly subdivided into two groups without regard to their previous EB or DES treatment. These subgroups were then treated with either testosterone or androsterone for 28 days (300 µg/day).

The results of these tests are summarized in Figure 8. In Phase 2, there was no significant difference between the mounting frequencies of EB or DES females. Therefore, to facilitate the comparisons of estrogen and androgen treatments, we pooled the scores for EB and DES. Figure 8 shows which groups received testosterone and which received androsterone in Phase 4. Androsterone failed to maintain mounting freqeuncy at a level significantly higher than in the control tests of Phase 1. For all groups, 6 µg daily of estrogen (EB or DES) were more effective in inducing and maintaining mounting than 300 µg of either androgen. That estrogen is more effective than testosterone was also shown by an additional group of androgenized females which were treated daily with 6 µg of testosterone propionate. These females failed to mount over a four-week test period.

These findings with the androgenized female contrast with those

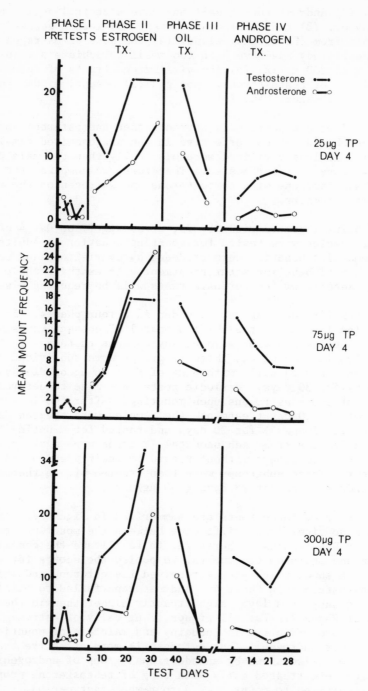

Fig. 8. Influence of estrogen and androgen treatment on mounting behavior of androgenized female hamsters; see text for details.

for the male. Although estradiol benzoate induced mounting in the
male hamster, EB treatment of the androgenized female induced higher
levels of mounting and was clearly more potent than testosterone or
androsterone. These data are consistent with the idea that in addition to any central neural action, androgen may facilitate mounting
in the male by acting on genital morphology. This theory is supported by the finding that male rats treated with 5 μg EB alone or
1 mg of DHT alone achieved mounts or intromissions only rarely;
whereas when these treatments were combined, high levels of male
sexual behavior resulted (Larsson et al., 1973). On the other hand,
this outcome might be the result of the increased amount of steroid
given when EB and DHT treatments were combined.

Another possibility is that estrogens and androgens are metabolized differentially by the male and female. Work with rats shows
that male rats inactivate estrogen more rapidly than females (Jel-
liniks and Lucieer, 1965) and that female rats inactivate testosterone more rapidly than males (Ota et al., 1972). Consequently, the
differences in behavior between males and females treated systemically with testosterone or estradiol may reflect sex differences in
the inactivation of estrogens and androgens rather than sensitivity
to these hormones per se. To control for these factors, other
techniques must be used in evaluating the relative contributions
of central versus peripheral hormone factors.

*COMPARISONS OF ESTRADIOL AND TESTOSTERONE BRAIN IMPLANTS
IN THE INDUCTION OF MALE SEXUAL BEHAVIOR IN THE RAT*

To circumvent the difficulties encountered by sex differences
in the hepatic inactivation of estrogens and androgens when these
steroids are injected systemically, we compared the effectiveness
of estradiol and testosterone after direct implantation into the
brain. This technique also limits the amount of hormone reaching
the genital tissues.

In our first study, Larry Christensen and I compared the effects
of brain implants of crystalline estradiol and testosterone on male
sexual behavior in long-term castrated rats. These animals were
selected for their readiness to copulate in the laboratory and then
castrated and allowed several weeks to rest until they no longer
copulated with a sexually receptive female. At this point, they
were implanted bilaterally with cannulae in the preoptic area (POA)
of the brain. This area was selected because testosterone propionate
implants here will reinstate the sexual behavior of castrated male
rats (Davidson, 1966; Lisk, 1967). Control implants were made in the
posterior hypothalamus. The double-walled cannulae that had been
implanted contained removable inner tubes (inserts) made of 26-gauge
stainless steel tubing. With these inserts, small pellets of crystalline hormone (30 μg testosterone: 15 μg from each cannula; 20 μg

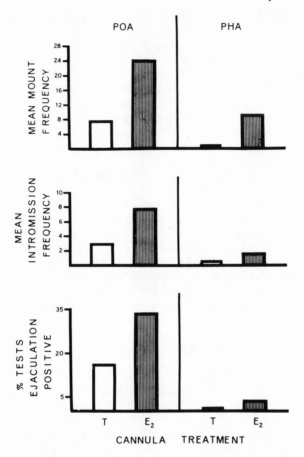

Fig. 9. Influence of bilateral implants of testosterone (T) or estradiol (E₂) in either the preoptic area (POA) or the posterior hypothalamus (PHA) or 3 measures of male sexual behavior in the rat.

estradiol: 10 µg from each cannula) were expelled into the implantation site every three days. Animals were then tested 7, 10, and 13
days after commencement of hormone treatment. The results of these
tests are shown in Figure 9. Estradiol implants in the POA resulted
in significantly greater frequencies of mounts, intromissions, and
ejaculations than implants of testosterone even though less estrogen
was implanted than testosterone. Implants in the posterior hypothalamus (PHA) were far less effective than POA implants.

That some hormones from these implants reached systemic circulation was shown by the fact that in testosterone-implanted
animals, the penile papillae showed signs of androgenic stimulation.

However, since there was no difference between the penes of POA and
PHA animals, and since systemic injections of 25 µg/day of testosterone failed to induce male sexual behavior, it is doubtful whether
the changes in behavior in testosterone-implanted animals were due
to systemic hormone influences.

The greater potency of estradiol over testosterone is consistent
with the hypothesis that testosterone must be converted to estrogen.
However, to challenge this idea more directly, we designed a second
experiment to block conversion of testosterone to estrogen in the
POA. Long-term castrated male rats were implanted with cannulae
in the POA and were twice daily infused with 300 µg of metapirone,
which blocks the conversion of testosterone to estrogen (Giles and
Griffith, 1968). One group of metapirone animals was treated systemically with 150 µg of testosterone, another with 50 µg of estradiol;
a control group was treated systemically with testosterone and
infused with sucrose. If conversion to estrogen is a necessary step
in the action of testosterone, metapirone should prevent testosterone
induction of male sexual behavior without affecting estradiol. The
results are summarized in Figure 10.

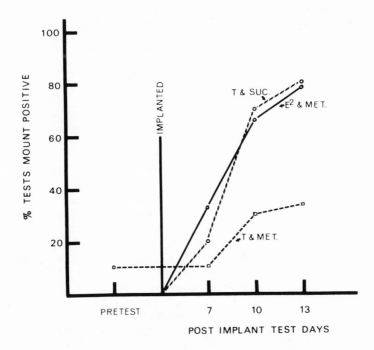

Fig. 10. Influence of brain implants of metapirone (MET) on male
sexual behavior in rats treated systemically with either testosterone
(T) or estradiol (E); SUC: sucrose.

The copulatory scores of the testosterone animals were reduced by metapirone; whereas the estradiol-treated males mounted at the same level as the testosterone-sucrose group. From these data, we suggest that the conversion of testosterone to estrogen within the POA is an important if not an essential step in the induction of male sexual behavior by testosterone. This conclusion is also supported by the work of Naftolin and his colleagues (Ryan et al., 1972; Naftolin et al., 1971a, b; Reddy et al., 1973) who demonstrated conversion of androstenedione to estrone in the central nervous system. The conversion hypothesis is further corroborated by Johnston and Davidson (1973), who found that POA implants of dihydrotestosterone failed to induce copulatory behavior in the castrated male rat. Despite its potency, dihydrotestosterone cannot be aromatized to estrogen.

PERIPHERAL ANDROGEN ACTION AND ITS INFLUENCE ON MALE SEXUAL BEHAVIOR IN THE RAT

Up to this point our work has been consistent with the theory that central activation and induction of male sexual behavior requires the conversion of testosterone to estradiol. However, we have been concerned that at least part of the total effect of androgen on male sexual behavior may result from action on genital tissues.

Historically, the theory that male sexual behavior is the result of peripheral excitatory stimuli is probably much older than our theories concerning central neurohormonal determinants of sexual behavior. A few quotations will supply a thumb-nail review of theories about peripheral determinants of sexual behavior.

> (The penis) confers with the human intelligence and sometimes has intelligence of itself, and although the will of the man desires to stimulate it it remains obstinate and takes its own course, and moving sometimes of itself without license or thought by the man, whether he be sleeping or waking, it does what it desires; and often the man is asleep and it is awake, and many times the man is awake and it is asleep; many times the man wishes it to practise and it does not wish it; many times it wishes it and the man forbids it. It seems therefore that this creature has often a life and intelligence separate from the man, and it would appear that the man is in the wrong in being ashamed to give it a name or to exhibit it, seeking rather constantly to cover and conceal what he ought to adorn and display with ceremony as a ministrant. (Leonardo da Vinci, 16th century)[1]

[1] Quotation taken from *The Literary Works of Leonardo da Vinci*, Oxford University Press (London, New York, 1939.

To the use of the sexual organ for the continuance of his
race, Man is prompted by a powerful instinctive desire,
which he shares with the lower animals. This instinct,
like the other propensities, is excited by sensations;
and these may either originate in the sexual organs themselves, or may be excited through the organs of special
sensation. (Carpenter, 1852, p. 956)

The original stimulus for sex responses is....an internal
excitant. In the male it is the gradual distention of
the seminal vesicles, a condition requiring a fairly
periodic discharge of their contents. The distention
produces an increase of tonicity in the wall of the vesicle, and this internal activity, combined, no doubt,
with similar glandular effects....stimulates the interoceptive end organs in these parts. In the female the
excitatory visceral changes are probably caused, not by
distention, but by some hormonic (glandular) process occurring about the time of menstruation. The response which
follows this stimulus consists of random and restless
activity quite analogous to that in the case of hunger.
(Allport, 1924, p. 69)

Some organ, other than the gonads, but in structural
and functional dependence on the testes, effects a tu-
mescense or tension in itself or in another tissue or
organ; this tension or tumescence initiates the afferent
impulses which stimulate sex activity. (Nissen, 1929,
p. 526)

We believe that the effects of castration upon sexual behavior in the male rat are due in some measure to lowering
of tactile sensitivity in the glans penis as a result of
deterioration in the genital papillae. We do not, however,
imagine that this is the only nor even the most important
avenue of hormonal action....It is obvious....that the
hormone (androgen) exerts its effects in a number of ways
and more central parts of the nervous system undoubtedly
are involved. Nevertheless, the alteration in peripheral
sensitivity may very well be one of the several hormonally
controlled changes which contribute to the behavioral function. (Beach and Levinson, 1950, p. 167)

....sexual differentiation of the neuromuscular substrate
for male and female sexual behavior in the rat, under the
influence of androgen, may be viewed as two distinct events.
Facilitation of the substrate for male sexual behavior
begins prenatally and presumably through an action on a
peripheral structure. Facilitation continues during early
postnatal development at the same locus. Inhibition of

female sexual behavior occurs primarily during the postnatal period at a central locus. (Nadler, 1969, p. 62)

        Statements such as these make it difficult to dismiss the pos-
sibility that hormones effect some interoceptive stimulus which
brings about or at least facilitates sexual activity. In a previous
section, we reported a dissociation between genital virilization and
adult male sexual activity in the female hamster as a result of neonatal treatment with androgen. However, the range of virilization
produced by such neonatal treatments was narrow; and the extent of
virilization is far short of that seen even in the male hamster castrated at birth. Hence failure to see a correlation between behavior
and morphology may reflect only the limited range of morphological
virilization achieved by our treatments.

        To estimate quantitatively peripheral androgenic influences on
male sexual behavior, we measured sexual activity in castrated male
rats that were being treated daily with the androgen fluoxymesterone
(FM). This androgen is reported not to influence mental erotic
imagery in Man, even though it exerts pronounced virilizing effects
upon genital tract tissues (Reilly and Gordon, 1961). This suggests
that its action is limited to peripheral genital tissues. In addition, fluoxymesterone does not reinstate sexual behavior in long-
term castrated male rats (Beach and Westbrook, 1968). In our experiments, Archie Vomachka and I castrated male rats and placed them
on TP replacement therapy. During this time, we measured their male
sexual activity and then switched half of the males to fluoxymesterone therapy. Over a nine-week test period, both intromission frequency and ejaculation frequency declined under the influence of FM.
However, this decline was not as rapid as that observed in castrated
males treated with the oil vehicle. The main question here is
whether the facilitatory effect of FM was the result of maintaining
penile integrity or of a central influence of the supposed peripheral
androgen. To determine the answer, we sectioned the two branches of
the dorsal nerve of the penis. If the effect of fluoxymesterone
was via the penis, then FM-treated males should show changes in
mounting similar to oil controls. (Sectioning the dorsal nerve of
the penis abolishes intromission and ejaculation and hence the possibility of directly assessing any further the influence of these
measures on FM activity.) When the dorsal nerve of the penis was
sectioned, mount frequency declined at the same rate in FM-treated
animals as in oil-treated intact males; but it declined at a slower
rate in intact FM males. Sectioning the dorsal nerve of the penis in
TP-treated castrates did not affect mounting behavior (Fig. 11).

        Androgens acting on the morphology of the penis apparently fa-
cilitate mount, intromission, and ejaculatory responses; however, this
effect can be eliminated by sectioning the dorsal nerve of the penis.
How this facilitation is effected--whether as a result of tonic feed-

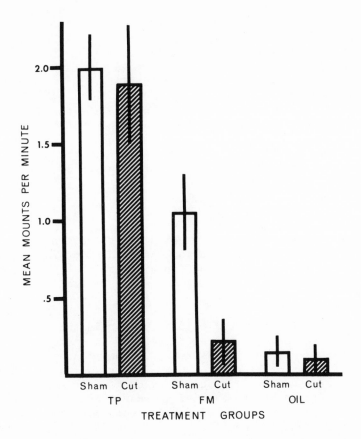

Fig. 11. Comparison of mounting behavior scores of castrated rats
after sectioning of the dorsal nerve of the penis and maintenance on
different androgens. Bars represent mean scores obtained on the last
of 10 tests given at 4-day intervals after inititation of FM (fluoxymesterone) treatment of one group.

back from androgen-stimulated nerve endings or of stimulation derived
from copulation--is not known.

Regardless of the precise mechanism involved, we can return to
our original question: Are estrogen metabolites the active agents
in the control of male sexual behavior? We have presented experiments
to show that prevention of testosterone metabolism in the preoptic
area of the brain does reduce its potency; however, the present experiment indicates that androgens like FM, which cannot be converted
to estrogen, facilitate male sexual behavior and that this facilitatory effect can be eliminated by removing the sensory input from

the penis. The idea that emerges from this work as well as from
studies conducted in other laboratories is that gonadal hormones
acting at several sites increase the probability of sexual responses.
Action in the preoptic, anterior hypothalamic region is a necessary
condition for male sexual behavior. At the same time, hormonal
action on the penis provides additional facilitation. According to
our results, the active hormonal agent may not be the same in each
case. So far, we have not been able to specify the essential characteristics of this activation. However, we suggest that facilitation
of male sexual behavior at the level of the preoptic area requires
a carbon ring structure with a saturated A ring and that facilitation
via genital structures demands a C 19 configuration.

SUMMARY AND CONCLUSIONS

The studies reported here were conducted to determine the effectiveness of various steroids in inducing and maintaining male
sexual behavior. They were designed to distinquish between hormone
effects on behavior which are mediated via the central nervous system
and those which are mediated via the genital structures.

Studies of sex differentiation of behavior in the hamster have
shown that testosterone, as well as estradiol and diethylstilbestrol,
given around the time of sexual differentiation induces the potential
for male sexual behavior. Despite their failure to establish a
similar potential, androsterone and dihydrotestosterone do induce
genital virilization equal to that of testosterone. These results
suggest that the changes in behavioral sex potential resulting from
testosterone treatment are not due entirely to changes in genital
morphology. Moreover, these data are consistent with, but do not
prove, the idea that to induce male sexual behavior, testosterone
must first be converted to estrogen. When testosterone was administered with compounds which prevent its hydroxylation, its masculinizing potential was markedly reduced.

In the adult hamster, male mounting behavior can be induced by
estrogens as well as by testosterone; in the androgenized female hamster estrogen is clearly more potent than testosterone for activating
this sexual response. Both androsterone and dihydrotestosterone facilitate male sexual behavior in the male hamster, but they are less
potent than testosterone and cannot maintain male sexual behavior in
the androgenized female. Whether they affect the adult male by direct central activity or via the genital morphology is not yet clear.

When the effects of intracerebral implants of estradiol and of
testosterone in the male rat were compared, the former were more
potent than the latter in promoting male sexual behavior. Moreover,
the effects of systemically administered testosterone were decreased
by the intracerebral infusion of compounds which block the conversion

of testosterone to estradiol. However, these same compounds did not decrease the effectiveness of estradiol.

When castrated male rats were treated with fluoxymesterone, male sexual behavior was facilitated. However, this compound probably does not have central action on male sexual behavior because, when the dorsal nerve of the penis was sectioned, the facilitatory effect of fluoxymesterone was abolished. Thus, androgens may affect male sexual behavior via genital morphology rather than via central activation.

REFERENCES

Allport, F., 1924, *Social Psychology*, Houghton, Mifflin, Boston.

Anders, M. W., 1971, Enhancement and inhibition of drug metabolism, Ann. Rev. Pharm. 11:37-56.

Barraclough, C., and Gorski, R. A., 1961, Evidence that the hypothalamus is responsible for androgen induced sterility in the female rat, Endocrinology 68:68-79.

Beach, F. A., 1948, *Hormones and Behavior*, Hoeber, New York.

Beach, F. A., 1971, Hormonal factors controlling the differentiation development, and display of copulatory behavior in the ramstergig and related species, in *Biopsychology of Development* (E. Tobach, L. Aronson, and E. Shaw, eds.), pp. 249-296, Academic Press, New York/London.

Beach, F. A., and Levinson, G., 1950, Effects of androgen on the glans penis and mating behavior of castrated male rats, J. Exp. Zool. 114:159-171.

Beach, F. A., and Westbrook, W. H., 1968, Dissociation of androgenic effects on sexual morphology and behavior in male rats, Endocrinology 83:395-398.

Carpenter, W. B., 1853, *Principles of Human Physiology*, 5th Amer. ed., Blanchard and Lea, Philadelphia.

Carter, C. S., Clemens, L. G., and Hoekema, D. J., 1972, Neonatal androgen and adult sexual behavior in the golden hamster, Physiol. Behav. 9:89-95.

Christensen, L., Coniglio, L. P., Paup, D.C., and Clemens, L. G., 1973, Sexual behavior of male golden hamsters receiving diverse androgen treatments, Horm. Behav. (in press).

Clemens, L. G., 1973, Development and behavior, in *Comparative Psychology: A Modern Survey* (D. Dewsbury, ed.), McGraw-Hill, New York.

Clemens, L. G., and Coniglio, L. P., 1971, Influence of prenatal litter composition on mounting behavior of female rats, Amer. Zool. 11:617

Coniglio, L. P., Paup, D. C., and Clemens, L. G., 1973, Hormonal factors controlling the development of sexual behavior in the male golden hamster, Physiol. Behav. 10:1087-1094.

Conney, A. H., and Burns, J. J., 1962, Factors influencing drug metabolism, Adv. Pharmacol. 1:31-58.

Conney, A. H., and Klutch, A., 1963, Increased activity of androgen hydroxylases in liver microsomes of rats pretreated with phenobarbital and other drugs, J. Biol. Chem. 238:1611-1617.

Davidson, J. M., 1969, Effects of estrogen on the sexual behavior of male rats, Endocrinology 84:1365-1372.

Dorfman, R. I., and Ungar, F., 1965, *Metabolism of Steroid Hormones*, Academic Press, New York/London.

Edwards, D., and Burge, K. G., 1971, Estrogenic arousal of aggressive behavior and masculine sexual behavior in male and female mice, Horm. Behav. 2:239-246.

Fels, E., and Bosch, L. R., 1971, Effect of prenatal administration of testosterone on ovarian function in rats, Amer. J. Obstet. Gynecol. 111:964-969.

Giles, C., and Griffiths, K., 1964, Inhibition of the aromatizing activity of human placenta by SU 4885 (Metapirone), J. Endocrin. 28:343-344.

Goy, R. W., Bridson, W. E., and Young, W. C., 1964, Period of maximal susceptibility of the prenatal guinea pig to masculinizing actions of testosterone propionate, J. Comp. Physiol. Psychol. 57:166-174.

Jellicnik, P. H., and Lucieer, I., 1965, Sex differences in the metabolism of oestrogens by rat liver microsomes, J. Endocrin. 32:91-98.

Johnston, P., and Davidson, J. M., 1972, Intracerebral androgens and sexual behavior in the male rat, Horm. Behav. 3:345-358.

Kuntzman, R.,1969, Drugs and enzyme induction, Ann. Rev. Pharm. 9:21-36.

Larsson, K., Sodersten, P., and Beyer, C., 1973, Induction of male
    sexual behavior by oestradiol benzoate in combination with
    dihydrotestosterone, J. Endocrin. 57:563-564

Lisk, R., 1967, Sexual behavior: Hormonal control, in *Neuroendocrin-
    ology* (L. Martini and W. F. Ganong, eds.), Vol II., pp. 197-239,
    Academic Press, New York/London.

McDonald, P. G., and Doughty, C., 1972, Inhibition of androgen-
    sterilization in the female rat by administration of an anti-
    estrogen, J. Endocrin. 55:455-456.

McDonald, P., Beyer, C., Newton, F., Brien, B., Baker, R., Tan, H.S.,
    Sampson, C., Kitching, P., Greenhill, R., and Pritchard, D.,
    1970, Failure of 5 α-dihydrotestoterone to initiate sexual
    behaviour in the castrated male rat, Nature (Lond.) 227:964-965.

Nadler, R., 1969, Differentiation of the capacity for male sexual
    behavior in the rat, Horm. Behav. 1:53-63.

Naftolin, F., Ryan, K. J., and Petro, Z., 1971a, Aromatization of
    androstenedione by limbic system tissue from human festuses,
    J. Endocrin. 51:795-796.

Naftolin, F., Ryan, K. J., and Petro, Z., 1971b, Aromatization of
    androstenedione by the diencephalon, J. Clin. Endocrin. Metab.
    33:368-370.

Neri, S., Florance, K., Kozial, P., and VanCleave, S., 1972, A
    biological profile of a nonsteroidal antiandrogen, SCH 13521
    (4'-nitro-3'-trifluoromethylisobutyranilide), Endocrinology
    91:427-437.

Nissen, H. W., 1929, The effects of gonadectomy, vasotomy and in-
    jection of placental and orchic extracts on the sex behavior
    of the white rat, Genet. Psychol. Monogr. 5:451.

Nucci, L. P., and Beach, F. A., 1971, Effects of prenatal androgen
    treatment on mating behavior of female hamsters, Endocrinology
    88:1514-1515.

Ota, M., Sato, N., and Obara, K., 1972, Sex differences in metabolism
    of testosterone by rat liver, J. Biochem. 72:11-20.

Paup, D. C., Coniglio, L. P., and Clemens, L. G., 1972, Masculiniza-
    tion of the female golden hamster by neonatal treatment with
    androgen or estrogen, Horm. Behav. 3:123-131.

Paup, D. C., Coniglio, L. P., and Clemens, L. G., 1974, Hormonal
    determinants in the development of masculine and feminine

behavior in the female hamster, Behav. Biol. (in press).

Pfaff, D. W., 1970, Nature of sex hormone effects on rat sex behavior: Specificity of effects and individual patterns of
response, J. Comp. Physiol. Psychol. 73:349-358.

Phoenix, C. H., Goy, R. W., Gerall, A. A., and Young, W. C., 1959,
Organizing action of prenatally administered testosterone
propionate on the tissues mediating mating behavior in the
female guinea pig, Endocrinology 65:369-382.

Price, D., and Ortiz, E., 1965, The role of fetal androgen in sex
differentiation in mammals, in Organogenesis (R. S. DeHaan and
H. Ursprung, eds.), pp. 629-652, Holt, Rinehart & Winston, New
York.

Reilly, W. A., and Gordon, G. S., 1961, Dissociation of growth-stim-
ulating and skeleton-maturing action of the synthetic androgen,
fluoxymesterone, J. Pediat. 59:188-193.

Reddy, V. V. R., Naftolin, F., and Ryan, K. J., 1973, Aromatization
in the central nervous system of rabbits: Effects of castration
and hormone treatment, Endocrinology 92:589-594.

Ryan, K. J., Naftolin, F., Reddy, V., Flores, F., and Petro, Z.,
1972, Estrogen formation in the brain, Amer. J. Obstet. Gynecol.
114:454-460.

Sutherland, S., and Gorski, R. A., 1972, An evaluation of the inhibition of androgenization of the neonatal female rat brain
by barbiturate, Neuroendocrin. 10:94-108.

Ward, I. L., and Renz, F. J., 1972, Consequences of perinatal hormone manipulation on the adult sexual behavior of female rats,
J. Comp. Physiol. Psychol. 78:349-355.

Whalen, R. E., Battie, C., and Luttge, W. G., 1972, Anti-estrogen
inhibition of androgen induced sexual receptivity in rats, Behav.
Biol. 7:311-319.

Whalen, R. E., and Edwards, D., 1967, Hormonal determinants of the
development of masculine and feminine behavior in male and
female rats, Anat. Rec. 157:173-183

Whalen, R. E., and Hardy, D. F., 1970, Induction of receptivity in
female rats and cats with estrogen and testosterone, Physiol.
Behav. 5:529-533.

Whalen, R. E., and Luttge, W. G., 1971, Testosterone, androstenedione
and dihydrotestosterone: effects on mating behavior of male

rats, Horm. Behav. 2:117-126.

Young, W. C., 1961, The hormones and mating behavior, in *Sex and Internal Secretions*, 3rd ed., (W. C. Young, ed.), Vol. 2., pp. 1173-1239, Williams & Wilkins, Baltimore.

RELATIONSHIP BETWEEN THE CENTRAL REGULATION OF

GONADOTROPINS AND MATING BEHAVIOR IN FEMALE RATS[1]

Robert L. Moss

Department of Physiology, Southwestern Medical School

The University of Texas Health Science Center at Dallas

INTRODUCTION

Elsewhere in this volume Drs. McCann (pp. 1-21) and Clemens (pp. 23-53) report on the baffling array of hormones and other biologically active substances produced by the hypothalamus, the gonads, and the pituitary gland which are involved in the control of reproductive activities. The causal relationship in the female rat between the discharge of pituitary gonadotropins, the release of ovarian hormones, and the subsequent onset of sexual receptivity is fairly well understood. In the normal cyclic female rat on the morning of proestrus, the secretion of estrogens from the developing graafian follicle reaches peak values and, in turn, acts on the hypothalamus to trigger the ovulatory surge of luteinizing hormone (LH) and follicle-stimulating hormone (FSH) via the release of the hypothalamic releasing factors, LH-releasing factor (LRF) and possible FSH-releasing factor (FRF). The surge in LH and FSH occurs on the afternoon of proestrus; ovulation takes place in the early hours of the following morning (estrus).

A preovulatory surge of progestin, i.e., progesterone and 20·α-dihydroxy-progesterone, occurs shortly after the peak in LH. To become sexually receptive and display normal sexual behavior patterns, the female rat must be under the influence of a high level of estrogen, immediately followed by a high level of progesterone. Thus, the ovarian-hypothalamic mechanism controlling reproductive

[1]This research was supported in part by a United States Public Health Service Grant No. NS-10434 END and in part by an Institutional Research Grant No. 72-1.

behavior is a dual system; not only do the ovarian hormones feedback on the hypothalamus to control the secretion of pituitary gonado- tropins via the releasing factor, but they initiate the overt display of sexual behavior.

With increasing knowledge of how the pituitary gonadotropins interact with the ovarian hormones to control reproductive activities, it has become more and more important to investigate the possibility that coitus exerts a stimulatory action on gonadotropin release and that the gonadotropins and their releasing factors act on sexual receptivity. The present communication was written with two basic objectives in mind: (1) to summarize our own systematic study of how coitus induces the release of gonadotropins in the female rat; and (2) to summarize our recent efforts to determine whether the gonadotropins or releasing factors act to induce lordosis behavior in the ovariectomized female rats.

STIMULATORY ACTION OF COITUS ON SERUM GONADOTROPIN CONCENTRATIONS

Coitus-induced release of gonadotropins in the normal cyclic, proestrous female rat[1]. Under certain conditions, the female rat, a spontaneous ovulator, displays coitus-induced ovulation and subse- quent pregnancy. Thus, reflex ovulation can be demonstrated when the ovulatory surge of LH and FSH is blocked before the critical period by either pharmacological, environmental, or endocrinological factors (Everett, 1952a; Everett, 1952b; Everett, 1964, Aron et al., 1965; Harrington et al., 1966; Zarrow and Clark, 1968; Kalra and Sawyer, 1970). On the afternoon of proestrus, coitus between 5:50 p.m. and 6:30 p.m. increased the number of ova detected on the morn- ing of estrus (Rodgers, 1971). These findings suggest that copula- tion is a sufficient stimulus to cause the release of LH in animals whose normal spontaneous ovulatory surge had been blocked.

A few attempts have been made to verify this hypothesis by measuring the effect of mating on serum gonadotropins in the normal cyclic, proestrous animal. Early investigations of the effects of mating on LH and FSH (Taleisnik et al., 1966; Spies and Niswender, 1971) used either biological or radioimmunoassays and evaluated the action of copulation on the morning of estrus after ovulation had presumably taken place. In subsequent studies, Rodgers and Schwartz (1973) investigated the effects of mating on LH levels on the after- noon of proestrus; Brown-Grant et al. (1973) studied coitus-induced release of LH and FSH in the constant estrous animal; and Moss and Cooper (1973) explored the action of copulation on plasma LH through- out the total sexual receptivity period in the normal cyclic, pro- estrous rat.

[1]Dr. K.J. Cooper was the principal collaborator during this phase of the research.

Here I summarize our work on this topic and include both recent
and previously reported data (Moss et al., 1973; Moss and Cooper,
1973). To investigate changes in the serum concentration of LH in
normal cyclic unmated and mated animals, female rats with at least
two consecutive, four-day cycles were decapitated via the guillotine
method at hourly intervals throughout the proestrous-estrous phase
of the estrous cycle. Before being sacrificed, these females had
been mated with a sexually active male for 15 minutes in a mating
chamber and then given 15 minutes isolation period. Blood samples
collected from unmated control and experimental mated animals were
assayed for LH content by radioimmunoassay according to the method
developed by Niswender et al. (1968)[1]. The reference preparation
was NIAMD-rat LH-RP-1, which has a biologic potency of 0.03 x NIH-
LH-S1 according to the ovarian ascorbic acid depletion assay.

Figure 1 clearly demonstrates that copulation elicits a discharge of LH. Coitus thus appears to be an adequate stimulus to evoke
an increase in serum LH concentration throughout most of the behavioral receptivity period. This finding is corroborated by the
observations of Taleisnik et al. (1966) and Spies and Niswender
(1971) who demonstrated that copulation between 3:00 and 5:00 a.m.
on the morning of estrus results in increased serum LH levels. In
addition, the present findings are in accord with the results of
Rodgers (1971), who showed that ova increased only if intromissions
were received during the hour preceding the onset of the dark cycle,
a time at which we observed a substantial coitus-induced increase in
LH. On the other hand, contrary to our findings (Moss and Cooper,
1973)[2] and those of Rodgers (1971), Rodgers and Schwartz (1973) reported that mating during the afternoon of proestrus failed to elevate serum LH. The discrepancy between these experiments may be due
to the method of collecting blood samples. In the Rodgers and
Schwartz study blood samples were collected by cardiac puncture, in
the others, by the guillotine technique.

Blood samples obtained immediately after copulation showed no

[1]We are indebted to Drs. Rees Midgley (University of Michigan) and
Gordon Niswender (Colorado State University) for supplies of anti-
ovine LH, and to Dr. Leo Reichert of Emory University for the supply
of purified ovine LH for radioiodination.

[2]The animals used in all the experiments reported in this chapter
were Sprague-Dawley rats purchased from Simonsen Laboratories, Gil-
roy, California. They were maintained in individual cages under a
standard laboratory diet, temperature, and light-dark schedule of
14 hours on and 10 hours off, with the dark phase beginning at 2:00
p.m. All times in Figure 1 have been converted to the conventional
light-dark schedule so that our 2:00 p.m. is shown as 7:00 p.m., its
conventional equivalent.

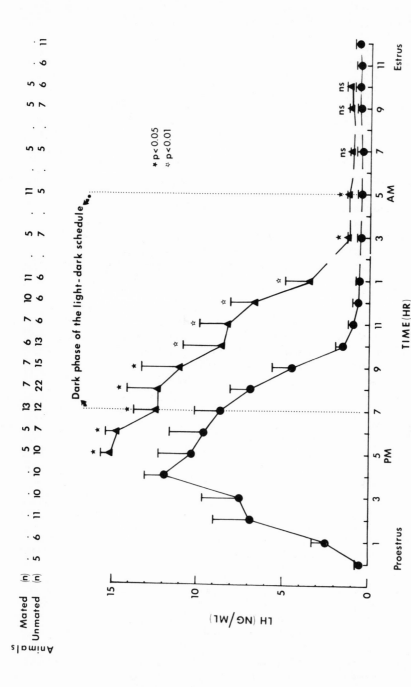

Fig. 1. Profile of serum LH concentration for control unmated (circles) and mated (triangles) cyclic female rats. The mated animals displayed a mean lordosis-to-mount ratio (L/M) 0.84; S.E. ±0.09. The vertical bars represent the standard error of the mean; * = p < 0.05; ✩ = p < 0.01; ns = nonsignificant. (Data modified from Moss and Cooper, 1973; see footnote 2, p. 57.)

significant rise in LH concentration over the levels in the unmated
control animals. A significant elevation was first observed at 10
minutes, and the levels remained elevated for 60 minutes after
coitus.

In a subsequent study, LH as well as FSH and prolactin[1] were
measured by radioimmunoassay and the results are shown in Figure
2. Copulation evoked a significant increase in the serum LH (p <
0.01) and prolactin (p < 0.05) concentrations. No difference in
serum FSH was observed between the unmated control and mated animals.
Serum LH increased at 15, 30, and 60 minutes, but by 90 minutes the
levels had significantly lowered. The serum prolactin concentration increased slightly at 15 minutes as a result of copulation. In

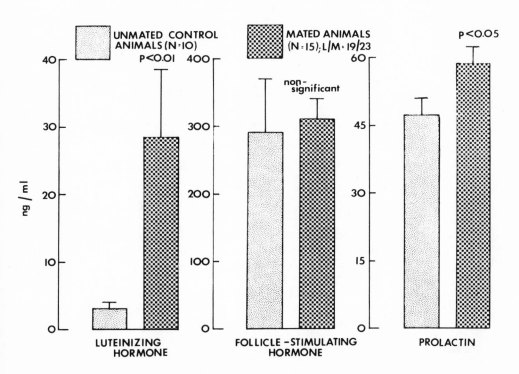

Fig. 2. Changes in serum LH, FSH, and prolactin levels after coitus.
Blood samples were collected between 9:00 and 10:00 p.m. The vertical bars represent the standard error of the mean. (R.L. Moss,
1973, unpublished observations.)

[1]FSH and prolactin were assayed by means of the kits supplied by
NIAMD and the results expressed in terms of the RP-1 rat FSH and
prolactin.

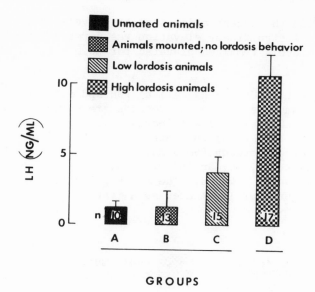

Fig. 3. Effects of varying the degree of sexual activity on serum
LH concentration. Group A: unmated animals; Group B: animals
mounted, no lordosis behavior; Group C: low lordosis group; Group
D: high lordosis group. The vertical bars represent the standard
error of the mean. All blood samples were collected at 10:00 to
11:00 p.m. Groups underlined by a common line do not differ;
groups not underlined by a common line do differ (p < 0.01). (Data
from Moss and Cooper, 1973.)

addition, prolactin concentrations progressively increased as the
time delay in blood sampling changed from 15 to 30, 60, and 90
minutes, the highest value occurring at 90 minutes.

Since the coitus-induced increase in LH and prolactin was obtained in the animals that had received high levels of coital stimulation, i.e., displayed a high lordosis-to-mount (L/M) ratio of
0.70 and above, the possibility could not be ruled out that levels
of coital stimulation could evoke a similar response.

Figure 3 summarizes the results of a series of experiments
investigating how varying the amount of sexual activity affects the
copulation-induced release of LH. High levels of coital stimulation, which resulted in a high L/M ratio, also resulted in a substantial increase in LH concentrations over the levels observed
after minimal or low-level sexual stimulation. Animals in the low
lordosis group were given a limited amount of sexual contact for
15 minutes with male sexual partners which were sexually exhausted
from previous testing.

Thus, the coitus-induced release of LH appears to be related
to penile intromission and the subsequent female lordosis behavior.
To test this hypothesis, we conducted experiments on sexually active
proestrous female rats whose vaginal area was taped to prevent intromission by the male sexual partner. The importance of penile insertion in the coitus-induced mechanism is demonstrated in Figure 4.
In the normal cyclic, proestrous female, mounting with intromissions
consistently induced a substantial increase in serum LH concentration. On the other hand, mounts without intromissions were much
less effective than mounts with intromissions. The occurrence
or nonoccurrence of ejaculation did not affect the concentrations
of serum LH. These findings are similar to those reported by Brown-
Grant et al. (1973), who used light-induced constant estrous animals,
and by Aron et al. (1968), who used normal cyclic four-day rats
primed with estrogen. As shown in Figure 4, vaginal probing by a
glass rod to stimulate the cervix did not produce an increase in
serum LH levels. Very few lordosis responses were observed in these
animals.

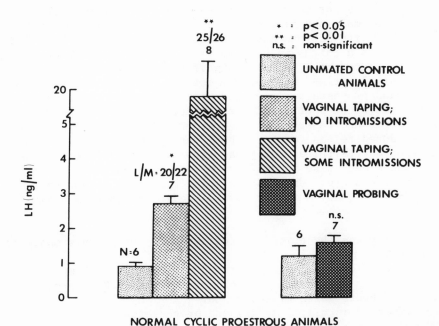

NORMAL CYCLIC PROESTROUS ANIMALS

Fig. 4. Effect of vaginal masking and probing on the coitus-induced
release of LH. Vertical bars represent the standard error of the
mean. Blood samples were collected between 9:00 and 10:00 p.m.
(R.L. Moss, 1973, unpublished data.)

*Coitus-induced release of LH in the ovariectomized, estrogen-
and reserpine-primed rat*[1]. A second set of experiments attempted to
dissociate the coitus-induced release from the spontaneous release
of LH. The basic assumption of these experiments was that a coitus-
induced mechanism for LH release exists in sexually active animals
primed with a known gonadotropin-blocking agent.

Numerous studies have demonstrated that after a period of
conditioning with estrogen, progesterone facilitates the induction
of sexual receptivity and lordosis behavior in the ovariectomized
female rat (Boling and Blandau, 1939; Beach, 1942; Young, 1961).
More recently, Meyerson (1964a, 1964b) found that the ovariectomized,
estrogen-primed rat became sexually receptive after the administration of reserpine. Besides being able to release progesterone
from the adrenals (Paris et al., 1971) and to substitute for
progesterone in the induction of lordosis behavior in the spayed
estrogen-primed rats, reserpine is also uniquely able to block the
normal, spontaneous release of LH and subsequent ovulation in the
normal cyclic, proestrous animal (Barraclough and Sawyer, 1957;
Ratner and McCann, 1971).

Ovariectomized animals primed[2] with either estrone[3] or estradiol benzoate[4] did not show an increase in serum LH concentrations
as functions of male mounting behavior (Fig. 5). On the other hand,
animals primed with estrone and progesterone[3] displayed a high L/M
ratio as well as a coitus-induced increase in the serum LH levels.
In addition, there was a slightly significant increase in serum LH
levels in animals primed with either estrone, estradiol benzoate,
or estrone and progesterone combined with reserpine[5] (Fig. 6).
In other words, when pretreated with either estrogen and progesterone or estrogen and reserpine, the ovariectomized animals displayed
a coitus-induced release of LH. Thus, reserpine, a known blocker
for the spontaneous release of gonadotropins, does not affect the
coitus-induced mechanism in ovariectomized, estrogen-primed animals.
The key questions are: What is reserpine's mode of action on the
coitus-induced mechanism in the normal cyclic, proestrous animal?

[1]Mr. G. Cearley was the principal collaborator during this phase of
the research.
[2]Ovariectomized female rats were primed with either 0.25 mg estrone
or 10 μg estradiol benzoate, then 42 hours later by 2 mg progesterone and/or 44 hours later by reserpine.
[3]Estrone (Theelin in oil; 1 mg/kg) and progesterone (Lipolutin in
oil; 50 mg/kg) were generously supplied through the courtesy of
Dr. S.N. Peterson, Parke-Davis Company.
[4] Estradiol benzoate was supplied by the generosity of Dr. P. Perlman,
Schering Corporation.
[5]Reserpine (Serpasil) was generously supplied by Dr. A. Plummer of
Ciba-Geigy Pharmaceutical Company.

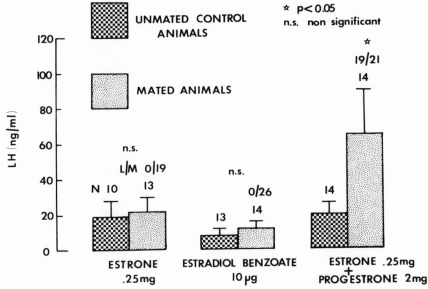

Fig. 5. Coitus-induced release of LH in ovariectomized, hormone-treated female rats. Vertical bars represent the standard error of the mean. (R.L. Moss and G. Cearley, 1973, unpublished data.)

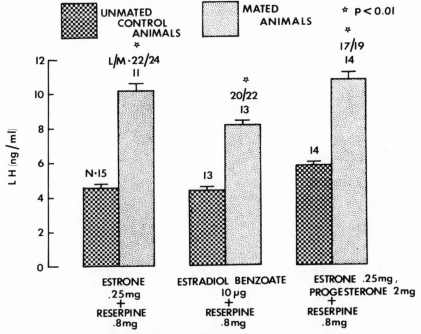

Fig. 6. Coitus-induced release of LH in ovariectomized, reserpine-steroid-primed female rats. Vertical bars represent the standard error of the mean. (R.L. Moss and G. Cearley, 1973, unpublished data.)

Is there a dual neural system, one involved in the spontaneous release and the other in the coitus-induced release of LH?

In trying to explain the action of reserpine, we have tested its effect on the spontaneous release and coitus-induced release in the normal cyclic, proestrous female rat. In these animals, reserpine administered before the critical period blocks the normal spontaneous release of LH (Fig. 7). This confirms the previous fidings of Barraclough and Sawyer (1957) and Ratner and McCann (1971) that reserpine is an effective blocker for the spontaneous release of gonadotropins and subsequent ovulation. The interesting aspect of the data is that reserpine does effectively block the coitus-induced release as well as the spontaneous release of LH. The blocking action of reserpine appears to be within the brain proper and not on the pituitary gland (Fig. 7). Fifty nanograms of luteinizing hormone releasing factor (LRF) injected intravenously substantially increased the serum LH concentration in animals pretreated with reserpine.

Reserpine acts on brain tissue to deplete brain monoamines. In addition, it stimulates adrenocorticotrophic hormone (ACTH) release which results in a release of adrenal progesterone (Grant et al., 1961; Munson, 1961; Westermann et al., 1962). Westermann et al. (1962) point out that the discharge of ACTH is not related to the change in stored norepinephrine since depletion of brain norepinephrine stores by means of α-MMT (α-methyl-m-tyrosine) does not cause ACTH hypersecretion. But, administering reserpine to animals pretreated with α-MMT produces a degree of adrenocortical activation identical with that in controls and closely related to the blockade of serotonin synthesis. Van Loon et al. (1971) recently hypothesized that catecholamines block ACTH secretion and that monoamine depletors like reserpine release the inhibitory influence of ACTH.

In a recent publication, Paris et al. (1971) suggested that reserpine influences lordosis not as Meyerson (1964a, 1964b) hypothesized by acting directly on the neural tissue that regulates sexual behavior, but indirectly by stimulating the release of ACTH and the subsequent release of progesterone. Thus, reserpine appears to have two possible effects: to evoke hypersecretion of ACTH as well as progesterone and to act as a sedative and thus decrease or minimize the aversive effects of environmental stimuli. In the present experiments, we have demonstrated coitus-induced release in normal cyclic, proestrous rats without reserpine and in estrogen-primed, ovariectomized females treated with progesterone and/or reserpine. In addition, reserpine injected into the normal cyclic rat has been shown to block the spontaneous release of LH as well as to diminish the coitus-induced release of LH. At this point, we decided to investigate the coitus-induced release of LH in the ovariectomized, adrenalectomized, estrone-, and reserpine-primed animal, i.e., in an animal whose progesterone levels are nil and whose ACTH levels

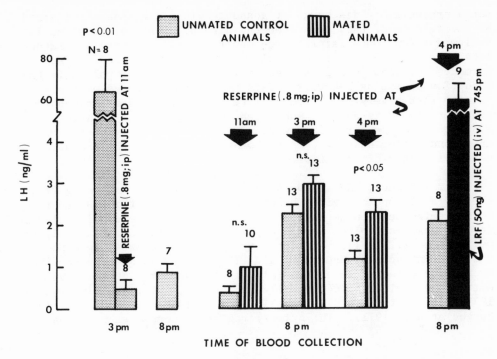

Fig. 7. Effect of reserpine on the coitus-induced release of LH in normal cyclic, proestrous female rats.

are high. Adrenalectomy in rats increases four-fold the toxicity of reserpine (Westermann et al., 1962) which is also four times more active in producing the characteristic reserpine syndrome.

Much to our surprise, copulation (mean L/M = 13/16) did not increase the serum concentration of LH in these rats (Fig. 8), either because of high levels of ACTH or because of little or no progesterone in the system. At this point, we believe that reserpine influences the coitus-induced release of LH, not by acting directly on the neural tissue that regulates this mechanism, but indirectly by stimulating the release of ACTH and progesterone. It must be emphasized that these are preliminary experiments; further research is necessary before any conclusions on the mechanism of reserpine on the coitus induced system can be made.

STIMULATORY ACTION OF LRF ON LORDOSIS BEHAVIOR[1]

Previous evidence indicates that in the normal cyclic female

[1]Dr. S. M. McCann was the principal collaborator during this phase of the research.

rat a preovulatory discharge of gonadotropins, presumably triggered
by the release of LRF, occurs on the afternoon of proestrus and is
followed a few hours later by the onset of heat (Ramirez and McCann,
1964; Everett, 1961; Monroe et al., 1969; Schwartz, 1969). The
hormones released during this surge may be directly involved in the
induction of mating behavior. This hypothesis is based on the
premise that a very close temporal relationship exists between the
gonadotropin surge and the subsequent onset of sexual behavior. In
addition, the areas regulating gonadotropin secretion, on the one
hand, and mating behavior, on the other, overlap in their central
neuronal system distribution.

For instance, LRF is found in a band of tissue extending from
the medial preoptic region through the ventral hypothalamus to the
median eminence-arcuate region from which LRF is presumably released
in juxtaposition to the portal capillaries (Creighton et al., 1970;
Quijada et al., 1971). LRF activity was first reported in hypothalamic extracts by McCann et al. (1960) and subsequently confirmed
by various authors (Fawcett et al., 1965; Guillemin et al., 1963;
Mittler and Meites, 1964; Schally and Bowers, 1964; Schally et al.,
1973; Matsuo et al., 1971; Schally et al., 1971; Amoss et al., 1971;
Burgus et al., 1971). The primary region concerned with mating

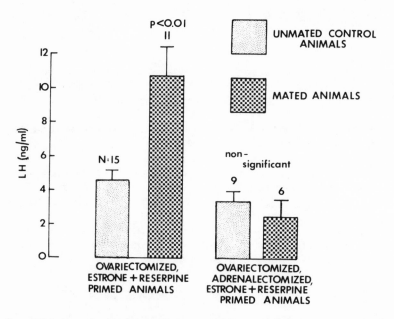

Fig. 8. Effect of adrenalectomy on the coitus-induced release of
LH in the ovariectomized, estrone- and reserpine-primed female rat.
Vertical bars represent the standard error of the mean. (R.L. Moss
and G. Cearley, 1973, unpublished data.)

behavior in the female rat appears to be the preoptic area since lesions in this area abolish this behavior (Law and Meagher, 1958; Goy and Phoenix, 1963; Sawyer, 1962) and implantation of estrogen evokes it (Lisk, 1962; Chambers and Howe, 1968). On the other hand, Napoli et al. (1972) and Moss et al. (1974) have found that electrical stimulation of the medial preoptic area reduced the intensity of sexual receptivity in the hormone-primed female rat. In addition, Powers and Valenstein (1972) have demonstrated that when lesioned, the medial preoptic area facilitates sexual receptivity in the rat. These findings suggest that the medial preoptic area exerts an inhibitory influence on copulatory behavior and that estrogen can reduce this inhibitory action.

Recently we examined the possible role of the gonadotropins and releasing factors in initiating and maintaining lordosis behavior in the ovariectomized female. The procedures have been described elsewhere (Moss and McCann, 1973). All hormones were injected subcutaneously (sc). All ovariectomized animals were injected first with estrone, then 42 hours later with progesterone or 48 to 50 hours later with either LRF[1], LH, FSH, or thyrotropin releasing

TABLE 1. *Summary of the Lordosis-to-Mount Ratio Data for Ovariectomized Female Rats under Various Hormonal Treatments Tested 50 Hours after the Initial Injection of Estrogen*

GROUPS	1	2	3	4	5	6
TREATMENT	E alone	E + P	E + LRF	E + LH	E + FSH	E + TRF
Total number of tests	28	18	21	14	13	16
Number of tests where at least one lordosis occurred	2	18*	18*	1	0	1
Percent responding	0.07%	100%	86%	0.07%	0.00%	0.06%
Cumulative number of mounts	273	223	364	153	147	160
Cumulative number of lordoses	8	193	283	2	0	3
Mean lordosis-to-mount ratio (L/M)	0.02	0.86	0.77	0.01	0.00	0.01
Dose Injected	.25 mg	.25 mg + 2 mg	.25 mg + 500ng	.25 mg + .03mg	.25 mg + .05mg	.25 mg + 500 ng

*Significantly different (P <.001) from all other groups by chi square test.

[1]Luteinizing hormone releasing factor (LRF) was generously supplied through the courtesy of Dr. Romano Deghenghi, Ayerst Ltd., Montreal.

factor (TRF). Results of the preliminary observation are summarized
in Table 1; data from subsequent experiments are shown in Figures 9
to 12. Ovariectomized females treated with estrone alone or estrone
with either LH, FSH, or TRF displayed little or no lordosis behavior
in response to male sexual contact (Table 1). On the other hand,
animals primed with estrogen combined with either progesterone or
LRF exhibited sexual behavior typified by a high L/M ratio (Fig. 9).
As expected, estrone-, progesterone-primed females diaplayed the
characteristic sexual behavior of normal cyclic, proestrous animals.
In contrast, the majority of the estrone-, LRF-primed animals exhibited lordosis behavior (Fig. 10); but some displayed some characteristics of a typical diestrous nonreceptive animal, i.e., hind kick-
ing and squealing.

At a high level of sexual receptivity, female rats display
characteristic behavior including a stiff-legged, hopping gait and
darting and rapid vibratory movements of the ears. In the present
experiment, the majority of animals displayed some hopping and darting behavior, not no ear wiggling. The possibility that progesterone
from the adrenals and not LRF is the predominant factor in initiating
lordosis behavior was considered. However, ovariectomized, adrenalectomized, estrone-primed female rats displayed good lordosis
behavior (L/M = 19/27) after the injection of LRF (Fig. 9, 10).

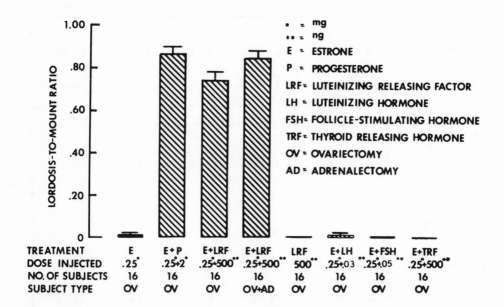

Fig. 9. The lordosis-to-mount ratio exhibited in response to estrone
alone or in combination with progesterone, LRF, LH, FSH, and TRF.
Vertical bars represent the standard error of the mean. (Data
modified from Moss and McCann, 1973.)

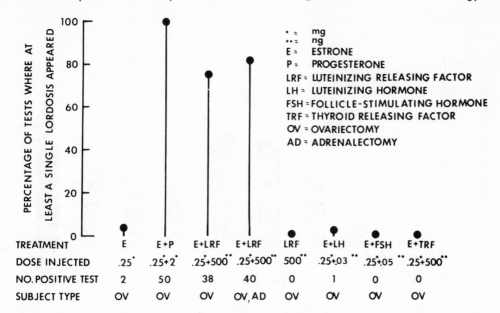

Fig. 10. Lordosis response to estrone alone or in combination with progesterone, LRF, LH, FSH, and TRF. (Data modified from Moss and McCann, 1973.)

To determine whether the response to LRF was specific, we tested releasing factor TRF. This one seemed particularly appropriate since it is localized in the bed nucleus of the stria terminals (Creighton et al., 1970; Quijada et al., 1971), a region near the LRF-containing and sexual behavior centers. Lordosis behavior was observed in only one of 50 tests with TRF.

In the initial study, the dose level of LRF (i.e., 500 ng, sc) was selected on the basis of experiments by Zeballos and McCann (unpublished data) which indicated that 500 ng given sc produced a substantial increase in LH serum content. A dose 10-fold smaller produced only a very slight effect on LH; thus, the initial dose reaching the hypothalamic-pituitary axis probably is within the physiological range. In terms of lordosis behavior, some animals responded to 50 ng of LRF, but the majority of the animals displayed consistent lordosis behavior only after an injection of 150 ng of LRF (Fig. 11). Thus, the dose response curve for LRF, measuring either L/M ratio or the percentage of lordosis evoked, increased with progressive increases in the amount of LRF injected.

We next concentrated our efforts on the time-course for the display of lordosis behavior. Estrone-, LRF-primed ovariectomized animals showed the first signs of lordosis behavior about one hour after the injection of LRF (500 ng, sc); but the best and most

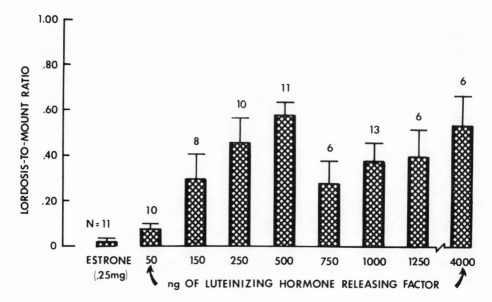

Fig. 11. Dose-response curve for LRF expressed in terms of its effect on the lordosis-to-mount ratio. Vertical bars represent the standard error of the mean. (R.L. Moss, S.M. McCann, and C. Dudley, 1973, unpublished data.)

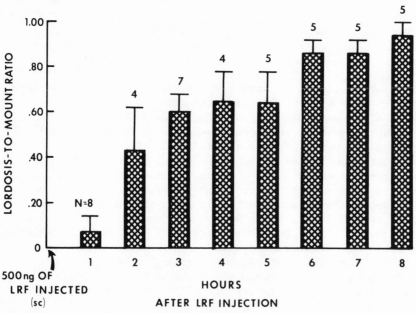

Fig. 12. Time-course for LRF-induced lordosis based on lordosis-to-mount ratios. Vertical bars represent the standard error of the mean. (R.L. Moss, S.M. McCann, and C. Dudley, 1973, unpublished data.)

consistent response occurred three to four hours after LRF injection (Fig. 12).

The present series of experiments demonstrates that, like pro-gesterone, LRF facilitates the induction of mating behavior. After an injection of LRF, ovariectomized female rats, pretreated with estrone, displayed a lordosis pattern which differed little from that produced by progesterone; whereas in response to estrogen alone or estrogen combined with LH, FSH, or TRF, only a few females exhibited the lordosis pattern. LRF injected into ovariectomized females without pretreatment with estrogen did not induce lordosis behavior. The fact that LRF potentiates lordosis behavior in ovariectomized, adrenalectomized, estrone-primed animals suggests that adrenal pro-gesterone is not involved in the LRF facilitation of lordosis be-havior. Independent confirmation of the ability of LRF to induce mating behavior in estradiol-primed, hypophysectomized, ovariectom-ized female rats has recently been reported (Pfaff, 1973).

LRF triggers the ovulatory discharge of LH and facilitates lordosis. If LRF is released locally into the preoptic area when the LRF-secreting neurons are active, then very high concentrations might reach the cells concerned with the mating response. It is tempting to postulate that if LRF is released on the afternoon of proestrus it will initiate mating behavior in female rats after a short delay. Next we must determine whether LRF can enhance mating behavior in males as well as females and determine whether these results can be duplicated in man.

I have outlined briefly the direction of our research on the interaction and interrelationship between pituitary gonadotropins and mating behavior in the female rat. We have also demonstrated that coitus is an adequate stimulus to trigger changes in plasma LH and prolactin levels but has no apparent effect on FSH concen-trations. In studying the mechanism for this release, we have found that coitus increases LH throughout the major portion of the behavioral receptivity cycle not only in the normal cyclic rat but also in the ovariectomized, estrone-, estrone-progesterone-, and estrone-reserpine-primed female rat. We have also shown that the prehormone LRF, a precursor for spontaneous release of LH, initiates sexual receptivity and lordosis behavior in the ovariectomized, estrogen-primed rat. Our next step is to inquire whether LRF released during the preovulatory discharge is essential in regula-ting mating behavior as well as gonadotropin secretion. Attempts to understand these relationships will, we hope, provide insight into the temporal relations between endocrine and behavioral mani-festations of estrus during the cycle of the normal female rat.

ACKNOWLEDGMENTS

The author is indebted to Dr. S.M. McCann, Mr. George Cearley, M. Kelly, and P. Riskind for their many useful discussions and valuable comments and to Drs. C.P. Fawcett and L. Krulich for their assistance and advice with assays. Grateful acknowledgment is also made to Mrs. Carol Dudley and Ms. Betty Turicchi for their excellent technical assistance. Bibliographic assistance was received from the UCLA Brain Information Service, which is part of the Neurological Information Network of NINDS and is supported under contract DHEWPH-43-66-59.

REFERENCES

Amoss, M., Burgus, R., Blackwell, R., Vale, W., Fellows, R., and Guillemin, R., 1971, Purification, amino acid composition and N-terminus of the hypothalamic luteinizing hormone releasing factor (LRF) of ovine origin, Biochem. Biophys. Res. Commun. 44:205-210.

Aron, C.L., Asch, G., Asch, L., Luxembourger, M., and Roos, J., 1965, Donnees nouvelles sur les facteurs neuro-hormonaux de la luteinisation chez la ratte. Mise en evidence de l'action ovulation du coit au cours du cycle oestral, Path. Biol. 13:603-614.

Aaron, C., Roos, J., and Asch, G., 1968, New facts concerning the afferent stimuli that trigger ovulation by coitus in the rat, Neuroendocrin. 3:47-54.

Barraclough, C.A., and Sawyer, C.H., 1957, Blockade of the release of pituitary ovulating hormone in the rat by chlorpromazine and reserpine: possible mechanisms of action, Endocrinology 61:341-351.

Beach, F.A., 1942, Importance of progesterone to induction of sexual receptivity in spayed female rats, Proc. Soc. Exp. Biol. Med. 51:369-371.

Boling, J.L., and Blandau, R.J., 1939, The estrogen-progesterone induction of mating responses in the spayed female rat, Endocrinology 25:359-364.

Brown-Grant, K., Davidson, J.M., and Grieg, F., 1973, Induced ovulation in albino rats exposed to constant light, J. Endocrin. 57:7-22.

Burgus, R., Butchner, M., Ling, N., Monahan, M., Rivier, J., Fellows, R., Amoss, M., Blackwell, R., Vale, W., and Guillemin, R.,

1971, Structure moleculaire du facteur hypothalamique (LRF) d'origine ovine contralant la secretion de l'hormone gonadotrope hypophysaire de luteinisation (LH), C.R. Acad. Sci. (D) (Paris) 273:1611-1613.

Chambers, W.R., and Howe, G., 1968, A study of estrogen sensitive hypothalamic centers using a technique for rapid application and removal of estradiol, Proc. Soc. Exp. Biol. Med. 128:292-300.

Crighton, D.B., Schneider, H.P.G., and McCann, S.M., 1970, Localization of LH-releasing factor in the hypothalamus and neurohypophysis as determined by *in vitro* assay, Endocrinology 87:323-329.

Everett, J.W., 1952a, New information concerning effects on hypophyseal function induced by coital stimuli in rats, Anat. Rec. 112:327.

Everett, J.W., 1952b, Presumptive hypothalamic control of spontaneous ovulation, Ciba Foundation Colloquia on Endocrinology 4:167.

Everett, J.W., 1961, The mamalian female reproductive cycle and its controlling mechanism, in *Sex and Internal Secretions* (W.C. Young, ed.), pp. 497-555, Williams & Wilkins, Baltimore.

Everett, J.W., 1964, Central neural control of reproductive functions of the adenohypophysis, Physiol. Rev. 44:373-431.

Fawcett, C.P., Harris, G.W., and Reed, M., 1965, The purification of the luteinizing hormone releasing factor (LRF), (abstr. only) in *Lectures and Symposia of the XXIIIrd International Congress of Physiological Sciences* (Tokyo, September 1962) (D. Noble, ed.), Int. Congr. Ser. No. 87, p. 300, Excerpta Medica, Amsterdam.

Gaunt, R., Chart, J.J., and Renzi, A.A., 1961, Endocrine pharmacology, Science 133:613-621.

Goy, R.W., and Phoenix, C.H., 1963, Hypothalamic regulation of female sexual behavior; establishment of behavioral oestrus in spayed guinea pigs following hypothalamic lesions, J. Reprod. Fertil. 5:23-40.

Guillemin, R., Jutisz, M., and Sakiz, E., 1963, Purification partille d'un facteur hypothalamique (LRF) stimulant la secretion de l'hormone hypophysaire de luteinisation (LH), C.R. Acad. Sci. (D) (Paris) 256:504-507.

Harrington, F.E., Eggert, R.G., Wilbur, R.D., and Lindenheimer, W.H, 1966, Effect of coitus on chlorpromazine inhibition of ovulation in the rat, Endocrinology 79:1130-1134.

Kalra, S.P., and Sawyer, C.H., 1970, Blockade of copulation-induced ovulation in the rat by anterior hypothalamic deafferentation, Endocrinology 87:1124-1128.

Law, O.T., and Meagher, W., 1958, Hypothalamic lesions and sexual behavior in the female rat, Science 128:1626-1627.

Lisk, R.D., 1962, Diencephalic placement of estradiol and sexual receptivity in the female rat, Am. J. Physiol. 203:493-496.

McCann, S.M., Taleisnik, S., and Friedman, H.M., 1960, LH-releasing activity in hypothalamic extracts, Proc. Soc. Exp. Biol. Med. 104:432-434.

Matsuo, H., Baba, Y., Nair, R.M.G., Arimura, A., and Schally, A.V., 1971, Structure of the porcine LH and FSH releasing hormone. I. The proposed amino acid sequence, Biochem. Biophys. Res. Commun. 43:1334-1339.

Meyerson, B.J., 1964a, Estrus behavior in spayed rats after estrogen or progesterone treatment in combination with reserpine or tetrabenazine, Psychopharmacologia 6:210-218.

Meyerson, B.J., 1964b, Central nervous monoamines and hormone in-duced estrus behavior in the spayed rat, Acta Physiol. Scand. 63:241-273.

Mittler, J.C., and Meites, J., 1964, *In vitro* stimulation of pituitary FSH release by hypothalamic extract, Proc. Soc. Exp. Biol. Med. 117:309-313.

Monroe, S.E., Rebar, R.W., Gay, V.L., and Midgley, A.R., 1969, Radioimmunoassay determination of luteinizing hormone during the estrous cycle of the rat, Endocrinology 85:720-724.

Moss, R.L., and Cooper, K.J., 1973, Temporal relationship of spon-taneous and coitus-induced release of luteinizing hormone in the normal cyclic rat, Endocrinology 92:1748-1753.

Moss, R.L., Cooper, K.J., and Danhof, I.E., 1973, Coitus-induced release of luteinizing hormone in the normal cyclic female rat, Fed. Proc. 32(3):239Abs. (abstr. #156).

Moss, R.L., and McCann, S.M., 1973, Induction of mating behavior in rats by luteinizing hormone releasing factor, Science 181:177-179.

Moss, R.L., Paloutzian, R.F., and Law, O.T., 1974, Electrical stim-
     ulation of forebrain structures and its effect on copulatory
     as well as stimulus-bound behavior in ovariectomized hormone-
     primed rats, Physiol. Behav. (in press).

Munson, P.L., 1961, Endocrine pharmacology: selected topics, Ann.
     Rev. Pharmacol. 1:315-350.

Napoli, A., Powers, J.B., and Valenstein, E.S., 1972, Hormonal
     induction of behavioral estrus modified by electrical stim-
     ulation of hypothalamus, Physiol. Behav. 9:115-117.

Niswender, G.D., Midgley, A.R., Monroe, S.E., and Reichert, L.E.,
     1968, Radioimmunoassay for rat luteinizing hormone with anti-
     ovine LH serum and ovine LH-   I, Proc. Soc. Exp. Biol. Med.
     128:807-811.

Paris, C.A., Resko, J.A., and Goy, R.W., 1971, A possible mechanism
     for the induction of lordosis by reserpine in spayed rats,
     Biol. of Reprod. 4:23-30.

Pfaff, D.W., 1973, Luteinizing hormone releasing factor (LRF) poten-
     tiates lordosis behavior in hypophysectomized ovariectomized
     female rats, Science 182:1148-1149.

Powers, B.J., and Valenstein, E.S., 1972, Sexual receptivity: facil-
     itation by medial preoptic lesions in female rats, Science
     175:1003-1005.

Quijada, M., Krulich, L., Fawcett, C.P., Sundberg, D.K., and McCann,
     S.M., 1971, Localization of TSH-releasing factor (TRF), LH-RF
     and FSH-RF in rat hypothalamus, Fed. Proc. 30(2):197Abs.
     (abstr. #21).

Ramirez, V.D., and McCann, S.M., 1964, Fluctuations in plasma
     luteinizing hormone concentrations during the estrous cycle
     of the rat, Endocrinology 74:814-816.

Ratner, A., and McCann, S.M., 1971, Effect of reserpine on plasma
     LH levels in ovariectomized and cycling proestrus rats, Proc.
     Soc. Exp. Biol. Med. 138:763-767.

Rodgers, C.H., 1971, Influence of copulation on ovulation in the
     cycling rat, Endocrinology 88:433-436.

Rodgers, C.H., and Schwartz, N.B., 1973, Serum LH and FSH levels in
     mated and unmated proestrous rats, Endocrinology 92:1475-1479.

Sawyer, C.H., 1962, Reproductive behavior, in *Handbook of Physiology-
     Neurophysiology* (Sect. 1; H.W. Magoun, ed.), Vol. II, Chapt.

XLIX, pp. 1225-1240, American Physiological Soc., Washington, D.C.

Schally, A.V., and Bowers, C.Y., 1964, Purification of luteinizing hormone releasing factor from bovine hypothalamus, Endocrinology 75:608-614.

Schally, A.V., Baba, Y., Arimura, A., Redding, T.W., and White, W.F., 1971, Evidence for peptide nature of LH- and FSH-releasing hormones, Biochem. Biophys. Res. Commun. 42:50-56.

Schally, A.V., Kastin, A.J., and Arimura, A., 1973, FSH-releasing hormone and LH-releasing hormone, Vitam. Horm. 30:84-164.

Schwartz, N.B., 1969, A model for the regulation of ovulation in the rat, in Recent Progress in Hormone Research (Proceedings of the 1968 Laurentian Hormone Conference, E.B) (Astwood, ed.), vol. 25, pp. 1-55, Academic Press, New York.

Spies, H.G., and Niswender, G.D., 1971, Levels of prolactin, LH and FSH in the serum of intact and pelvic-neurectomized rats, Endocrinology 88:937-943.

Taleisnik, S., Caligaris, L., and Astrada, J.J., 1966, Effect of copulation on the release of pituitary gonadotropins in male and female rats, Endocrinology 79:49-54.

Van Loon, G.R., Scapagnini, U., Moberg, C.P., and Ganong, W.F., 1971, Evidence for central adrenergic neural inhibition of ACTH secretion in the rat, Endocrinology 89:1464-1469.

Westermann, E.O., Maickel, R.P., and Brodie, B.B., 1962, On the mechanism of pituitary-adrenal stimulation by reserpine, J. Pharmacol. Exp. Ther. 138:208-217.

Young, W.C., 1961, The hormones and mating behavior, in Sex and Internal Secretions (W.C. Young, ed.), pp. 1173-1239, Williams & Wilkins, Baltimore.

Zarrow, M.X., Clark, J.H., 1968, Ovulation following vaginal stimulation in a spontaneous ovulator and its implications, J. Endocrinol. 40:343-352.

SOME MECHANISMS GOVERNING THE INDUCTION, MAINTENANCE, AND SYNCHRONY OF MATERNAL BEHAVIOR IN THE LABORATORY RAT[1]

Howard Moltz

The University of Chicago

INTRODUCTION

As long ago as 1925, Stone advanced the hypothesis that the endocrine changes accompanying the termination of pregnancy underlie the onset of maternal behavior in the rat. In a sense, this is an obvious assumption because such maternal activities as nursing, nest building, and retrieving are initiated in close temporal association with the birth process. However, more than 40 years elapsed before Stone's hypothesis was actually demonstrated and before some of the endocrine agents responsible for the induction of maternal behavior were finally identified. Specifically, these agents include progesterone, estrogen, and prolactin.

Progesterone, which has been assayed in both peripheral plasma (Grota and Eik-Nes, 1967; Wiest et al., 1968) and ovarian vein blood (Eto et al., 1962; Fajer and Barraclough, 1967; Hashimoto et al., 1968), begins to increase on Day 4 of pregnancy, reaching maximal concentrations on Day 14. Thereafter, the output begins to fall slowly until, on Day 20, the decline becomes abrupt. This decline is apparently due to the sharp increase of 20α-hydroxy-steroid dehydrogenase, the enzyme largely responsible for the conversion of progesterone to 20α-hydroxypregn-4-en-3-one (Kuhn, 1969; Wiest, 1968; Wiest et al., 1968). That progesterone is taken up in rat brain during most of pregnancy can reasonably be inferred from data on the accumulation of the tritiated preparation in nonpregnant

[1] This paper was written during the tenure of NIH Research Grant HD-06872. The author wishes to thank Ms. Gail Orgelfinger for her careful preparation of the manuscript.

females (Hamburg, 1966; Lumas and Farooq, 1966; Raisinghani et al., 1968).

Virtually coincident with the onset of progesterone withdrawal is a characteristic increase in plasma estrogen (Yoshinaga et al., 1968). Prevailing at low concentrations before mid-pregnancy, the secretory rate of estrogen begins to increase about Day 15. At first slowly and then more rapidly, it rises to peak values on or shortly before the day of parturition. That it is selectively bound in different brain regions, at least in the nonpregnant female, has been shown by the work of Eisenfeld and Axelrod, (1965, 1966), Pfaff (1965, 1968), and Stumpf (1968).

Prolactin, the third hormone essential for the induction of maternal behavior, increases shortly after coition (Dilley and Adler, 1968), then quickly recedes and is maintained at low plasma concentrations until about Day 20 of pregnancy (Grindeland et al., 1969, Kwa and Verhofstad, 1967). On Day 22, however, the day of parturition, it increases sharply (Amenomori et al., 1970). To my knowledge, there is no evidence of prolactin uptake in the central nervous system nor of the effect of prolactin on the uptake of other hormones.

Thus the role of estrogen, progesterone, and prolactin in maternal behavior is as follows.

The neuronal system mediating maternal behavior is initially insensitive to arousal by estrogen and prolactin, which at even relatively high titers are characteristically unable to exert an activational effect. Before these critical hormones can become effective, the mediating system must first be primed, i.e., made to sustain lower-than-normal thresholds of estrogen and prolactin. Progesterone, or rather, the withdrawal of progesterone, is thought to function at this point.

Existing at high concentrations during most of pregnancy, progesterone probably acts at first to maintain or even to increase activation thresholds within the neuronal system. Beginning on Day 15, however, the blood levels of progesterone typically start to decline. This decline, occurring from the high progesterone titers of pregnancy, effects within the mediating system a condition functionally akin to what Kawakami and Sawyer (1959) called the "rebound from progesterone dominance." Such a rebound not only decreases selected thresholds within the mediating system but carries these thresholds to lower-than-normal levels. In other words, at a time more or less coincident with the birth process, the mediating system is made sensitive to endocrine influence, particularly of estrogen and prolactin. At this point, presumably, estrogen and prolactin, functioning synergistically and at concentrations increased by near-term secretory rates, excite the mediating

substrate to a level at which it becomes acutely responsive to the
sight, sound, and odor of the young. Thus affected, the puerperal
female responds as soon as the young emerge from the birth canal.

To determine whether this triad of hormones actually does induce maternal behavior, we attempted, on the one hand, to disrupt
the estrogen and progesterone balance of parturition and, on the
other, to induce maternal behavior in nulliparous females by injecting them with these hormones.

In our first study (Moltz and Wiener, 1966), we tried to reverse the characteristic near-term increase in estrogen by ovari-
ectomizing primiparous females which were between 12 and 15 hours
from parturition. Fully 50% failed to behave maternally, with the
result that their young died of neglect. We next attempted to
manipulate near-term progesterone levels by injecting 2 mg of the
steroid twice daily beginning on Day 19 of pregnancy (Moltz et al.,
1969a). Here again, fully 50% of the primiparous females neglected
their litters. We then turned to nulliparous females, hoping to
induce maternal behavior through selected injections of estrogen,
progesterone,and prolactin (Moltz et al., 1970).

At between 90 and 110 days of age, nulliparous rats were ovariectomized and three weeks later subjected to the following hormone
regimen. Estradiol was injected once daily from Day 1 through Day
11 at a dosage level of 12μg per injection; progesterone twice
daily on Days 6 through 9 at a dosage level of 3 mg per injection;
and prolactin on the evening of Day 9 and the morning of Day 10,
each at a dosage of 50 IU. On the afternoon of Day 10--approxi-
mately 20 hours after the progesterone had been withdrawn--six
foster pups, 10 to 15 hours old, were given to each female. The
pups remained until the following morning, at which time a fresh
litter was substituted. This procedure continued until maternal
behavior was displayed or, failing maternal behavior, until seven
days had elapsed. To be scored maternal, a female had not only to
retrieve, but to build a nest, assume a nursing posture, lick the
young, and keep them warm. She was, in other words, obliged to
exhibit the full spectrum of nurtural attachment.

All of the 10 nulliparous females subjected to this schedule
showed full maternal behavior at between 35 and 40 hours from the
time the pups were first proffered. This represents not only a significant reduction in latency from the average of six to seven days
characteristic of untreated nulliparae but a uniformity in the time
of onset closely approaching that of the puerperal female. Control
females, on the other hand, which had been given only two of the
three inductor substances (the vehicle in each case having been substituted for the hormone omitted), showed marked variability in time
of onset and, of course, a significantly higher median latency
(Fig. 1).

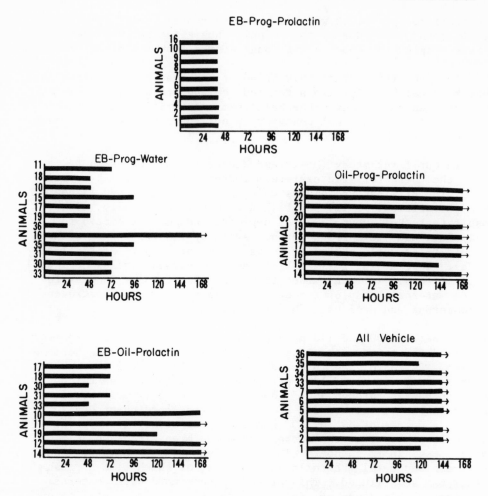

Fig. 1. Latency and variability in time of onset for the display of maternal behavior in experimental and control females; arrow at end of bar indicates that the animal had failed to act maternally at the conclusion of the observation period (from Moltz et al., 1970).

If there is a rebound from progesterone dominance during the latter half of pregnancy--a rebound that renders the maternal mediating system sensitive to the influence of estrogen and prolactin--then it should be possible to replace progesterone with an agent capable of producing the same central effect. The psychoactive phenothiazines seemed promising since, like progesterone, they first depress and then, upon withdrawal, heighten neuronal activity (Doyle et al., 1968). Because of this parallel,

we replaced progesterone with perphenazine in the hormone schedule
we had used to induce maternal behavior in the nulliparous females
(Moltz et al., 1971). Of the 12 experimental animals, eight became fully maternal at between 35 to 40 hours from the time the
foster young were first proffered; and four responded at between
67 to 72 hours. Control females, i.e., those in whom the vehicle
was injected in place of perphenazine and those that received none
of the inductor agents, exhibited more variability in the time of
onset to maternal behavior and an appreciably higher median latency.
Figure 2 represents these data and includes for comparison a group

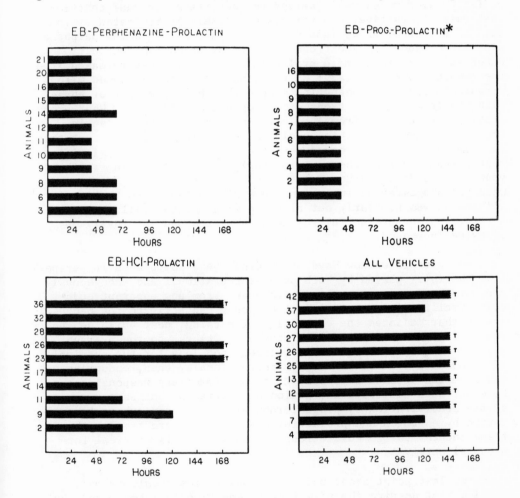

Fig. 2. Latency and variability in time of onset for the display of
maternal behavior in experimental and control females; broken bar
(marked with T) indicates that the animal had failed to act maternally at the conclusion of the observation period (fig. from Moltz et
al., 1971; *--data taken from the study by Moltz et al., 1970).

of ovariectomized nulliparous females taken from our earlier study. As indicated, these females had received progesterone rather than perphenazine.

No drug has only a single physiological effect. Thus, in addition to influencing neuronal activity, perphenazine and the other phenothiazines act characteristically to release large amounts of prolactin from the adenohypophysis (Ben-David, 1968; Ben-David et al., 1965). In the present experiment, the resulting concentration of prolactin, rather than a threshold change in the maternal mediating system, may have served to facilitate maternal behavior. That this alternative, however, is not likely is indicated by the data of a control group (not shown in Figure 2), in which perphenazine was extended an additional day in our injection schedule. Under such a regimen, more prolactin should have been released from the female's own pituitary; on the other hand, the presumed withdrawal effects of perphenazine should then have been out of phase with the terminal injections of estrogen and prolactin. What proved critical was the withdrawal of perphenazine, not its prolactin-promoting properties; this conclusion is indicated by the fact that the females of this extended-perphenazine group differed significantly from those of the experimental group by responding to their young not in 35 to 40 hours but in six to seven days. Moreover, when the progesterone injections were also extended so that their withdrawal was similarly out of phase with the terminal injections of estrogen and prolactin, the latency to respond maternally was likewise increased.

The picture we now have of maternal behavior places in perspective the roles played by both pup-related stimuli and endogenously produced hormones. Clearly, quite apart from any apparent endocrine facilitation, the presence of young can activate the neuronal system that mediates the expression of nursing, nest building, and retrieving. But activation by pups alone takes, on the average, six or seven days, an interval entirely too long for the survival of a neonatal litter. In the puerperal female, then, endocrine facilitation becomes necessary to ensure immediate responsiveness. Here our data elucidate the action of a triad of hormones: substrate sensitization by progesterone withdrawal and substrate excitation by estrogen and prolactin. In the near-term female, this triad induces the immediate responsiveness that is critical for neonatal survival.

One last point about our injection studies should be made, namely, that despite the significant reduction in latency and the marked uniformity in the time of onset to maternal behavior, the nulliparous females still took as long as 35 to 40 hours to respond. This latency may indicate that 1) the dosage level of one or other of the hormones administered was too low, or perhaps too high; 2) progesterone, estrogen, and prolactin can effect only a limited reduction in latency to respond, and to induce immediate maternal behavior additional hormones must be administered; or 3) the time lapse between the injection of one or other of our inductor agents was not optimal. With respect to this last condition, we already know that the interval between the last progesterone injection and the first prolactin injection is critical. Perhaps a different scheduling of our hormone regimen, for example, allowing for complete sensitization by progesterone withdrawal, would have reduced our obtained latency to 15 to 20 hours. However, whether discrete injections can ever duplicate the effects of endogenous release and thereby induce immediately attention to young remains to be determined.

MAINTENANCE

Thus far we have spoken only of the initiation of maternal behavior. However, once initiated, the behavior has to be maintained for three to four weeks until weaning. The problem, then, is to ascertain how these care-taking activities are sustained once they become established.

In the rat and in other mammals as well, the onset of lactation occurs at about the time of parturition so that when the young emerge the mammae are already distended with milk (Kuhn, 1968, 1969; Kuhn and Lowenstein, 1967; Shinde et al., 1964). In this condition, the female begins to nurse and thereafter, for a period of several weeks, regularly promotes suckling by approaching the nest, licking the young, and adjusting her posture to permit nipple attachment. Several investigators (e.g., Lehrman, 1961; Rosenblatt and Lehrman, 1963) have suggested that mammary engorgement, or more precisely, the peripheral tension accompanying that engorgement plays a role in maintaining nursing, retrieving, and nest building.

One way of determining the importance of mammary engorgement in maintaining maternal behavior is to show whether the postparturient female continues to display maternal behavior even when deprived of all mammary tissue. Therefore, we (Moltz et al., 1967) subjected weanling rats to a total mammectomy and, in addition, to a total thelectomy (surgical removal of the nipples). Briefly, the operation consisted of removing adipose tissue from both the anal region and the dorsomedial aspect of the thigh, ligating and removing the median branch of the thoracic artery, and, finally, removing each nipple and either excising or cauterizing the surrounding tissue. When these animals matured and were impregnated and allowed to give birth normally, they not only persisted in their attempts to nurse but in all respects behaved toward their young exactly like normal postparturient animals. Mammary engorgement, evidently, is not essential for the maintenance of maternal behavior.

But nursing and the presumed peripheral relief it affords are
not the only avenues through which the young can influence the
postparturient female. A litter, after all, provides visual,
auditory, and olfactory stimuli, which, even when unaccompanied by
actual suckling, may be sufficient to sustain nurtural responsiveness. Indeed, perhaps these pup-related stimuli to maternal behavior act directly on the mother. Or, on the other hand, perhaps
they first induce in the mother a particular endocrine profile
which in turn mediates the continued appearance of the behavior.
Rosenblatt (1965), for example, has suggested that high levels of
prolactin are essential for the continued postpartum display of
each of the several response items of the nurtural complex.

Prolactin is normally released by suckling and is maintained
at elevated titers during at least the first two weeks after parturition (Amenomori et al., 1970). But despite its prominent role in
the mother-litter interaction, suckling is not essential for the
characteristic postpartum prolactin response. Puerperal females
in contact with young but surgically deprived of the opportunity
to nurse have virtually the same pattern of hypophyseal output as
that of intact puerperae (Moltz et al., 1969b).

The data just cited do not indicate how prolactin acts in
maintaining maternal behavior. To ascertain this role, we used
ergocornine (Numan et al., 1972), an ergot alkaloid that interferes
with prolactin release both in vivo and in vitro (Wuttke et al.,
1971; Pasteels and Ectors, 1970; Lu et al., 1971). The ergocornine
was injected daily for a period of 21 days, beginning on the day
of parturition. Those that received the drug did not differ significantly from control females on any measure of maternal behavior,
the control females themselves having received either ergocornine
plus prolactin or simply the vehicle. Since both lactation and the
decidual cell response to uterine traumatization were inhibited
after ergocornine and at least partially reversed by exogenous prolactin, it was evident that ergocornine interfered with the postpartum release of prolactin. The fact that maternal behavior remained unaffected warrants the conclusion that prolactin is essential in initiating but not in maintaining nursing, nest building,
and retrieving.

Estrogen likewise is necessary for inducing but not for maintaining maternal behavior. Ovariectomy performed about 12 hours
before parturition has a disruptive effect on maternal behavior
(Moltz and Wiener, 1966) but performed shortly after parturition
is entirely without effect (Rosenblatt, 1969). Thoman and Levine
(1970) reported that adrenal steroids are not essential for sustaining maternal responsiveness, and we observed (unpublished
observations) that adrenalectomy and ovariectomy performed soon

after parturition were without effect. Moreover, Moltz and co-workers (Moltz et al., 1969a) showed that progesterone administered to postpartum female rats had no influence whatever on already established maternal behavior; but the same treatment before parturition led to neglect of the litters.

We have not, of course, investigated every possible hormonal change which characterizes the postpartum period but believe that the maintenance of maternal behavior does not require the mediation of endocrine agents. In other words, once the neuronal mechanisms underlying maternal behavior are activated, they are governed, solely and directly, by the stimulative properties of the young.

But where in the central nervous system do these mechanisms lie? Recently, Michael Numan, a graduate student working in the author's laboratory, attempted to explore this question (Numan, 1973). Taking females five days after parturition and the initiation of maternal behavior, he inflicted bilateral electrolytic lesions to the medial preoptic area of the hypothalamus (MPOA). Such lesions terminated completely nursing, nest building, and retrieving. To determine whether MPOA neurons rather than merely fibers of passage were critically involved in this behavioral disruption, Numan severed the stria terminalis and the medial cortico-hypothalamic tract (MCHT). The stria arises in the amygdala and sends components which synapse in and pass through the MPOA (Heimer and Nauta, 1969); the MCHT arises in the ventral hippocampus, negotiates the caudal region of the MPOA, and terminates in both the anterior hypothalamus and the arcuate nucleus (Nauta, 1956). Interrupting these major fibers of passage had no effect whatever on the continued display of maternal behavior.

Numan then made knife cuts which severed the connections between the medial and lateral preoptic areas; and as a result, his postpartum females ceased attending to their litters. Thus the MPOA may mediate the expression of maternal behavior through its connections with the lateral hypothalamus. Figure 3 shows photomicrographs of a representative MPOA lesion and a representative medio-lateral knife cut.

One final point should be mentioned here. In any lesion study the question of specificity of effect arises; and here we may ask whether the inhibition that followed MPOA insult was restricted to maternal behavior, or whether the animal was so invaded that unrelated response systems were affected as well. There is simply no conclusive answer to such a question, for it is obviously impossible to examine *all* behaviors in an effort to determine whether just *one* was disrupted. However, Numan did report that his animals appeared in good health after surgery and that their eating and drinking habits were similar to those of sham operates. Moreover, their

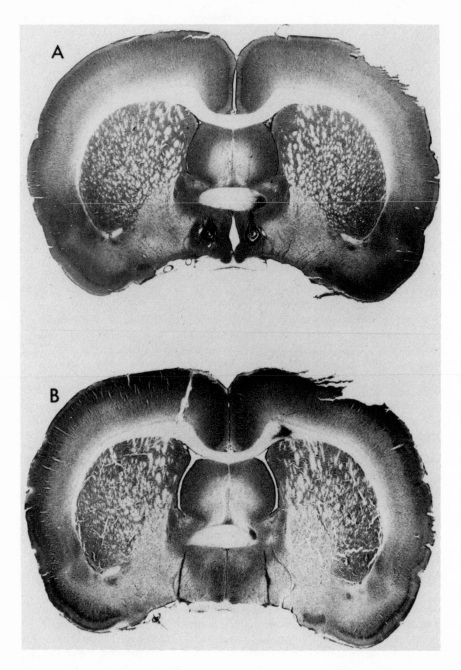

Fig. 3. Photomicrographs of a representative MPOA lesion (A) and
a representative medio-lateral knife cut (B) (from Numan, 1973).

sexual behavior, induced through estrogen and progesterone priming, was normal. All that can be said conclusively is that the MPOA, which is evidently critical for the continued display of maternal behavior, may also be critical for the continued display of other response systems as well.

SYNCHRONY

Maternal behavior does not remain static but changes as the pups themselves change from largely immobile neonates into active, adult-looking young. Thus retrieving, which occurs frequently during the first nine or 10 days after parturition, begins to decrease soon thereafter just about the time the young begin to leave the nest (Beach and Jaynes, 1956; Moltz and Robbins, 1965; Rosenblatt and Lehrman, 1963; Wiesner and Sheard, 1933). By Day 14 postpartum, retrieving is rarely displayed. Nest building follows the same functional course: a high compact nest is seen for some nine or 10 days postpartum, after which it progressively disintegrates and nest-building behavior declines (Moltz and Robbins, 1965; Rosenblatt and Lehrman, 1963). Nursing, however, does not decline until 20 or 21 days postpartum (Moltz and Robbins, 1965); but once the young begin to leave the nest and the nest in turn begins to disintegrate, it is the young that are likely to initiate nursing rather than the mother (Rosenblatt and Lehrman, 1963). Only by the end of the fourth week does nursing disappear altogether.

What we have just described is the behavior characteristically displayed toward a single litter as it progresses to weaning age. It contrasts with the behavior displayed toward litters repeatedly replaced by newborn pups each time they become 10 or 12 days old (Bruce, 1961; Nicoll and Meites, 1959; Wiesner and Sheard, 1933). Under such conditions of stimulation, maternal attachment is prolonged, with the mother continuing to respond for as long as three or four months and in a manner consistently appropriate to each new litter.

These postpartum changes in maternal behavior, apparently synchronized with the nurtural demands and response attainments of a growing litter, present a more complex picture than the one considered so far, which included only the maintenance and not these changes in the expression of maternal behavior. Moreover, we have viewed the young merely as passive recipients of care, not as agents which influence the kind of care they receive. What we must now do is consider the interaction between mother and litter. We believe we have discovered a mechanism whereby an important dimension of that interaction is expressed.

Young rats during the first 10 to 12 days of life remain in the nest; and it is the mother, for the most part, who characteristically

approaches to initiate nursing and other care-taking activities.
Beginning, however, at about 14 days postpartum, the pups become
sufficiently mobile to leave the nest; and it is they, in turn,
who approach the mother for nursing (Rosenblatt, 1965; Wiesner and
Sheard, 1933). Since it would be maladaptive for freely moving
young to seek out a nonlactating adult for a nursing bout, the
question arose whether the young can discriminate between a lactating and a nonlactating female and if so, on what basis the discrimination is made.

We observed that a nulliparous female housed in the same cage
as a mother was rarely approached by young and that the lactating
mother had an odor noticeably different from her nulliparous coun-
terpart. It seemed likely, therefore, that the pups' discrimination was based on olfactory cues. To test this assumption, we
(Leon and Moltz, 1971) built a discrimination apparatus, pictured
in Figure 4, designed to permit approach across an open field from
a start box to either of the two goal boxes. A 16-day lactating
mother, a nulliparous female, or no stimulus animal at all were presented in the goal boxes in different paired combinations. Opaque
plexiglas separated these boxes from the rest of the apparatus so
that their contents were not visible from the open field. The pups,
however, could enter either goal box by descending a small cliff.
Forced air was introduced into each goal compartment from a central
valve to pass, from there, up through the cliff-opening to the
larger area of the apparatus and then to the start box.

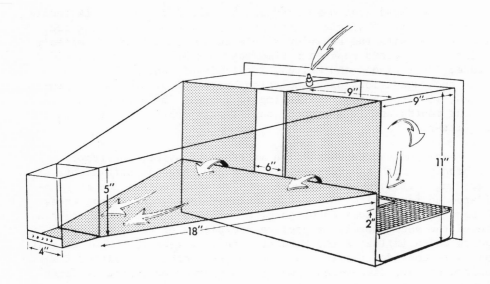

Fig. 4. Olfactory discrimination apparatus. Arrows indicate direction of airflow; dimensions are shown (from Leon and Moltz, 1971).

When 16-day-old young were given a choice between their own 16-day lactating mother and a nulliparous female, they overwhelmingly chose their own mother. And when a strange 16-day lactating female, rather than their own mother, was opposed to a nulliparous female, the young again chose the lactating female; thus each lactating mother does not emit a unique odor to which only her own litter will respond.

To demonstrate that the preference was directed by odor and not by other cues which might have come from the goal compartments, we used two other experiments. In the first, a strange 16-day lactating female and a nulliparous female were placed individually in the goal compartments for three hours and then removed. Immediately afterwards, the pups were required to choose between the now soiled but otherwise empty compartments. In the second experiment, a nulliparous female and a 16-day lactating female were placed in the goal compartments; but the airflow through the central valve was reversed so that the air within the goal compartments flowed *away* from rather than *towards* the pups. Under both of these conditions, the preference for the lactating female was odor-based; required to choose between two soiled boxes, the pups chose the box previously occupied by the lactating animal; but when the airflow was reversed, they showed no preference for the lactating female.

The aim of our next study (Leon and Moltz, 1972) was two-fold: 1) to establish when this maternal pheromone appears during the postpartum period and for how long it is effectively emitted; and 2) to ascertain when the young begin to respond to the maternal pheromone and for how long they maintain this response. We hoped that the data would provide insight into what role, if any, the pheromone actually plays in synchronizing the mother-young relationship.

Accordingly, with our olfactory discrimination apparatus, we systematically tested young of different chronological ages against mothers of different lactational ages. A striking synchrony between mothers and young in the development and dissolution of what we call the "pheromonal bond" emerges. This synchrony is expressed in two ways: 1) the pheromone was first released by the mother at about 14 days postpartum, at exactly the same time that the young first became responsive to the pheromone; and 2) the mother ceased to release the pheromone at about 27 days postpartum, again at the time that the young ceased to be attracted by the pheromone.

Our next step was to determine what role the pheromonal bond plays in synchronizing the mother-young relationship. We have already mentioned that the young begin to leave the nest at about 14 days of age, at the same time that the incidence of maternal retrieving declines sharply. Since nursing continues for another two weeks, mother and young have still to be reunited periodically.

To assure such a reunion, the release of a maternal pheromone and
a responsiveness to it by the young would seem eminently suited.
On the other hand, eventually the young must be weaned; and their
attraction to the mother and to her pheromone must be terminated
accordingly. Weaning generally occurs at about 27 days postpartum
at precisely the time when the mother ceases to emit the pheromone
and the young cease to be attracted to it.

We next investigated the extent to which this synchrony between
emission and attraction is governed by the stimulus properties of
the young. Two findings, in particular, emerged. First, pheromonal
emission was clearly inhibited when our females were made to experience only neonatal pups, i.e., pups which through litter substitution were not allowed to advance beyond Day 1 of age (Moltz and
Leon, 1973). Second, pheromonal emission was just as clearly prolonged by repeated substitution of young between 16 and 21 days of
age, beginning when the female's own litter reached 21 days (Moltz
et al., 1974). Some of our animals continued to release the
pheromone for more than 100 days when new litters were introduced.
Thus, pheromonal emission would appear to be governed by the stimulus characteristics of the young, not by some endogenous condition
of the postparturient female specifically programmed to mediate
release at 14 days and cessation at 27 days.

But what of the physiological mechanisms underlying this pup-
controlled maternal pheromone? We speculated (Leon and Moltz,
1973) about their hormonal nature, since the period just before the
initiation of emission is characterized by high endocrine output
during which ovarian and adrenal steroids are actively discharged,
as is prolactin from the adenohypophysis. Prolactin, in particular,
seemed a likely candidate for a governing mechanism. Not only is
it characteristically present at high titers throughout the first
half of the postpartum period (Amenomori et al., 1970); but it is
released in response to pup stimulation (Grosvenor, 1965; Moltz et
al., 1969b; Sar and Meites, 1969), a condition evidently essential
for pheromonal emission. Accordingly, we investigated the role of
prolactin along with the role of ovarian and adrenal steroids. One
group of females was ovariectomized on the day of parturition, a
second adrenalectomized, and a third subjected to the combined operation. The most likely possibility, that prolactin alone is critical for pheromonal emission, was explored through daily injections
of ergocornine, which inhibits the release of hypophyseal prolactin
in vivo. At 16 days postpartum, each female was then paired in our
discrimination apparatus with a nulliparous female and tested with
colony young, 16 days old, for the presence of the pheromone. The
data (Table 1) confirm the importance of prolactin and the apparent
unimportance of adrenal and ovarian hormones in the emission of
the maternal pheromone. Ergocornine inhibited the pheromone, which
was restored by prolactin replacement. Removal of the adrenals
and ovaries, on the other hand, had no effect on the release

of the attractant.

We have in the maternal pheromone what seems to be an instrument of adaptation, one that is essential for synchronizing the behavior of mother and litter during the latter half of the maternal episode. However, caution must be exercised, for we do not know how, if at all, the pheromone operates outside the laboratory, particularly when mother and young are free to withdraw completely from each other and when species members of all ages and both sexes are present. For such a situation, we have only the observations of Calhoun (1962) that after long periods of separation, wild, pre-weanling rats are often found reunited with their mothers in various areas of the burrow system. Calhoun was unaware of the possibility of a pheromonal attraction, but such attraction now seems obvious.

TABLE 1.   Choice Behavior of 16-Day-Old Young

Condition of Adult Females	Number of Young Choosing Mothers	Number Choosing Nulliparous Females	No Choice	Significance
Intact Control	110	4	6	< 0.001
Ergocornine	53	65	7	> 0.05
Ergocornine and Prolactin	88	13	19	< 0.001
Adrenalectomized	101	17	2	< 0.001
Ovariectomized	106	8	6	< 0.001
Adrenalectomized-Ovariectomized	82	24	14	< 0.01
Adrenalectomized-Ovariectomized and Prolactin	95	8	17	< 0.001

(from Moltz and Leon, 1973)

REFERENCES

Amenomori, Y., Chen, C.L., and Meites, J., 1970, Serum prolactin levels in rats during different reproductive states, Endocrinology 70:506-510.

Beach, F.A., and Jaynes, J., 1956, Studies of maternal retrieving in rats, II. Effects of practice and previous parturitions, Amer. Naturalist 90:103-109.

Ben-David, M., 1968, Mechanism of induction of mammary differentiation in Sprague-Dawley rats by perphenazine, Endocrinology 83:1217-1223.

Ben-David, M., Dikstein, S., and Sulman, F.G., 1965, Production of lactation by non-sedative phenothiazine derivatives, Proc. Soc. Exp. Biol. Med. 118:265-270.

Bruce, H.M., 1961, Observations on the suckling stimulus and lactation in the rat, J. Reprod. Fertil. 2:17-34.

Calhoun, J.B., 1962, *The Ecology and Sociology of the Norway Rat*, Public Health Service Pub. No. 1008, U. S. Public Health Service, Bethesda, Md.

Dilley, W.G., and Adler, N.T., 1968, Postcopulatory mammary gland secretion in rats, Proc. Soc. Exp. Biol. Med. 129:964-967.

Doyle, C., Shimuzu, A., and Himwich, H.E., 1968, Effects of chronic administration of some psychoactive drugs on EEG arousal on rabbit, Int. J. Neuropharm. 7:87-95.

Eisenfeld, A.J., and Axelrod, J., 1965, Selectivity of estrogen distributions in tissues, J. Pharm. Exp. Ther. 150:469-475.

Eisenfeld, A.J., and Axelrod, J., 1966, Effect of steroid hormones, ovariectomy, estrogen pre-treatment, sex and immaturity on the distribution of ^{3}H-estradiol, Endocrinology 79:38-42.

Eto, T., Masuda, H., Suzuki, Y., and Hsai, T., 1962, Progesterone and pregn-4-en-20α-ol-3-one in rat ovarian venous blood at different stages in the reproductive cycle, Jap. J. Anim. Reprod. 8:34-40.

Fajer, A.B., and Barraclough, C.A., 1967, Ovarian secretion of progesterone and 20α-hydroxy-pregn-4-en-3-one during pseudo-pregnancy in rats, Endocrinology 81:617-622.

Grindeland, R.E., McCulloch, W.A., and Ellis, S., 1969, Radio-immunoassay of rat prolactin (Unpub. paper, 51st meeting,

Endocrine Soc., New York, June, 1969).

Grota, L.J., and Eik-Nes, K.B., 1967, Plasma progesterone concentrations during pregnancy and lactation in the rat, J. Reprod. Fertil. 13:83-91.

Grosvenor, C.E., 1965, Evidence that exteroceptive stimuli can release prolactin from the pituitary gland of the lactating rat, Endocrinology 76:340-342.

Hamburg, D., 1966, Effects of progesterone on behavior, in Endocrines and the Central Nervous System (R. Levine, ed.), pp.251-263, Williams & Wilkins, Baltimore.

Hashimoto, I., Henricks, D.M., Anderson, L.L., and Melampy, R.M., 1968, Progesterone and pregn-4-en-20α-ol-3-one in ovarian venous blood during various reproductive stages in the rat, Endocrinology 82:333-341.

Heimer, L., and Nauta, W.J.H., 1969, The hypothalamic distribution of the stria terminalis in the rat, Brain Res. 13:284-297.

Kawakami, M., and Sawyer, C.H., 1959, Neuroendocrine correlates of changes in brain activity thresholds by sex steroids and pituitary hormones, Endocrinology 65:652-668.

Kuhn, N.J., 1968, Lactogenesis in the rat. Metabolism of uridine diphosphate galactose by mammary gland, Biochem. J. 106: 743-748.

Kuhn, N.J., 1969, Progesterone withdrawal as the lactogenic trigger in the rat, J. Endocrinol. 44:39-54.

Kuhn, N.J., and Lowenstein, J.M., 1967, Lactogenesis in the rat. Changes in metabolic parameters at parturition, Biochem. J. 105:995-1002.

Kwa, H.G., and Verhofstad, F., 1967, Prolactin levels in the plasma of female rats, J. Endocrinol. 39:455-456.

Lehrman, D.S., 1961, Hormonal regulation of parental behavior in birds and infra-human mammals, in Sex and Internal Secretions, 3rd ed. (W.C. Young, Ed.), Vol. 2, pp.1268-1382, Williams & Wilkins, Baltimore.

Leon, M., and Moltz, H., 1971, Maternal pheromone: Discrimination by pre-weanling albino rats, Physiol. Behav. 7:265-267.

Leon, M., and Moltz, H., 1972, The development of the pheromonal bond in the albino rat, Physiol. Behav. 8:683-688.

Lu, R.H., Koch, F., and Meites, J., 1971, Direct inhibition by ergo-
    cornine of pituitary prolactin release, Endocrinology 89:229-
    233.

Lumas, K.R., and Farooq, A., 1966, The uptake *in vivo* of 1, 2-^{3}H
    progesterone by the brain and genital tract of the rat, J.
    Endocrinol. 36:95-96.

Moltz, H., and Leon, M., 1973, Stimulus control of the maternal
    pheromone, Physiol. Behav. 10:69-71.

Moltz, H., and Robbins, D., 1965, Maternal behavior of primiparous
    and multiparous rats, J. Comp. Physiol. Psychol. 60:417-421.

Moltz, H., and Wiener, E., 1966, Effects of ovariectomy on maternal
    behavior of primiparous and multiparous rats, J. Comp. Physiol.
    Psychol. 62:382-387.

Moltz, H., Geller, D., and Levin, M., 1967, Maternal behavior in
    the totally mammectomized rat, J. Comp. Physiol. Psychol.
    64:225-229.

Moltz, H., Leidahl, L., and Rowland, R., 1974, Prolongation of
    pheromonal emission in the maternal rat, Physiol. Behav.
    (in press).

Moltz, H., Leon, M., Numan, M., and Lubin, M., 1971, Replacement
    of progesterone with a phenothiazine in the induction of
    maternal behavior in the ovariectomized nulliparous rat,
    Physiol. Behav. 6:735-737.

Moltz, H., Levin, R., and Leon, M., 1969a, Differential effects of
    progesterone on the maternal behavior of primiparous and
    multiparous rats, J. Comp. Physiol. Psychol. 67:36-40.

Moltz, H., Levin, R., and Leon, M., 1969b, Prolactin in the post-
    partum rat: Synthesis and release in the absence of suckling
    stimulation, Science 163:1083-1084.

Moltz, H., Lubin, M., Leon, M., and Numan, M., 1970, Hormonal in-
    duction of maternal behavior in the ovariectomized nulliparous
    rat, Physiol. Behav. 5:1373-1377.

Nauta, W.J.H., 1956, An experimental study of the fornix in the rat,
    J. Comp. Neurol. 104:247-273.

Nicoll, C.S., and Meites, J., 1959, Prolongation of lactation in
    the rat by litter replacement, Proc. Soc. Exp. Biol. Med.
    101:81-82.

Numan, M., 1973, The role of the medial preoptic area in the regu-
    lation of maternal behavior (Unpub. Ph.D. diss., Univ. of
    Chicago).

Numan, M., Leon, M., and Moltz, H., 1972, Interference with pro-
    lactin release and the maternal behavior of female rats,
    Horm. Behav. 3:29-38.

Pasteels, J.L., and Ectors, F., 1970, Mode d'action de l'ergocor-
    nine sur la secretion de prolactine, Arch. Int. Pharm. Ther.
    186:195-196.

Pfaff, D.W., 1965, Cerebral implantation and autoradiographic
    studies of sex hormones, in Sex Research: New Developments
    (J. Money, ed.), pp.219-234, Holt, Rinehart and Winston, New
    York.

Pfaff, D.W., 1968, Uptake of ^{3}H-estradiol by the female rat brain:
    An autoradiographic study, Endocrinology 82:1149-1155.

Raisinghani, K.H., Dorfman, R.I., Forchielli, E., Gyermek, L., and
    Genther, C., 1968, Uptake of intravenously administered pro-
    gesterone, pregnanedione, and pregnanolone by the rat brain,
    Acta Endocrin. (Copenhagen) 57:395-404.

Rosenblatt, J.S., 1965, The basis of synchrony in the behavioral
    interaction between the mother and her offspring in the lab-
    oratory rat, in Determinants of Infant Behavior, (B.M. Foss,
    ed.), Vol. 3, pp.3-41, John Wiley & Sons, New York.

Rosenblatt, J.S., 1969, The development of maternal responsiveness
    in the rat, Amer. J. Orthopsychiat. 39:3-45.

Rosenblatt, J.S., and Lehrman, D.S., 1963, Maternal behavior in
    the laboratory rat, in Maternal Behavior in Mammals (H. Rhein-
    gold, ed.), pp.8-58, John Wiley & Sons, New York.

Sar, M., and Meites, J., 1969, Effects of suckling on pituitary
    release of prolactin, GH, and TSH in postpartum lactating
    rats, Neuroendocrin. 4:25-31.

Shinde, Y., Ota, K., and Yokoyama, A., 1964, Lactose content of
    mammary glands of pregnant rats near term: Effect of removal
    of ovary, placenta, and foetus, J. Endocrinol. 31:105-114.

Stone, C.P., 1925, Preliminary note on maternal behavior of rats
    living in parabiosis, Endocrinology 9:505-512.

Stumpf, W.E., 1968, Estradiol-concentrating neurons: Typography
    in the hypothalamus by dry-mount autoradiography, Science

$\underline{162}$:1001-1003.

Thoman, E.B., and Levine, S., 1970, Effects of adrenalectomy on maternal behavior in rats, Devel. Psychobiol. $\underline{3}$:237-244.

Wiesner, B.P., and Sheard, N.M., 1933, *Maternal Behaviour in the Rat*, Oliver and Boyd, Edinburgh.

Wiest, W.G., 1968, On the function of 20α-hydroxypregn-4-en-3-one during parturition in the rat, Endocrinology $\underline{83}$:1181-1184.

Wiest, W.G., Kidwell, W.R., and Balogh, K., 1968, Progesterone catabolism in the rat ovary: A regulatory mechanism for progestational potency during pregnancy, Endocrinology $\underline{82}$:844-860.

Wuttke, W., Cassell, E., and Meites, J., 1971, Effects of ergocornine on serum prolactin and LH, and on hypothalamic content of PIF and LRF, Endocrinology $\underline{88}$:737-741.

Yoshinga, K., Hawkins, R.A., and Stocker, J.F., 1969, Estrogen secretion by the rat ovary *in vivo* during the estrous cycle and pregnancy, Endocrinology $\underline{85}$:103-112.

NEURAL AND HORMONAL INTERACTIONS IN THE REPRODUCTIVE

BEHAVIOR OF FEMALE RATS[1]

Barry R. Komisaruk

Rutgers University, The State University of New Jersey

Institute of Animal Behavior, Newark

I. INTRODUCTION

The lordosis posture in female rats (elevation of the rump and head and depression of the back) is a hormone-sensitive reflexive response to tactile stimulation of the flank region. It is normally elicited by the rapid pelvic thrusting movements of the male. After the male dismounts, the female maintains the lordosis posture longer if an intromission occurred than if one did not (Kuehn and Beach, 1963; Diakow, 1974). This finding suggests that stimuli provided by genital tract stimulation facilitate the lordosis response. If the genital tract is denervated, the prolongation of lordosis resulting from intromission does not occur (Diakow, 1970). The following studies, which focus on the effect of genital tract stimulation on lordosis, demonstrate not only that genital tract stimulation (a visceral stimulus) facilitates the effect of flank stimulation (a somatic stimulus) on lordosis, but also that this facilitating effect persists for several hours after the cessation of the stimulus. In contrast to the facilitating effect, the same genital tract stimulation *abolishes* different skeletal reflex responses (e.g., leg withdrawal to foot pinch). Furthermore, it blocks behavioral and neurophysiological responses to painful stimulation.

Genital tract stimulation has been shown to affect neuroendocrine

[1]This article is Contribution No. 184 of the Institute of Animal Behavior. These studies were supported by Research Grant NIMH-MH 13279, Research Scientist Development Award Type II S-K02-MH 14711, and Training Grant GM 1135 and by funds from the Alfred P. Sloan Foundation. The excellent technical assistance of Ms. E. Azevedo, Ms. M. Buntin, and Ms. R. Harris is gratefully acknowledged.

processes, by inducing: 1) ovulation (Zarrow and Clark, 1968; Aron et al., 1968; Harrington et al., 1967); 2) pregnancy and pseudopregnancy (Kollar, 1953; Carlson and DeFeo, 1965; Adler, 1969); and 3) milk letdown (Cross, 1959; Hays and Vandemark, 1953). The present findings demonstrate unexpected and potent effects of this stimulation on sensorimotor processes as well.

II. INFLUENCE OF INTROMISSION AND VAGINAL STIMULATION ON THE LORDOSIS RESPONSE

A. The Role of Vaginal Stimulation in Facilitating Lordosis

Previous investigators showed that when the vaginal cervix was probed with a glass rod in intact (Carlson and DeFeo, 1965) or decorticate female rats (Beach, 1944, and personal communication) held in the hand, the rats showed a lordosis response. We examined this effect in greater detail in intact rats during the estrous cycle, three weeks after ovariectomy, and after treatment with selected dosages of estrogen (Komisaruk, 1971; Komisaruk and Diakow, 1973). Lordosis intensity was measured photographically as the elevation in millimeters of rump and head above the lowest portion of the back. When appropriate, a subjective estimate of lordosis intensity was noted on a seven-point intensity scale. All studies were performed with blind procedures.

1. Lordosis in ovariectomized, hormonally untreated females. Probing the vaginal cervix of 98 ovariectomized, hormonally untreated rats held by the base of the tail did not elicit lordosis in any of them; nor did palpating the flanks and perineum. However, probing the cervix while the flanks and perineum were being palpated elicited lordosis in 48% of these females. Lordosis was defined as elevation of both the head and rump at least above the lowest point on the back; such responses were scored at least in the upper three of the seven-point scale (the cut-off point when a dichotomy into lordosis and no lordosis was used). Of the 98 rats tested, 9.2% showed an "excellent" lordosis, i.e., marked elevation of the rump and head. (All the rats used in these lordosis experiments were Sprague-Dawley derivatives obtained from Camm Research Laboratories, Wayne, New Jersey. In unpublished studies with Sprague-Dawley derivatives obtained from Charles River Laboratories, North Wilmington, Massachusetts, almost all ovariectomized, hormonally untreated animals showed lordosis in response to cervical probing with flank-perineum palpation; the basis of this difference is not known.)

2. Lordosis in relation to the estrous cycle. Before ovariectomy, each rat was tested for lordosis in response to cervical probing in conjunction with flank-perineum palpation at one selected stage of its estrous cycle. Comparison of the group mean lordosis intensities, showed that during proestrus (P) and the transitional stages from

diestrus to proestrus (DP) and proestrus to estrus (PE), the lordosis
intensities were significantly higher than during estrus (E), metestrus (M), and diestrus (D). Although the intensity of lordosis and
the proportion of females showing lordosis during DP, P, or PE
(61.9%) were significantly greater than during the rest of the estrous cycle, 44.0% showed lordosis during E, M, or D, when females
are normally unreceptive to males (Young et al., 1941).

3. Facilitatory effect of estrogen on lordosis; relative sensitivity
of mating, vaginal smear, and the lordosis reflex. Forty to 46 hours
before being tested for mating responses to males, the ovariectomized
rats were treated with a single subcutaneous injection of 0.01, 0.1,
1.0, or 10.0 µg estradiol benzoate (EB)/100 g bw and no progesterone.
Even at the highest dosage, only three of nine females showed spontaneous lordosis responses to the males, an indication of the desired low level of estrogenic stimulation (Fig. 1). In contrast, 23
of 26 females in the three highest estrogen-dosage groups showed
lordosis in response to cervical probing in conjunction with palpation of the flanks and perineum.[1] Lordosis was not induced at any
estrogen dosage by either cervical probing alone or palpation alone.
The sensitivity of this behavioral response to estrogen is at least
as great as the sensitivity of cornification of the vaginal smear to
estrogen. The criterion for cornification was selected to permit e-
ven weak effects of estrogen to be considered positive. Thus, a smear
was considered cornified if more than about 10% of the cells were of
the squamous type. The smears were rated in a blind procedure. At
the two highest estrogen dosages, cornification was observed in all
18 females and lordosis in 16 of 18. At the next lower dose (0.1 µg
EB/100 g bw), cornification occurred in only three of eight females;
whereas lordosis occurred in seven of the same eight females. Thus,
this reflex response appears to be somewhat more sensitive, and
certainly not less sensitive, to estrogen than cornification of the
vaginal smear (Komisaruk and Diakow, 1973).

4. Differential blockage of estrogenic effects by the antiestrogen,
MER-25. Our findings with the antiestrogen, MER-25 (ethamoxytri-
phetol) (Komisaruk and Beyer, 1972), also support this conclusion that
the lordosis response to cervical probing is hypersensitive to estrogen. Ovariectomized rats were treated daily for 10 days with
either 2 µg EB or sesame oil vehicle. MER-25 in suspension was injected subcutaneously two hours earlier each day. A high (75 mg/kg)
or low (25 mg/kg) dosage of MER-25 or an oil control was administered. In contrast to the oil control group, the group that received
EB without MER-25 showed a significant increase in uterine weight,
number of fully cornified smears, number of females showing lordosis

---

[1]Unless otherwise specified, the phrase "in conjunction with palpation of the flanks and perineum" will be abbreviated to "with palpation."

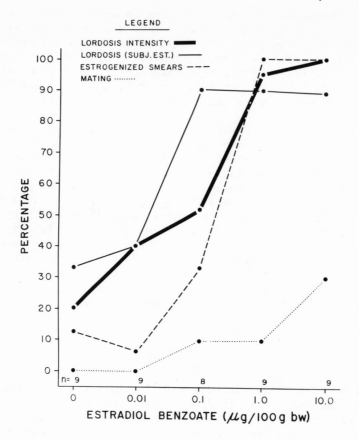

Fig. 1. Relative responses of several estrogen indicators at increasing estradiol benzoate (EB) dosage levels. Lordosis intensity was converted to percentage by comparing each group mean to the group mean at the highest EB dosage. The other indicators represent the percentage of rats in each dosage group (n's shown along abscissa) showing: (a) at least good lordosis on the basis of a subjective estimate; (b) at least 10% of the cells in each vaginal smear to be cornified; or (c) at least one lordosis in response to ten male mounts.

in response to mounting by males, and the number of females showing lordosis in response to cervical probing (with palpation). Administration of either the high or the low dosage of MER-25 blocked the stimulatory effects of estrogen on each of the indicators except lordosis in response to cervical probing with palpation, evidence

that this response is more sensitive than the other indicators to
estrogen. There was actually some evidence of an estrogenic effect
of MER-25 alone: four of eight females in the high dosage MER-25
group showed lordosis in response to cervical probing with palpation;
whereas only one of eight in the low dosage group showed lordosis to
this stimulus. Thus, by competing with estrogen, MER-25 blocks estrogenic effects on relatively insensitive indicators, but on the
relatively sensitive measures, such as partial cornification of the
vaginal smear or lordosis in response to cervical probing with palpation, it acts as a weak estrogen by itself.

5. Synergism between estrogen and progesterone in facilitating lordosis. Although progesterone is not essential for the lordosis response
in rats (estrogen is sufficient: Davidson et al., 1968; Edwards et
al., 1968), it strongly facilitates the lordosis response to mating
attempts by males (e.g., Edwards et al., 1968). When we added a constant dosage of progesterone to graded dosages of estradiol benzoate,
we observed a synergistic effect on lordosis intensity induced by
cervical probing with palpation (Diakow et al., 1973). Mean lordosis
intensities (in millimeters of dorsiflexion; cf. Komisaruk, 1971, for
measurement technique) at EB dosages (single subcutaneous injection
40 hours before testing) of 10.0, 1.0, 0.01, or 0 µg/100 g bw were
40.2, 18.8, 6.7, and 10.5, respectively. When 0.5 mg progesterone
was added subcutaneously four hours before testing to four subgroups
at the same EB dosages, the mean lordosis intensities were 41.0, 43.1,
13.5, and 16.8, respectively. Without progesterone, the lordosis
reflex intensity increased significantly only at the highest estrogen
dosage. However, when progesterone was added to the same estrogen
dosages, there was a significant increase in lordosis reflex intensity
at the next lower estrogen dosage as well as in the highest estrogen-
dosage groups. The lordosis response induced either by cervical
probing with palpation or by normal mating attempts by males has the
properties of sensitivity to estrogen and the synergistic action
of progesterone.

6. Induction of lordosis responsiveness by previous vaginal stimulation. A previous study (Komisaruk and Diakow, 1973) indicated that
ovariectomized rats showed lordosis in response to palpation after,
but not before, they had received cervical probing. This finding
implied that the effects of cervical probing could persist beyond
the actual application of this stimulus. In a subsequent systematic
investigation (Rodriguez-Sierra et al., 1974), we confirmed this
effect and demonstrated that lordosis in response to flank-perineum
palpation alone can be induced by previous probing of the cervix.
That is, even though cervical probing by itself was almost completely
ineffective in inducing lordosis, it enabled a previously ineffective
stimulus (flank-perineum palpation alone) to become effective within
seconds after the cessation of the probing stimulus (the rats were
first tested within 10 to 15 seconds of the start of application of
the probing stimulus, which lasted about two seconds). The effect

Barry R. Komisaruk

of the previous application of the probing stimulus persisted for
several hours: at two hours, 50% of 12 ovariectomized females not
treated with estrogen were still showing lordosis in response to
flank-perineum palpation alone; and five hours after the single application of the cervical probing, 33% were still showing lordosis in
response to flank-perineum palpation alone (Fig. 2). We are currently
investigating the nature of this "trigger" process.

7. Induction of sexual receptivity by vaginal stimulation. In a
further test of the lordosis-facilitating effect of cervical probing,
we determined whether unreceptive females could be made receptive to
sexually active males by cervical probing. Using ovariectomized,
estrogen-primed (5 μg EB/kg x 2) females (Sprague-Dawley from Charles
River), we selected 36, each of which showed no lordosis to five
mounts by vigorous males. Eighteen were subjected to cervical probing
in conjunction with flank-perineum palpation, and the others were only
handled for a comparable period. Immediately after these treatments,
each female was returned to the male's cage and allowed to receive
five additional mounts. One of 18 handled females then showed lordosis in response to the male; whereas eight of the 18 probed females
showed lordosis in response to mounting attempts by the male (X^2 =
5.33; p< 0.05). The lordosis quotient of the responding females was
50. Initially, 11 of the 18 probed females vs. one of the 18 non-
probed females (X^2 = 12.50; p< 0.001) kicked away the males but ceased
rejecting them within several minutes.

8. Role of the adrenals in mating-induced facilitation of lordosis.
Recently we have found (Larsson et al., 1974) that repeated testing
of ovariectomized female rats with males every two hours resulted in
a significant increase in lordosis quotient (LQ) by the sixth hour
(mean LQ: 0.20, 0.40, and 0.80 at 2, 4, and 6 hr, respectively); but
this increase did not occur in females which were also adrenalectomized (mean LQ: 0.11, 0.34, and 0.17 at 2, 4, and 6 hr, respectively).
This facilitatory effect of repeated mating attempts by males may
be due to the secretion of adrenal progestins which could facilitate
lordosis since ACTH induces receptivity in ovariectomized estrogen-
primed rats (Feder and Ruf, 1969). However, this type of facilitation
mediated by the adrenal is probably different from that involved in
the prolonged facilitation of lordosis resulting from cervical probing. That is, even if cervical stimulation released progestin (perhaps as in the induction of pseudopregnancy), this alone without estrogen would not facilitate lordosis (Diakow et al., 1973). Therefore, prolonged facilitation of lordosis by cervical probing (section
II-A,6) in ovariectomized, estrogen-untreated rats is probably not
due to the reflex release of progestin.

B. *Neural Factors in Lordosis*

1. Enlargement of the genital sensory field by estrogen. Estrogen
and progesterone could facilitate lordosis, by increasing motor res-

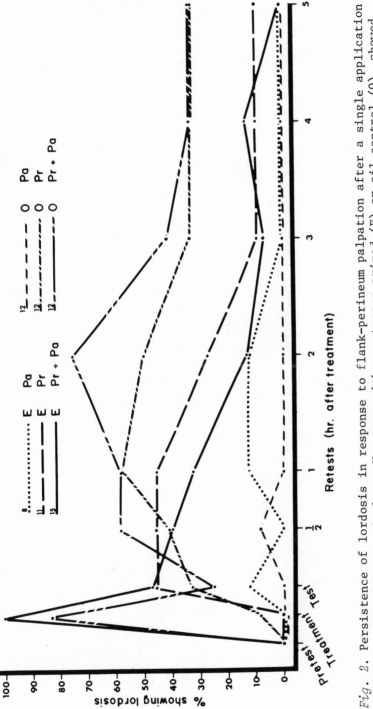

Fig. 2. Persistence of lordosis in response to flank-perineum palpation after a single application of the vaginal probing stimulus. No rats, either estrogen-primed (E) or oil control (0), showed lordosis in response to flank-perineum palpation in the Pretest. At Treatment rats received either cervical probing only (Pr), flank-perineum palpation only (Pa), or both together (Pr + Pa). On Tests, administered 10 sec, 30 min, and 1, 2, 3, 4, and 5 hr after Treatment, the rats were tested for lordosis in response to flank-perineum palpation only (no cervical probing).

ponsiveness, sensitivity to adequate peripheral stimulation, and/or
responsiveness of integrative processes. Various hormone-induced
modifications of neural activity have been described (for reviews,
cf. Beyer and Sawyer, 1969; Komisaruk, 1971); but they do not identify sensory, motor, or integrative mechanisms as the primary site(s)
of hormone action.

To determine whether estrogen modifies sensory activity, we recorded neuronal activity in the pudendal nerve, which receives input
from the perineal region (Komisaruk, Adler, and Hutchison, 1972, 1973).
Treating ovariectomized rats with EB (2 µg/day every other day for
several weeks) significantly increased the size of the genital sensory
field (increase in length: 22.0%; width: 26.3%; area: 31.9%; field
of the clitoral region: 75.0%). The sensory field was measured by
determining the limits of the body surface which, when brushed or
scratched mechanically, generated action potentials in the nerve.
The estrogen-induced enlargement of field size persisted after the
pudendal nerve had been cut, eliminating its efferent activity; only
the activity from the distal portion was recorded. (We cannot, however, exclude the possibility of efferent control by other nerves
innervating the same region). This finding indicates that estrogen
may modify sensory activity by acting directly at the periphery. The
sensitivity of the genital sensory field was lowest at the borders
of the field and increased toward the midline. Increasing the size
of the field may assist the female to orient to the copulatory thrusts
by the male. Kow and Pfaff (1973) have confirmed this increase in
the size of the genital sensory field measured by recording from the
pudendal nerve and demonstrated an increase in the tactile sensitivity
of the field.

2. Intense lordosis in ovariectomized hormonally untreated rats as
a result of ablation of the septum. How would the estrogen-sensitive
lordosis response be affected if the responsiveness of the rats to
sensory stimulation were increased by nonhormonal means? Hyper-
responsiveness to sensory stimulation is one of the major characteristics of rats with lesions of the septum. For example, septal-lesioned rats show increased numbers of responses to foot shock over a range
of suprathreshold shock intensities (Lubar et al., 1970; Doty and
Forkner, 1971). However, the current intensity required to elicit
jump in 50% of the trials may (Doty et al., 1971) or may not (Lubar
et al., 1970) be substantially lower in the septal-lesioned rats than
in the controls. Thus, septal lesions appear to increase the responsiveness, but not necessarily the sensitivity, to sensory stimuli.
We had already shown that lordosis can be induced in 48% of 96 ovariectomized, untreated rats by adding cervical probing to flank-perineum
palpation. When we applied this stimulation to ovariectomized, hormonally untreated rats sustaining complete suction ablation of the
septum, we observed an intense lordosis response, the median intensity
of which was not less than that induced by optimal estrogen plus progesterone treatment (Fig. 3) (Komisaruk, Larsson, and Cooper, 1972).

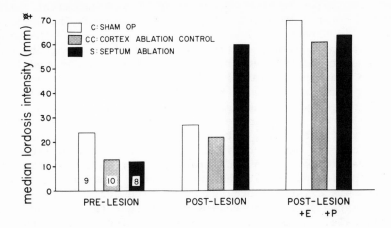

Fig. 3. Intensification of lordosis as a result of septal ablation
in ovariectomized rats in the absence of estrogen. Lordosis was
elicited by cervical probing in conjunction with flank-perineum palpation.

This was a marked increase over their preoperative levels and those
of sham-operated or cortex-lesioned controls. These nonhormonally
treated females did not show lordosis in response to mounting attempts
by vigorous males, but they did show a significant increase in lordosis responsiveness after daily injections of 1 μg EB/100 g bw/day.
In the sense that the septal-ablated rats showed this increase after
three rather than four days of estrogen treatment, they were more
sensitive to the estrogen than were the controls. Powers and Valenstein (1972) have shown a decrease in the amount of estrogen necessary
to induce receptivity in rats with medial preoptic area lesions.
These findings suggest the existence of a septo-preoptic system which
is normally inhibitory to the lordosis response. This system may be
inhibited by estrogen and thus release the lordosis response.

3. Lordosis in males after septal ablation in the absence of estrogen.
The lordosis-facilitating effect of cervical probing is mediated by
the pelvic nerve because pressure on the cervix activates neuronal
activity in the nerve (Komisaruk, Adler, and Hutchison, 1972) and
transection of the nerve abolishes the lordosis-facilitating effect
of cervical probing (C. Diakow, in preparation). The pelvic nerve
is also activated by mechanical stimulation of the rectum (Komisaruk,
Adler, and Hutchison, 1972). Rectal probing is almost as effective
as cervical probing in facilitating lordosis in female rats (Komisaruk, 1971; Komisaruk and Diakow, 1973). In intact males not treated
with estrogen, rectal probing with palpation did not induce lordosis;
but after a single injection of 10 or 100 μg EB/100 g bw subcutaneous-

ly, four of five males in each dosage group showed excellent lordosis
about 42 hours later (Komisaruk, 1971). We (Komisaruk, K. Larsson,
and R. Cooper, in preparation) removed the septum in six otherwise
intact males and performed a control procedure of ablating the cortex
overlying the septum in six others. None of the males showed lordosis
in response to flank-perineum palpation alone; but when we added the
rectal probing stimulus, five of the six septal males and none of the
cortical controls showed lordosis. Thus, it appears that the septum
normally inhibits the lordosis response in males as well as females,
an indication that the neural circuitry underlying the lordosis response is present and may be activated in males as well as females.
Since high levels of estrogen facilitate the lordosis response in
intact males (Davidson, 1969), perhaps the same central neural circuits for lordosis exist in both sexes; and the sex difference lies
in the lower degree of responsiveness rather than the absence of the
system in the male. In both sexes, input from the pelvic nerve
facilitates the lordosis reflex.

III. INFLUENCE OF VAGINAL STIMULATION ON SENSORY-MOTOR ACTIVITY

A. Blockage of Spinal and Cranial Withdrawal Reflexes and Simultaneous Facilitation of the Lordosis Reflex by Vaginal Stimulation

When pressure was exerted against the vaginal cervix with a glass
rod, we observed that rats became immobile and could be slid stiff-
legged along a table top without any restraint (Komisaruk and Larsson,
1972). This immobilization also resulted in a marked suppression or
total abolition of reflex leg withdrawal in response to foot pinch and
of facial movements in response to ear pinch. There was also a partial suppression of the righting reflex and of the eye blink response
to touching the cornea. Transection of the spinal cord at the mid-
thoracic level still enabled cervical probing to block leg withdrawal
to foot pinch, even though the cranial reflexes were no longer suppressed by cervical probing. Thus the blocking effect occurs at both
the lower spinal cord level and at the level of the cranial nuclei.

The fact that the same cervical probing stimulus that blocks
these withdrawal reflexes simultaneously facilitates the lordosis
response to flank-perineum stimulation suggests a visceral sensory
process which reflexly shifts central nervous mechanisms from flexor
to extensor activation. This suggestion is discussed in section IV-D.

B. Reflex Blockage: A Motor Effect?

To determine whether reflex blockage occurs at the motor level,
we stimulated the pyramidal tract directly through electrodes in the

medulla of six rats (J. Wallman, J. Hassenbey, and Komisaruk, in
preparation). Each showed movements of the face or fore- or hindlimb.
Cervical probing never suppressed these movements even though the
current levels were just suprathreshold for observing the movements
(Fig. 4). Thus, cervical probing blocks the withdrawal reflexes by
acting elsewhere than at the final common pathway from the pyramidal
tract to the motor neurons.

C. *Reflex Blockage: A Sensory Effect? Selective Blockage of*
   Responses to Painful Stimulation by Probing the Vaginal Cervix

1. Neurophysiological evidence. To determine whether the blockage
occurs closer to the sensory level, we recorded neuronal activity
in the lateral thalamic (sensory) nuclei in rats anesthetized with
urethane (Komisaruk and Wallman, 1973). The neurons selected for
study showed responses to light brushing of the fur as well as to
noxious stimulation (pinching the tail, a paw, or an ear). When
the cervix was probed, responses to brushing the fur lightly were
unaffected; but responses to noxious stimulation were markedly suppressed or almost entirely abolished (Fig. 5). This result demonstrated a selective blockage of responses to intense, painful, but
not to mild tactile stimulation. Thus the cervical probing acts as
an analgesic but not anesthetic stimulus. When pressure was applied
continuously to the cervix, the blockage of responsiveness to painful
stimulation persisted for several minutes; recovery of responsiveness
to painful stimulation began within 30 seconds after cessation of the
stimulus (Fig. 6).

It could be argued that cervical probing is itself a painful
stimulus which "distracts" the rat from painful stimulation applied
to the body surface This does not seem to be the case, however,
because noxious stimulation applied to one part of the body surface
did not block the response to noxious stimulation applied to another
part of the body surface. For example, thalamic neurons were activated by pinching the tail and the activity recorded; but pinching
the toe, although the rat responded behaviorally by withdrawing its
foot, did not activate the neurons at the particular recording site.
When a toe pinch was applied concurrently with the tail pinch, the
thalamic neurons were still activated by the tail pinch. However,
when the cervix was probed concurrently with the tail pinch, those
neurons were not activated. Thus, cervical probing specifically
blocked the neuronal response to tail pinch; but other "distracting"
stimuli did not. Similar specificity was observed in behavioral
observations in which "distracting" the rats by pinching the tail did
not block leg withdrawal to the foot pinch; whereas probing the cervix
did.

Another explanation for the blocking effect of cervical probing
might be that it induces a generalized EEG sleeplike condition (Rami-

Barry R. Komisaruk

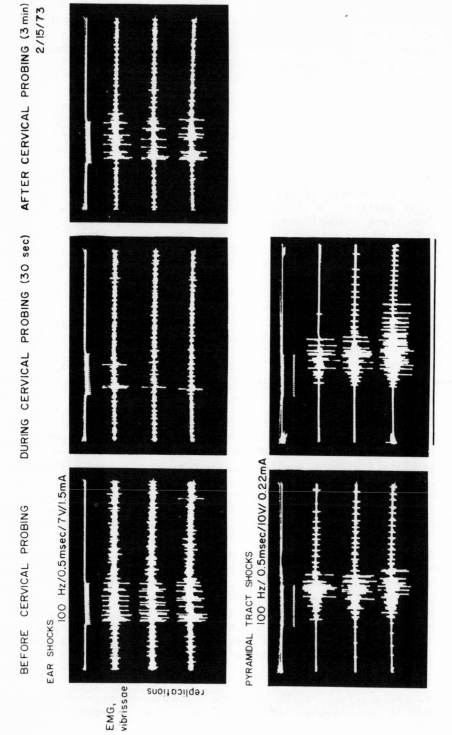

Fig. 4. Differential effects of cervical probing; blockage of reflex response to noxious stimulation (top) but not of response to direct electrical stimulation of the pyramidal tract (bottom).

a

A 4.6 / L 2.7 / V - 5.555   rat 647

tap forearm   thalamic neurons

                                                         cortical EEG

probe cervix (30 sec)
& tap forearm

pinch forearm

                                                         100 uV

probe cervix (30 sec)
& pinch forearm

—— I sec

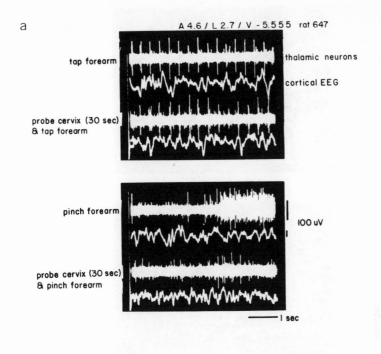

rat 646

A 4.2/L 2.7/V-5.430

b

BEFORE CERVICAL PROBING   DURING CERVICAL PROBING (30 sec)

brush fur (thorax)   thalamic
                                                         neurons

                                                         cortical
                                                         EEG

pinch hind foot

                                                         —— I sec

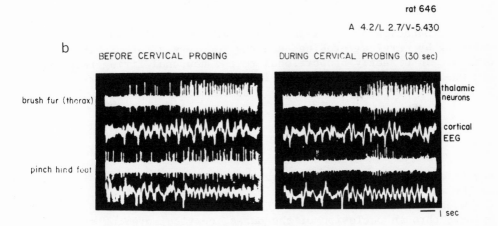

Fig. 5.   a) Suppression of thalamic neuronal response to pinch but
not touch by cervical probing. Repetitive tapping throughout the
top set of records generated corresponding neuronal responses. Pinch
was applied only during the middle portion of the lower set of records. b) Similar results as in (a); but in this slightly different
format, pinch and repetitive tapping were applied only during the
second half of each record.

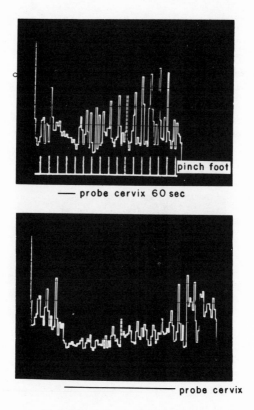

Fig. 6. Duration of blockage of thalamic neuronal response to pinching by cervical probing, and the recovery after probing. The ordinate is directly proportional to the neuronal firing rate. Pinching was applied repetitively as indicated by vertical marks adjacent to "pinch foot" (these marks apply to both records).

rez et al., 1967), which would reduce the responsiveness of neurons to the arousing noxious stimulation. This seems unlikely, however; for when we elicited a state of EEG arousal by inducing a temporary hypoxia through occlusion of the nares, cervical probing still blocked the response to painful stimulation (Fig. 7).

We can still not definitively exclude the possibility that the "pain-blocking" effect of cervical probing is related to the arousal condition of the brain. In previous studies (Komisaruk et al., 1967; Ramirez et al., 1967), we observed that neurons show marked changes in relation to the "sleep-arousal" state of the EEG. The great majority of neurons showed an increase of firing rate when the EEG

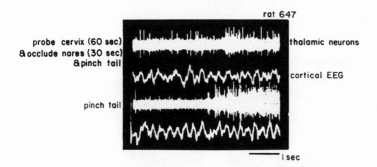

Fig. 7. Arousal by temporary hypoxia does not counteract suppressive effect of cervical probing; for explanation see text.

changed from a sleep-like (high amplitude slow waves) to an aroused (either low amplitude fast waves or low amplitude theta-like waves at 3-4 Hz) condition, either spontaneously or when painful stimulation was applied. In the current study, cervical probing tended to prevent the appearance of theta-like waves in response to pinching (Fig. 8). However, theta waves occasionally appeared even while the thalamic neuronal response to pinching was still blocked by the cervical probing (e.g., Fig. 5b); and conversely, thalamic neuronal responsiveness recovered while the theta-like response was still blocked. Thus, although the neuronal response may be linked with the blockage of the theta-like response, the two effects can be dissociated. Therefore, the cervical probing probably induces a blockage of responsiveness to painful stimulation not by inducing a general depression of brain activity, but rather by selectively altering the responsiveness of thalamic neurons to painful as opposed to tactile stimulation.

2. Behavioral evidence. To test the pain blockage hypothesis derived from these neurophysiological findings, we determined, using a behavioral endpoint, whether cervical probing could block responsiveness to painful stimulation (J. Wallman and Komisaruk, in preparation). We have established that behavioral responsiveness to tactile stimulation is not blocked universally since cervical probing markedly facilitates the effectiveness of light touch and pressure on the flanks in eliciting lordosis (e.g. Komisaruk and Diakow, 1973). The behavioral endpoint for the pain response we selected was vocalization (a squeak) in response to a 60 Hz sine wave stimulus, 50 msec in duration, applied to the tail. Similar vocalization was elicited by the presumably nonpainful procedure of suddenly lifting the rat off a table surface by placing one's open hands under the rat's abdomen and thorax. When cervical probing was applied by a second observer

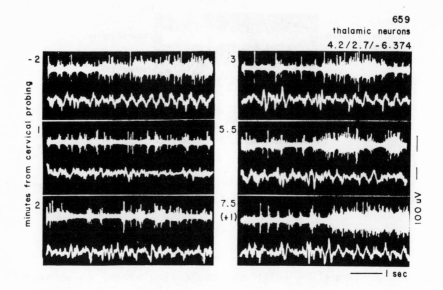

Fig. 8. Blockage of neuronal activation and EEG theta response to foot pinch by cervical probing. Neuronal activation and theta-like activity were induced by pinching the foot. Cervical probing was then applied continuously for 6.5 min, and the pinching stimulus was re-applied repeatedly during the probing. One minute after the initiation of cervical probing, there was no neuronal activation in response to the pinch; and the EEG showed no change, remaining in low amplitude fast activity. Neuronal responsiveness started to recover by 2 min without a clear change in the EEG. By 3 min, neuronal responsiveness was greater; and the theta-like response was beginning to recover. However, at 5.5 min, the neurons responded even though there was no clear change in the EEG. By 7.5 min (1 min after cessation of cervical probing), neuronal responsiveness had clearly recovered and was accompanied by the theta-like activity. Thus, cervical probing can suppress both thalamic neuronal responsiveness and the theta-like EEG response to noxious stimulation.

just before and during lifting the rats in this manner, the mean number of vocalizations *increased* in response to the lifting. Thus, cervical probing did not block the ability of rats to vocalize. On the other hand, when we induced vocalization by shocking the tail at a just suprathreshold constant current intensity which was adjusted for each individual, the vocalization response was abolished in each of 14 rats within 5 - 20 sec after the application of the cervical probing (800 g force applied with the tip of a plunger from a 1 cc tuberculin syringe) (Fig. 9).

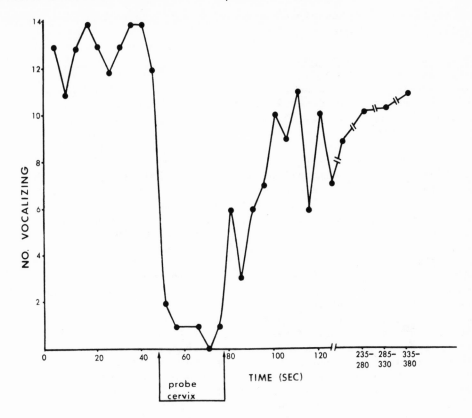

Fig. 9. Suppression of vocalization in response to tail shocks by probing the vaginal cervix in unanesthetized rats. Current intensity was maintained constant for each individual.

        This effect was confirmed with an alternate method: measuring the current threshold required to induce vocalization before, during, and after the cervical probing. A mean increase of 57% and a maximum (of each animal) mean increase of 157% in current intensity over pre-probing intensity was required to induce vocalization during the cervical probing. The vocalization threshold was still elevated significantly three minutes after the start of cervical probing when the probing was applied continuously. Thus cervical probing is a potent stimulus for blocking behavioral responses to painful stimulation. The observation that the ability of the rats to vocalize is not blocked by cervical probing (as in the nonpainful lifting procedure), but is by tail shock strongly suggests that the probing blocks responsive-

ness to pain. This behavioral study thus supports the earlier neurophysiological findings.

3. Blockage of pain responses by somatic as well as visceral stimulation: convergent systems? Since cervical probing potentiates the
effect of flank-perineum palpation in inducing lordosis, in a preliminary experiment we determined whether a similar principle operates
in pain blockage; that is, can flank-perineum palpation alone also
induce pain blockage?

       Estradiol benzoate (30 μg/kg x 2) was administered to 12 ovari-
ectomized rats and sesame oil to another 13. Pain thresholds were
then determined in all rats by determining the current intensity
necessary to elicit vocalization in response to tail shock under four
different sensory stimulation conditions: 1) gentle handling, 2)
vigorous flank scratching, 3) flank-perineum palpation (applying pressure), 4) cervical probing. Stimulation of the flanks (condition 2)
had no significant effect on the vocalization threshold in either
group. However, when lordosis was induced in the estrogen-treated
rats by flank-perineum palpation (condition 3), the mean vocalization
threshold was elevated significantly (26.9%; $\underline{t}$ test for correlated
means: $\underline{t}$ = 3.15, p.< 0.01) over the control mean. The oil-treated
controls showed no lordosis responses to this stimulation and no elevation of these thresholds (actually a decrease of 6.6%). Cervical
probing elevated the vocalization threshold significantly in both
groups, a greater increase occurring in the estrogen-treated group
(82.0%; $\underline{t}$ = 5.24, p < 0.001) than in the oil-treated controls (43.3%;
$\underline{t}$ = 2.40, p < 0.05). Thus, the apparent convergence between somatic
and visceral stimuli observed in the elicitation of lordosis appears
to occur in the blockage of pain responses as well. The implications
of this apparent convergence are discussed in section IV-B.

                              IV.   DISCUSSION

        A.   Possible Adaptive Significance of Pain Blockage
                   Induced by Vaginal Stimulation

       The suppression of pain responses by vaginal stimulation may be
significant in diminishing possibly aversive aspects of parturition
and copulation. In parturition, passage of fetuses through the cervix and vaginal canal applies sustained mechanical stretching and
thereby could diminish the pain resulting from dilatation and stretching of the birth canal and the surrounding skeletal structures.

       Several studies have suggested that during copulation, there
may be an aversive component to vaginal stimulation. (1) Receptive
females given the opportunity to escape from males into a separate
chamber stayed in the escape chamber for the longest period after
ejaculations, and for progressively shorter periods after intromissions, mounts without intromissions, and no sexual contacts (Pierce

and Nuttal, 1961). (2) Similarly, female rats given the opportunity to press a bar and thereby to allow a male to enter their cage
waited longest after ejaculation, less after intromission, and the
least after mounts without intromissions. Furthermore, the prolonged
delay in allowing the male to re-enter the female's cage in the post-
ejaculatroy interval was abolished by anesthetization of the vaginal
canal before copulation (Bermant, 1961; Bermant and Westbrook, 1966).
(3) When females were given a prolonged series of 50 intromissions,
the lordosis quotient was lower in the last 10 than the first 10 intromissions; but this decline did not occur if intromissions were prevented with a vaginal "mask" made of tape (Hardy and DeBold, 1972).

These possibly aversive properties of vaginal stimulation may
be diminished by the concomitant pain-blocking effect of vaginal
stimulation. Since multiple intromissions before ejaculation, and
not simply ejaculation alone, are required for pregnancy in rats
(Adler, 1969), this pain-blocking effect may increase the willingness
of the female to accept multiple intromissions.

B. Lordosis and Pain Blockage

1. Two aspects of the same process? Are the phenomena of facilitation of the effects of somatic stimulation by visceral stimulation
in eliciting lordosis and the blockage of pain responses by visceral
or somatic stimulation related? Both the principle of referred pain
(Ruch, 1961) and the "gate-control" theory of pain (Melzack and Wall,
1965) can account for both these phenomena, indicating that they
may be variations of the same process: convergence of visceral and
somatic input on the same central (tactile) neurons.

Referred pain has been accounted for (Ruch, 1961) by postulating
the existence of individual neurons in the central nervous system
responsive to somatic as well as to visceral noxious stimulation
which originates at the same dermatomal levels. Such convergence of
visceral and somatic input on single central neurons has been demonstrated in the spinal cord (Selzer and Spencer, 1969a, 1969b;
Duda, 1964a, 1964b; Fields et al., 1970; Aidar et al., 1952; Pomeranz
et al., 1968), thalamus (Aidar et al., 1952; McLeod, 1958), and cortex
(Amassian, 1952; Downman, 1951). Thus, when visceral sensory activity
initiated by visceral irritation activates a central neuron, the individual might misinterpret the true origin of the input as being in
the soma of the corresponding dermatome since the central neuron could
also be activated by that somatic input.

Some examples of pain referred to the soma and its associated
visceral disorders are: 1) angina pectoris (pain on the left side
of the chest, shoulder, arm, and elbow) arising from primary arterial
hypertension; 2) chest and scapular pain from lobar pneumonia and
pleuritis; 3) chest pain from esophageal carcinoma; 4) pain in the
epigastric region from stomach ulcer or inflammation of the gall-

bladder; 5) pain in the lower right quadrant of the abdomen from
appendicitis or inflammation of the fallopian tubes; 6) flank pain
from kidney stones (Weiss and Davis, 1958); and 7) testicular pain
from renal disease (Cohen, 1947).

Since the pelvic nerve, which innervates the vagina, projects
to the same spinal levels (Clegg and Doyle, 1967) as the somatic nerves
which innervate the lower back, both sets of input probably impinge
on the same spinal neurons. Indeed, convergence has been demonstrated between input from the bladder and cutaneous input from the
flanks, the former, like the cervix, receiving innervation from the
pelvic nerve (Fields et al., 1970). In the absence of ovarian hormone, stimulation of either set of inputs (visceral or somatic) is
almost always inadequate to elicit lordosis, but simultaneous stimulation of both sets often provides a suprathreshold stimulus for
generating lordosis (Komisaruk and Diakow, 1973). Somatic stimulation in the lower back region can be itself induce lordosis after
estrogen treatment, and we have occasionally observed lordosis in
response to cervical probing by itself. Thus, these two stimuli are
to some extent equivalent; perhaps the two inputs converge on the
same spinal neurons. Simultaneous stimulation of the cervix and
lower back could then summate and thereby increase the likelihood of
the occurrence and the intensity of the lordosis response.

To account for the pain blockage induced by cervical probing or
lower back stimulation, let us assume that these two pathways converge on central spinal neurons in the tactile pathway (convergence
of visceral with tactile input on spinal neurons does occur at least
between urinary bladder input and lower trunk input in cats (Fields
et al., 1970). According to the "gate-control" theory of Melzack
and Wall (1965), tactile stimulation blocks input originating in
pain pathways in the following way. Tactile input synaptically
stimulates (a) an ascending "action-system" neuron and (b) a neuron
of the substantia gelatinosa (SG). The SG neuron, activated by the
tactile input, exerts presynaptic inhibition on a tactile sensory
neuron at its synapse with the action-system neuron. A parallel
"pain-system" sensory neuron also directly activates (a) the same
action-system neuron and (b) *inhibits* the same SG neuron (in contrast to the excitatory effect which the tactile input exerts on the
SG neuron). This SG neuron, when active, exerts presynaptic inhibition on the pain-system sensory neuron at the synapse of the
latter with the action-system neuron. Thus, in contrast to the
tactile system, activation of the pain-system neuron *inhibits* the
SG neuron and thereby *reduces* its inhibition of the sensory pain-
system neuron. This allows the activity in the pain-system to
build gradually by positive feedback. Pain blockage occurs when
tactile stimulation is applied because it activates the SG neuron
and thereby increases its *inhibitory* influence at *both* the "tactile
system-action system" synapse and the "pain system-action system"

synapse.[1] If we assume that the input from both cervical probing and lower back stimulation activates the tactile system neurons, then either or both would block activity originating in the pain system. We observed a significant increase in pain threshold resulting from either stimulation of the back in estrogen-treated rats which showed lordosis or cervical probing (with or without estrogen). Of possible relevance in this regard is the finding of selective uptake of radio-active estrogen in neurons of the SG (Keefer et al., 1973; cf. Rexed, 1952), which might alter their threshold of activation.

2. Possible relation to acupuncture. This discussion has emphasized the similarity of effects induced by visceral or somatic stimulation at the same dermatomal levels. We have described a system in which visceral stimulation (cervical probing) facilitates a somatic res-ponse (lordosis) to somatic stimulation (flank-perineum palpation). There is a case in which the converse process occurs. Beach (1966) has reported that neonatal guinea pigs defecate and urinate only when the mother licks the perineal region. (This somatic stimulation induces lordosis in both males and females.) If the mother does not provide this tactile stimulation, the neonates die. The somatic stimulation may converge and summate with visceral input, which is apparently underdeveloped, and thereby generate suprathreshold stim-ulation for activating the visceral elimination reflex. This could be a nonpainful reverse "referred-pain" process. The effectiveness of acupuncture procedures may be accounted for by this process of somatic stimulation changing visceral activity at the corresponding dermatomal level. Somatic stimulation apparently can influence vis-ceral activity by direct reflex pathways and/or by reflexive alter-ations in the circulatory supply to the viscera (cf. Mann, 1972, pp. 5-26). An altered visceral circulatory supply could be one means by which the healing of visceral lesions could be accelerated. We are currently investigating ways in which somatic stimulation alters genital tract activity in rats.

C. Prolonged Effects of Vaginal Stimulation

In the present studies, we have found that the effects of vag-inal stimulation persist after the stimulation has been terminated.

---

[1]One prediction, based on the Melzack and Wall hypothesis, has not found support when tested (Vyklicky et al., 1969; Burke et al., 1971). These investigators found that input from pain fibers de-creased the input from tactile fibers through presynaptic inhibition; whereas Melzack and Wall predicted that input from pain fibers should increase rather than decrease the input from tactile fibers. Never-theless, our findings are consonant with the Melzack and Wall hypo-thesis in that tactile stimulation which elicits lordosis suppresses the vocalization response to painful stimulation.

Responses to painful stimulation were suppressed for several minutes after stimulation. Furthermore, females which initially did not show lordosis in response to flank-perineum palpation but were induced to show lordosis in response to cervical probing with palpation showed a persistence of lordosis in response to flank-perineum palpation alone two or more hours after the single application of the probing stimulus. Cervical probing alone, without concomitant lordosis, was adequate to induce lordosis responding to flank-perineum palpation alone hours after the probing stimulus.

Various studies by others also show a long-lasting influence of vaginal stimulation. While the nature of this persisting effect is not necessarily the same in each case, it may be worthwhile to review these effects. Hardy (personal communication) has found that a single intromission significantly increases the lordosis quotient in female rats tested immediately after the intromission. Repeated mating has been reported to increase female receptivity (Anderson, 1936; Koster, 1943; Clemens et al., 1969). An increase in neuronal activity which persists about 2.5 hours in the medial preoptic area and median eminence was induced by vaginal stimulation during the evening of proestrus but not diestrus in urethane-anesthetized rats (Blake and Sawyer, 1972). Similarly treated rats showed an increase in plasma luteinizing hormone 30 but not 90 minutes after vaginal stimulation.

These studies do not distinguish between the possibility that 1) a minimum amount and duration of prolactin or other hormone release is necessary to initiate a self-sustaining process which is not necessarily neurally mediated; or 2) sustained hormone (e.g., prolactin) secretion results from sustained neural activity after the copulatory stimuli. The former has been proposed by Everett (1967); the latter is supported (a) by the observations by Goldfoot and Goy (1970) that sexual receptivity can be terminated by vaginal stimulation, even in hypophysectomized guinea pigs, (suggesting a persistent neural effect) and also (b) by our finding of prolonged (several hours) induction of lordosis responsiveness even in ovariectomized, hormonally untreated rats after vaginal stimulation. However, we cannot exclude the possibility that releasing factors (Moss and McCann, 1973; Pfaff, 1973) or other neural secretions could be released reflexively by copulatory stimuli and act directly on the brain for prolonged periods. Edmonds, Zoloth, and Adler (1972) have shown that induction of a progestational state resembling pseudopregnancy requires multiple intromissions, but these intromissions can be spaced several hours apart. It is not possible to conclude from these findings, however, whether this "storage" of copulatory stimulation consists of summation of the effects of persisting hormonal (e.g. prolactin) levels in the blood and/or of prolonged neural activity.

Other effects of vaginal stimulation which persist long after

the termination of the stimulus have been observed. (1) Intact
female rats showed significantly elevated prolactin levels as long
as eight hours post coitum. In contrast, pelvic-neurectomized rats
showed elevated prolactin levels in the blood at 20 mintues post
coitum but not later; they did not become pregnant. If prolactin
was given to pelvic-neurectomized females during the first three
days after coitus, pregnancy was established. Thus, the normal
induction of pregnancy may require elevated serum prolactin levels
for several days after coitus (Spies and Niswender, 1971).

(2) Everett (1967) postulated that in female rats, copulation
can stimulate prolactin secretion which persists for many days after
the termination of the stimulus. Barbiturate was administered during
proestrus, and the females were allowed to mate later the same day;
but ovulation failed to occur. Within a few days, they returned to
full estrous cyclicity with attendant spontaneous ovulation and
formation of corpora lutea (based on sample autopsies). The females
then entered a phase of pseudopregnancy, as indicated by 12 to 14
days of diestrous vaginal smears and the ability to form a deciduoma.
Everett concluded that the corpora lutea resulting from the delayed
spontaneous ovulation responded to persistent high levels of prolactin induced by the original copulatory stimuli and thus initiated
pseudopregnancy.

(3) Diamond (1972) has shown in hamsters that vaginal stimulation at the time of artificial insemination (AI) profoundly influences the proportion of pups born alive 17 days later. AI alone did
not induce pregnancy. If progesterone (5 or 10 mg) was administered
in addition to AI, the females became pregnant; but no more than 25%
of the births were live. In contrast, if vaginal stimulation was
administered with AI, all births were live. If progesterone (15 or
10 mg) was administered with AI and vaginal stimulation, 43-86% of the
births were live. The mechanism underlying this very long lasting
effect of vaginal stimulation is unknown.

D.   *Lordosis as an Extensor-Dominated, Flexor-Suppressed Reflex:
An Attempt at a Unifying Hypothesis*

Lordosis is customarily considered to be an estrogen- and pro-
gesterone-controlled behavior pattern, but its neural basis is poor-
ly understood. As a clue to a better understanding, it may be useful to consider whether lordosis is a motor pattern which occurs in
contexts other than mating behavior and in the absence of ovarian
hormones.

In immature rabbits (e.g., at 10 days of age) the righting reflex which requires coordinated flexor activity is not yet developed
(Carmichael, 1954; Warkentin and Carmichael, 1939). Rabbits of this
age were dropped from a position in which they were initially suspended upside down by the four legs. During the fall, the fore- and

hind legs were extended and the back was arched, so that the table was hit first with the head and rump and then with the middle of the back. By 20 days, flexor activity matured to the stage in which the rabbits could twist around, flex the legs, and right themselves. Thus, in a situation in which extensor activity was activated, the body posture that emerged closely resembled lordosis. In the following discussion, I propose that lordosis in the rat results from selective activation of the extensor system and simultaneous suppression of the flexor system.

The lordosis posture in rats is characterized by an extension of all four legs, elevation of the rump and head, and depression of the middle of the back (dorsiflexion). Probing the vaginal cervix, which facilitates lordosis in response to palpation of the flanks and perineum (Komisaruk, 1971; Komisaruk and Diakow, 1973), also elicits partial extension of the hind legs when the rats are lifted by the tail off their hind feet and support themselves on their forefeet (unpublished observations).

Probing the vaginal cervix has a profound *inhibitory* effect on a variety of other reflexes simultaneously (Komisaruk and Larsson, 1971). Thus, it can completely eliminate the leg withdrawal response to intense foot pinch in awake, unanesthetized female rats. It can completely block a facial twitch response (vibrissa retraction) to ear pinch (probably a trigeminal nerve to facial nerve reflex, similar to the eye-blink reflex) in lightly anesthetized rats. The tail-flick response to pinch or heat can be completely abolished in awake rats by this stimulus (unpublished observations). When the vaginal cervix is probed, awake, unanesthetized rats can be turned upside down off their feet by slowly twisting the tail, their ability to right themselves being markedly diminished. The eye blink response to touching the cornea is attenuated in awake and lightly anesthetized rats by this stimulus.

How can we reconcile these two sets of observations? The same stimulus, probing the vaginal cervix, abolishes a wide variety of nociceptive reflexes, but simultaneously exerts a marked potentiation of the lordosis response to flank and perineum palpation. Thus, the cervical probing does not simply depress nervous system excitability, since lordosis responding is markedly enhanced. As indicated above, each of the components of lordosis appears as a constellation of activity before flexor responses have matured. These components of the lordosis response are certainly not inhibited by the cervical probing. On the other hand, each of the nociceptive responses we tested *are* inhibited by the cervical probing. Although leg withdrawal and righting reflexes are predominantly flexor in nature, the vibrissa-retraction, eye-blink, and tail-flick reflexes are not so clearly flexor. However, the characteristic of these responses is that they are responses to noxious stimulation (whereas the extensor reflexes are not). Thus, perhaps all the nociceptive reflexes which we tested and which were inhibited by cervical probing are predominantly controlled by a "flexor system."

To test this hypothesis, we carried out preliminary studies (with A. Naggar) on the effect of cervical probing on an extensor reflex independent of lordosis. We selected the knee-jerk as the extensor reflex. Female rats were lightly anesthetized with ether and placed on their sides. When the patellar tendon was struck lightly with the edge of a metal spatula, a leg extension was elicited reliably. Cervical probing abolished the leg-withdrawal and vibrissa-retraction reflexes. However, there was no suppression whatever of the leg-extension response. Aside from the lordosis response, the knee-jerk reflex is thus the only reflex we have tested which is not suppressed by cervical probing. Since it is an extensor reflex, this finding lends strong support to the hypothesis that in rats lordosis is an activation of the extensor system and a simultaneous suppression of the flexor system.

One of the implications of this hypothesis is that factors which facilitate extensor activity should also facilitate the lordosis reflex. The Moro reflex is an extensor reflex in normal infants, which is characterized by extension of legs and arms. It is elicited by rapid passive movements of the head, and is therefore considered to result from stimulation of the vestibular system (Peiper, 1963, p. 152). One of the behavioral patterns shown by female rats in a state of high sexual receptivity is "ear-wiggling" ("rapid head movements which produce vibration of the ears;" Beach, 1942). Its function is unknown. By means of high-speed motion picture analysis (Komisaruk and K. Larsson, unpublished observations), the head can be seen to shake around the nose-to-neck axis so that each eye is higher, then lower, then higher, than the other eye. This shaking movement is performed at rates up to about 25 times per second. This head shaking movement must generate a potent vestibular input which could potentiate the extensor system, as in the Moro reflex, and thereby "self-stimulate" an intensification of the lordosis posture.

The jump response to foot shock or a loud noise is significantly enhanced by septal lesions in rats (Lubar et al., 1970; Doty and Forkner, 1971). It is noteworthy that the sensory threshold of rats with septal lesions is not necessarily lowered, but the likelihood that they jump in response to the adequate stimulus is significantly increased by this lesion. This suggests that septal lesions potentiate reactivity of the extensor system. We have demonstrated that septal ablation in rats results in a striking intensification of the lordosis reflex, even in the absence of ovarian hormones. Perhaps the septum normally exerts an inhibitory effect on the extensor system and the septum itself is inhibited by estrogen; thereby activity of the extensor system may be facilitated.

Other effects of probing the vaginal cervix are blockage of pain

responses, which are recorded in the thalamus, and blockage of vocalization in response to tail shocks. As noted in these experiments, responses to touch were not blocked. This indicates that the
nociceptor response system is blocked by cervical probing but the
non-nociceptor system is not blocked. Cervical probing thus blocks
nociceptive sensory input and flexor (nociceptive response) activity.
Indeed, perhaps one means by which flexor reflexes are blocked by
cervical probing is via a blockage of the nociceptive input. Since
our recordings were all made in the thalamus, we cannot exclude the
possibility that the blockage of nociceptive activity occurred at
lower synaptic levels, e.g., the spinal cord. This may in fact be
the site of blockage, for cervical probing blocked hind-leg withdrawal reflexes even in rats sustaining mid-thoracic spinal transection (Komisaruk and Larsson, 1971).

     This discussion has emphasized as a fundamental property of the
nervous system its division into extensor and flexor systems and has
suggested how these two components could function in lordosis. The
point of emphasis is that this division is a fundamental property of
the nervous system which is functional in the absence of ovarian
hormones.

     Beach (1966) has demonstrated that lordosis is shown by neonatal
guinea pigs, males as well as females, in response to perineal licking by the mother. The licking stimulates urination and defecation.
The lordosis response persists throughout the nursing period; it re-
appears only in the females after puberty, in response to the appropriate levels of ovarian hormones. Beach (1967) has suggested that
the disappearance of the reflex is due to the ontogeny of inhibitory
processes which are themselves inhibited by the ovarian hormones, a
process that is further evidence that the neural substrate for lordosis is functional in the absence of ovarian hormones.

     It may thus be helpful to view the lordosis circuitry as an
estrogen- and progesterone-sensitive component of a more fundamental
extensor system of the nervous system, one which can be activated
by adequate sensory stimuli in the absence of the hormones but whose
activity is markedly enhanced by these hormones.

REFERENCES

Adler, N.T., 1969, Effects of the male's copulatory behavior on
     successful pregnancy of the female rat, J. Comp. Physiol.
     Psychol. 69:613-622.

Adler, N.T., Resko, J.A., and Goy, R.W., 1970, The effect of cop-
     ulatory behavior on hormonal change in the female rat prior
     to implantation, Physiol. Behav. 5:1003-1007.

Aidar, O., Geohegan, W.A., and Ungewitter, L.H., 1952, Splanchnic afferent pathways in the central nervous system, J. Neurophysiol. 15:131-138.

Amassian, V.E., 1952, Interaction in the somatovisceral projection system, Res. Publ. Assoc. Nerv. Ment. Dis. 30:371-402.

Anderson, E.E., 1936, Consistency of tests of copulatory frequency in the male albino rat, J. Comp. Psychol. 21:447-459.

Aron, Cl., Roos, J., and Asch, G., 1968, New facts concerning the afferent stimuli that trigger ovulation by coitus in the rat, Neuroendocrin. 3:47-54.

Beach, F.A., 1942, Importance of progesterone to induction of sexual receptivity in spayed female rats, Proc. Soc. Exp. Biol. Med. 50:369-371.

Beach, F.A., 1944, Effects of injury to the cerebral cortex upon sexually-receptive behavior in the female rat, Psychosom. Med. 6:40-55.

Beach, F.A., 1966, Ontogeny of "coitus-related" reflexes in the female guinea pig, Proc. Nat. Acad. Sci. 56:526-533.

Beach, F.A., 1967, Cerebral and hormonal control of reflexive mechanisms involved in copulatory behavior, Physiol. Rev. 47:289-316.

Bermant, G., 1961, Response latencies of female rats during sexual intercourse, Science 133:1771-1773.

Bermant, G., and Westbrook, W.H., 1966, Peripheral factors in the regulation of sexual contact by female rats, J. Comp. Physiol. Psychol. 61:244-250.

Beyer, C., and Sawyer, C.H., 1969, Hypothalamic unit activity related to control of the pituitary gland, in Frontiers in Neuroendocrinology (W.F. Ganong and L. Martini, eds.), pp. 255-287, Oxford University Press, New York.

Blake, C.A., and Sawyer, C.H., 1972, Effects of vaginal stimulation on hypothalamic multiple-unit activity and pituitary LH release in the rat, Neuroendocrin. 10:358-370.

Blandau, R.J., Boling, J.L., and Young, W.C., 1941, The length of heat in the albino rat as determined by the copulatory response, Anat. Rec. 79:453-463.

Burke, R.E., Rudomin, P., Vyklicky, L., and Zajac, F.E., 1971,

Primary afferent depolarization and flexion reflexes produced
by radiant heat stimulation of the skin, J. Physiol. 213:185–
214.

Carlson, R.R., and DeFeo, V.J., 1965, Role of the pelvic nerve vs.
the abdominal sympathetic nerves in the reproductive function
of the female rat, Endocrinology 77:1014–1022.

Carmichael, L., 1954, The onset and early development of behavior,
in Manual of Child Psychology, 2nd ed. (L. Carmichael, ed.),
pp. 60–185, John Wiley, & Sons, New York.

Carter, C.S., and Schein, M.W., 1971, Sexual receptivity and ex-
haustion in the female golden hamster, Horm. Behav. 2:191–
201.

Clegg, M.T., and Doyle, L.L., 1967, Role in reproductive physiology
of afferent impulses from the genitalia and other regions, in
Neuroendocrinology (L. Martini and W.F. Ganong, eds.), Vol. II,
pp. 1–17, Academic Press, New York/London.

Clemens, L.G., Hiroi, M., and Gorski, R.A., 1969, Induction and
facilitation of female mating behavior in rats treated neonatally with low doses of testosterone propionate, Endocrinology
84:1430–1438.

Cohen, H., 1947, Visceral pain, Lancet: 6487–6488.

Cross, B.A., 1959, Neurohypophyseal control of parturition, in
Recent Progress in the Endocrinology of Reproduction (C.W.
Lloyd, ed.), pp. 441–445, Academic Press, New York/London.

Davidson, J.M., 1969, Effects of estrogen on the sexual behavior
of male rats, Endocrinology 84:1365–1372.

Davidson, J.M., Smith E.R., Rodgers, C.H., and Bloch, G.J., 1968,
Relative thresholds of behavioral and somatic responses to
estrogen, Physiol. Behav. 3:227–229.

Diakow, C., 1970, Effects of genital desensitization on the mating
pattern of female rats as determined by motion picture analysis
(unpub. Ph.D. thesis, New York Univ.), Amer. Zool. 10:486
(abstr. #55).

Diakow, C., 1974, J. Comp. Physiol. Psychol. (in press).

Diakow, C., Pfaff, D.W., and Komisaruk, B.R., 1973, Sensory and
hormonal interactions in eliciting lordosis, Fed. Proc. 32:
241 Abs (abstr #166).

Diamond, M., 1972, Vaginal stimulation and progesterone in relation to pregnancy and parturition, Biol. Reprod. 6:281-287.

Doty, B.A., and Forkner, M.R., 1971, Alterations in pain thresholds and avoidance conditioning in rats with septal lesions, Neuropsychologia 9:325-330.

Downman, C.B.B., 1951, Cerebral destination of splanchnic afferent impulses, J. Physiol. 113:434-441.

Downman, C.B.B., 1955, Skeletal muscle reflexes of splanchnic and intercostal nerve origin in acute spinal and decerebrate cats, J. Neurophysiol. 18:217-235.

Duda, P., 1964a, Facilitatory and inhibitory effects of splanchnic afferentation on somatic reflexes, Physiol. Bohemoslovenica 13:137-141.

Duda, P., 1964b, Localization of the splanchnic effect on somatic reflexes in the spinal cord, Physiol. Bohemoslovenica 13:142-146.

Edmonds, S., Zoloth, S.R., and Adler, N.T., 1972, Storage of copulatory stimulation in the female rat, Physiol. Behav. 8:161-164.

Edwards, D.A., Whalen, R.E., and Nadler, R.D., 1968, Induction of estrus: estrogen-progesterone interactions, Physiol. Behav. 3:29-33.

Everett, J.W., 1967, Provoked ovulation or long-delayed pseudopregnancy from coital stimuli in barbiturate-blocked rats, Endocrinology 80:145-154.

Feder, H.H., and Ruf, K.B., 1969, Stimulation of progesterone release and estrous behavior by ACTH in ovariectomized rodents, Endocrinology 84:171-174.

Fields, H.L., Partridge, L.D., Jr., and Winter, D.L., 1970, Somatic and visceral receptive field properties of fibers in ventral quadrant white matter of the cat spinal cord, J. Neurophysiol. 33:827-837.

Goldfoot, D.A., and Goy, R.W., 1970, Abbreviation of behavioral estrus in guinea pigs by coital and vagino-cervical stimulation, J. Comp. Physiol. Psychol. 72:426-434.

Green, R., Luttge, W.G., and Whalen, R.E., 1970, Induction of receptivity in ovariectomized female rats by a single intravenous injection of estradiol-17b, Physiol Behav. 5:137-141

Hardy, D.F., and DeBold, J.F., 1972, Effects of coital stimulation upon behavior of the female rat, J. Comp. Physiol. Psychol. 78:400-408.

Harrington, R.E., Eggert, R.G., and Wilbur, R.D., 1967, Induction of ovulation in chlorpromazine-blocked rats, Endocrinology 81:877-881.

Hays, R.L., and Vandemark, N.L., 1953, Effect of stimulation of the reproductive organs of the cow on the release of an oxytocin-like substance, Endocrinology 52:634-637.

Keefer, D.A., Stumpf, W.E., and Sar, M., 1973, Estrogen-topographical localization of estrogen-concentrating cells in the rat spinal cord following ^{3}H-estradiol administration, Proc. Soc. Exp. Biol. Med. 143:414-417.

Kollar, E.J., 1953, Reproduction in the female rat after pelvic nerve neurectomy, Anat. Rec. 115:641-658.

Komisaruk, B.R., 1971a, Induction of lordosis in ovariectomized rats by stimulation of the vaginal cervix: hormonal and neural interrelationships, in *Steroid Hormones and Brain Function* (C.H. Sawyer and R.A. Gorski, eds.), pp. 127-141, University of California Press, Berkeley.

Komisaruk, B.R., 1971b, Strategies in neuroendocrine neurophysiology, Amer. Zool. 11:741-754.

Komisaruk, B.R., and Beyer, C., 1972, Differential antagonism, by MER-25, of behavioral and morphological effects of estradiol benzoate in rats, Horm. Behav. 3:63-70.

Komisaruk, B.R., and Diakow, C., 1973, Lordosis reflex intensity in rats in relation to the estrous cycle, ovariectomy, estrogen administration and mating behavior, Endocrinology 93:548-557.

Komisaruk, B. R., and Larsson, K., 1971, Suppression of a spinal and a cranial nerve reflex by vaginal or rectal probing in rats, Brain Res. 35:231-235.

Komisaruk, B.R., and Wallman, J., 1973, Blockage of pain responses in thalamic neurons by mechanical stimulation of the vagina in rats, Program and Abstracts, Society for Neuroscience, 3rd Annual Meeting, p. 315.

Komisaruk, B.R., Adler, N.T., and Hutchison, J., 1972, Genital sensory field: enlargement by estrogen treatment in female rats, Science 178:1295-1298.

Komisaruk, B.R., Larsson, K., and Cooper, R., 1972, Intense lordosis in the absence of ovarian hormones after septal ablation in rats, Program and Abstracts, Society for Neuroscience, 2nd Annual Meeting, p. 230 (abstr. #51.10).

Komisaruk, B.R., McDonald, P.G., Whitmoyer, D.I., and Sawyer, C.H., 1967, Effects of progesterone and sensory stimulation on EEG and neuronal activity in the rat, Exp. Neurol. 19:494-507.

Koster, R., 1943, Hormone factors in male behavior of the female rat, Endocrinology 33:337-348.

Kuehn, R.E., and Beach, F.A., 1963, Quantitative measurement of sexual receptivity in female rats, Behaviour 21:282-299.

Larsson, K., Feder, H.H., and Komisaruk, B.R., 1974, Role of the adrenal glands, repeated matings, and monoamines in lordosis behavior of rats, Pharm. Biochem. Behav. (in press).

Lubar, J.F., Brener, J.M., Deagle, J.H., Numan, R., and Clemens, W.J., 1970, Effect of septal lesions on detection threshold and unconditioned response to shock, Physiol. Behav. 5:459-463.

Mann, F., 1972, *Acupuncture*, Vintage Books/Random House, New York.

McLeod, J.G., 1958, The representation of the splanchnic afferent pathways in the thalamus of the cat, J. Physiol. 140:462-478.

Melzack, R., and Wall, P.D., 1965, Pain mechanisms: a new theory, Science 150:971-979.

Moss, R.L., and McCann, S.M., 1973, Induction of mating behavior in rats by luteinizing hormone-releasing factor, Science 181:177-179.

Peiper, A., 1963, *Cerebral Function in Infancy and Childhood*, 3rd rev. ed. (B. Nagler and H. Nagler, trans.), International Behavioral Science Series, Consultants Bureau, New York.

Peirce, J.T., and Nuttall, R.L., 1961, Self-paced sexual behavior in the female rat, J. Comp. Physiol. Psychol. 54:310-313.

Pfaff, D.W., 1973, Luteinizing hormone-releasing factor potentiates lordosis behavior in hypophysectomized ovariectomized female rats, Science 182:1148-1149.

Pomeranz, B., Wall, P.D., and Weber, W.V., 1968, Cord cells responding to fine myelinated afferents from viscera, muscle and skin, J. Physiol. 199:511-532.

Powers, B., and Valenstein, E.S., 1972, Sexual receptivity: fa-
    cilitation by medial preoptic lesions in female rats, Science
    175:1003-1005.

Ramirez, V.D., Komisaruk, B.R., Whitmoyer, D.I., and Sawyer, C.H.,
    1967, Effects of hormones and vaginal stimulation of the EEG
    and hypothalamic units in rats, Amer. J. Physiol. 212:1376-
    1384.

Rexed, B., 1952, The cytoarchitectonic organization of the spinal
    cord in the cat, J. Comp. Neurol. 96:415-495.

Rodriguez-Sierra, J.F., Crowley, W.R., and Komisaruk, B.R., 1974,
    Vaginal stimulation in rats induces prolonged lordosis re-
    sponsivness and sexual receptivity, J. Comp. Physiol. Psychol.
    (in press).

Ruch, T.C., Pathophysiology of pain, 1961, in *Neurophysiology* (T.C.
    Ruch, H.D. Patton, J.W. Woodbury, and A.L. Towe, eds.), pp.
    350-368, W.B. Saunders Co., Philadelphia.

Selzer, M., and Spencer, W.A., 1969a, Convergence of visceral and
    cutaneous afferent pathways in the lumbar spinal cord: Brain
    Res. 14:331-348.

Selzer, M., and Spencer, W.A., 1969b, Interaction between visceral
    and cutaneous afferents in the spinal cord: reciprocal pri-
    mary afferent fiber depolarization, Brain Res. 14:349-366.

Spies, H.G., and Niswender, G.D., 1971, Levels of prolactin, LH,
    and FSH in the serum of intact and pelvic-neurectomized rats,
    Endocrinology 88:937-943.

Vyklicky, L., Rudomin, P., Zajac, F.E., and Burke, R.E., 1969,
    Primary afferent depolorization evoked by a painful stimulus,
    Science 165:184-186.

Warkentin, J., and Carmichael, L., 1939, A study of the development
    of the air-righting reflex in cats and rabbits, J. Genet.
    Psychol. 55:67-80.

Weiss, S., and Davis, D., 1928, The significance of the afferent
    impulses from the skin in the mechanism of visceral pain.
    Skin infiltration as a useful therapeutic measure, Amer. J.
    Med. Sci. 176:517-536.

Young, W.C., Boling, J.L., and Blandau, R.J., 1941, The vaginal
    smear picture, sexual receptivity, and time of ovulation in

the albino rat, Anat. Rec., 80:37-45.

Zarrow, M.X., and Clark, J.H., 1968, Ovulation following vaginal
    stimulation in a spontaneous ovulator and its implications,
    J. Endocrinol. 40:343-352.

PHEROMONES IN PRIMATE REPRODUCTION AND SOCIAL BEHAVIOR[1]

Gisela Epple

Monell Chemical Senses Center

University of Pennsylvania, Philadelphia

INTRODUCTION

During recent years research on invertebrate and mammalian pheromones has aroused considerable interest, both within the scientific community and among the general public. This interest may be one of the reasons why the word *pheromone* became so widely accepted that it is now used by some authors for any kind of chemical communication among members of the same species.

Bronson (1971), however, has recently challenged the general applicability of the term pheromone to all forms of intraspecific chemical communication in mammals. He points out that many mammalian species seem to communicate information, such as identification of the species and individual, sex, sexual state, and age, by using highly complex mixtures of chemical substances from various sources (e.g., general skin odor, urine, feces, sex accessory glands, specialized skin glands, etc.). It seems difficult, if not impossible, to relate most of these messages to only one, or at least a limited number of chemicals present in a body product whose pheromonal activity is suspected or confirmed. Bronson (1971) therefore proposes to limit the term pheromone to "situations where there seems a reasonable probability of isolating one or at least a restricted mixture of compounds that could, in turn, be synthesized and whose actions could be confirmed experimentally."

[1]The author's studies were supported by NSF grants GB-12-660 and GB-33104X, and in part by a biomedical fellowship from the Population Council of The Rockefeller University.

Bronson's arguments have to be considered in a discussion of primate "pheromones." In many primate species, there is a large body of behavioral and anatomical evidence for the existence of chemical communication. The majority of all infrahuman primates shows olfactory and gustatory investigation of the bodies of conspecifics in different social and especially in sexual situations (cf. Michael, 1969, and many others). In those species which show scent-marking behavior, the marks are sniffed and sometimes also licked by the animal who produced them and by conspecifics. Many species possess specialized skin glands in various body regions which often produce odoriferous secretions. Moreover, many species show highly stereotyped behavioral patterns of scent marking during which feces, urine, saliva, vaginal secretions, and the secretions of scent glands are applied to their bodies, those of conspecifics, and to the environment.

Although these facts strongly suggest the importance of chemical signals in primate communication, it is not possible, on the basis of our present knowledge, to determine which of the chemical signals are pheromones in the stricter sense of the term proposed by Bronson (1971) and which are produced by a mixture of odors from various sources. Therefore, whenever the term pheromone is used in this review, it is with the understanding that this is an oversimplification which has to be revised as our knowledge of chemical communication increases.

Within the framework of this article, it is impossible to present a detailed review of the sometimes quite sketchy evidence for pheromones and their many possible functions in the sexual and social life of primates. I shall therefore limit the discussion to two areas of behavior where chemical signals have been shown to play a role: 1) the regulation of sexual behavior; 2) the production of chemical messages on the physiological, social, and emotional state of the individual.

THE REGULATION OF SEXUAL BEHAVIOR

Sexual Attraction and Arousal

One of the most widespread functions of primate pheromones seems to be the transmission of information of the reproductive condition of the female, that is, the production of female sexual attractants which not only inform the male about the sexual state of the female but also arouse his sexual activities. Only a few examples out of the many reports on chemical communication involved in sexual behavior can be mentioned here. Among the prosimians, for instance, *Galago senegalensis*, *Loris tardigradus*, and *Nycticebus coucang* females produce a vaginal discharge at the time of estrus which in *Loris* and *Nycticebus* is mixed with urine during marking. In all three species it seems to function as a strong sex attractant (Doyle et al., 1967; Sauer and Sauer, 1963; Seitz, 1969). Female *Lemur catta* show a peak of genital marking during and after estrus, applying vaginal mucus and probably urine to the substrate (Evans and Goy, 1968; Jolly, 1966). In higher primates, female *Callimico goeldii* impregnate their tails with urine and the secretions of the circumgenital scent glands with increased frequency during estrus. In these permanently mated primates, the male apparently keeps a check on the reproductive condition of his female by regularly smelling and licking her genital scent glands and urine. Chemical cues from the female seem to initiate courtship and mating during estrus (Lorenz, 1972). In spider monkeys, the males frequently sniff, lick, and even drink the urine of females. This behavior is shown in captive as well as in free living *Ateles* during all phases of the females' reproductive cycles, including pregnancy. The males also sniff their own hands after handling the clitora of females and investigate places where females sat. Since females do not show any visual signs of cycling, olfactory clues might be the main transmitters of information on the reproductive condition of the female (Klein, 1971).

Even Old World monkeys and apes frequently show olfactory investigation of the females' genitalia (cf., Michael and Zumpe, 1972). Before copulation, for instance, *Macaca arctoides* males insert their fingers into the vaginas of the females, sniffing them afterwards (Michael and Zumpe, 1972).

However, among all inferred primate sexual attractants, the only one whose behavioral function, hormonal control, and chemical nature has been analyzed in detail is the sex pheromone of the rhesus monkey. Michael and Keverne (1968, 1969, 1970) demonstrated that in *Macaca mulatta* the sexual attractiveness of the female and the sexual arousal of the male are, to a great extent, dependent upon chemical signals. These pheromones, called *copulins* by Michael and Keverne, are produced in the vagina of the female under estrogenic stimulation and strongly control the sexual behavior of males under laboratory conditions. When males were trained to gain access to females in a free cage situation, they frequently worked to obtain ovariectomized females who were rendered receptive by injections of estrogen,but did not do so to obtain ovariectomized, nontreated females. When the males were rendered anosmic by plugging their nasal olfactory area, they were no longer able to discriminate between receptive and nonreceptive females. Reversal of anosmia, however, restored their ability to do so.

In an attempt to localize the source of these olfactory cues, Michael and Keverne (1969, 1970) applied vaginal washings from estrogen-treated donor females to the sexual skin of untreated, unreceptive females. Although the males had shown little interest in the nonreceptive females previously, the application of vaginal

secretions from receptive females immediately caused them to show
a significant increase in sexual behavior. This occurred in spite
of the fact that the ovariectomized recipient females were totally
unreceptive and did not at all encourage the males.

Curtis and co-workers (1971) recently identified the pheromonally active constituents of vaginal washings from estrogenized
females as a mixture of five aliphatic acids. A synthetic mixture of these acids, applied to the sexual skin of unreceptive
females, mimicked the behavioral effects of the natural pheromone.
A similar, although not identical, content of aliphatic acids to
those of the rhesus monkey was found in vaginal washings of *Papio
anubis, Enythrocebus patas, Macaca nemestrina, Macaca fascicularis,
Saimiri sciureus,* and *Homo* (Michael et al., 1972). Unreceptive
rhesus females, treated with vaginal washings of intact anubis
baboons, stimulated the sexual activity of male rhesus monkeys
significantly, an indication that there is some interspecific
activity of the sex attractants.

Bonsall and Michael (1971) provide evidence which indicates
that pheromone production depends on bacterial action within the
vagina. After incubation at 37°C, the amount of volatile acids in
vaginal washings from rhesus monkeys increased. It was inhibited
by autoclaving or the addition of penicillin before incubation.
In vivo, gonadal hormones might regulate acid production by determining the availability of the substrate or the pH of the vagina
(Bonsall and Michael, 1971; Michael et al., 1972).

Reproductive Synchronization

Another area of reproductive biology in which pheromones have
been suspected of playing a role is the synchronization of breeding activities. Several authors have speculated that the phenomena of behavioral and physiological synchronization of reproductive activity in seasonal primates are, in part, controlled by
pheromones. In *Lemur catta,* Jolly (1966) reported that under
natural conditions the breeding season of each troop lasts for
only about two weeks or less and that each female seems to be
receptive for only one day. During the breeding season, the lemur
groups show a very high degree of behavioral synchronization,
correlated with a sharp increase in sexual and social interactions
and also in the various patterns of olfactory marking and olfactory investigation. Jolly (1966) pointed out that the great
variety of scent-marking patterns and the large amount of sniffing
each other's bodies and scent marks may be the major means of
communicating the rising excitement of the breeding season among
the members of lemur groups. She suggests, therefore, that synchrony of reproductive activities and of female estrus depends, in
part, on olfactory and other social communications within the

group. Evan and Goy's (1968) laboratory study of scent marking
and reproductive behavior in *Lemur catta* fully supports her views.

In a semi-free-ranging group of squirrel monkeys, Baldwin
(1970) noted that males frequently sniffed females and were attracted to branches where females had urine washed. This behavior was
shown all year round, but female odors seemed to become particularly attractive to males during the breeding season. Moreover,
Latta et al. (1967) reported that the females of their laboratory
colony urine washed with increased frequency during estrus and
that males sniffed the genitalia and urine of the females but
never sniffed male genitalia or male urine. In free-ranging
groups, adult males travel separately from the females except
during the breeding season, when they are spermatogenic (Baldwin,
1968). Baldwin (1970) suggested seasonal changes in the quality
of female odors. As female odors become more attractive to males
during the breeding season, they motivate them to interact with
the females. On the other hand, besides producing attractive
scents, females may also produce pheromones which stimulate the
males into a reproductively active condition and thus correlate
physiological as well as behavioral breeding activities. In
laboratory groups, where males and females are caged together,
the males are permanently exposed to female odors. This would
explain why the males of some laboratory colonies are more or less
sexually active all year round (Baldwin, 1970).

Stimulation of male sexual activity by the presence of females
was also observed in the rhesus monkey. Vandenberg (1969) induced
sexual activity in quiescent male rhesus monkeys by exposing them
to ovariectomized, estrogenized females during the nonbreeding
season. The monkeys were captured from a free-ranging group on
an island near Puerto Rico and remained exposed to their normal
environmental conditions during the experiments. At the same
time, no sexual behavior was recorded in the free-ranging group
mates of the experimental males. Moreover, Rose and co-workers
(1971, 1972) have recently shown that plasma testosterone levels
in male rhesus monkeys increase with social rank and sexual activity. Although no direct evidence is available at this point, it
is tempting to speculate that the stimulation of male reproductive
activity through sexual interactions and through the presence of
receptive females results in part from pheromones produced by the
females. The female sex attractants might well play a role here.

A synchrony of reproductive cycles related to the social
environment seems to exist even in man. In a population of college
girls living in an all-girl dormitory, McClintock (1971) demonstrated a synchronized onset of menstruation in pairs and groups
of girls who spent considerable time with each other (close
friends and roommates) that was lacking in pairs of girls randomly
formed from the population of subjects. Subjects who reported

that they spent time with males three or more times per week had
significantly shorter menstrual cycles than girls who saw males
less than three times per week. McClintock's (1971) study does
not elucidate the physiological mechanisms that might cause the
effects she has demonstrated. She speculates, however, that
pheromones are involved in the correlation of menstruation and in
the apparent effect of males on the female cycles, pointing out
that pheromonally mediated parallels to these phenomena exist in
mice (cf. Bronson, 1971).

CHEMICAL MESSAGES ON THE PHYSIOLOGICAL, EMOTIONAL, AND SOCIAL STATUS OF THE INDIVIDUAL

Many species of mammals use chemical signals to communicate
information on the identity of the species and of the individual
as well as its sex, age, physiological and emotional conditions,
and even its social status (cf. Bronson, 1971; Müller-Schwarze,
1974). We do not know how widespread the production of chemical
messages of this kind is among primates nor how detailed the information contained in the massages. That scent-marking behavior and
the development of specialized skin glands are so widespread among
prosimians and South American monkeys and are even present in some
Old World monkeys and apes (cf., Hill, 1944, 1954; Poirier, 1970;
Schaffer, 1940) suggests, however, that chemical signals of this
kind are much more important in the communication systems of primates than we realize. Detailed chemical messages on the physiological and psychological condition of the animal that produces
them may have important functions in several different areas of
behavior, e.g., in the demarcation and defense of home ranges,
orientation within the home range, friendly and aggressive interactions within the group and between groups, and parental behavior.
All of these behaviors, of course, may in turn affect the individual's reproductive behavior and success.

The evidence for the existence of chemical messages in the
physiological and emotional status of individuals and their biological roles will be discussed in the following pages. It is,
so far, limited to a very few species. The studies on sexual
attractants which communicate female receptivity have already been
reviewed above. Seitz (1969) and Epple (1970) showed that
Nycticebus coucang and *Callithrix jacchus* recognize species specific odors. Evidence for olfactory recognition of sex and individual identity in *Nycticebus coucang* and *Lemur fulvus* is provided
by Seitz (1969) and Harrington (in press), and Evans (cf., Evans
and Goy, 1968) reported that *Lemur catta* males preferred the scent
of conspecifics to their own odors. According to Evans and Goy's
(1968) study of the various scent-marking patterns of *Lemur catta*,
"ringtails may prove to possess an olfactory repertoire whose
complexity rivals the more sophisticated visual and acoustic systems

of larger-brained primates."

In our own laboratory, we have recently started to analyze the messages contained in the complex scent marks of a species of South American marmoset, the saddle backed tamarin (*Saguinus fuscicollis*) (Epple, 1971, 1972a, 1973). Male and female tamarins possess large scent glands in the circumgenital, suprapubic, and sternal areas (Perkins, 1966). Scent marking is performed by rubbing the glandular areas against items of the environment or against the body of a conspecific. When the combined circumgenital and suprapubic glands are used, not only skin secretions but also a few drops of urine are applied to the substrate. At times, the females also seem to be mixing vaginal discharge into the marks.

In a series of experiments, we attempted to obtain information on the types of messages which are contained in the scent marks of marmosets (Epple, 1971, 1972a, 1973). The results are reviewed below and are supplemented with unpublished data. The function of the signals in some areas of behavior is discussed and compared with findings in other species.

Odors Identifying the Sex of an Individual

On theoretical grounds, one would expect the scent marks of adult *Saguinus fuscicollis* to carry information on the sex of the animal that produced them. To test this possibility, the marmosets were given a series of preference tests during which they could chose between two perches of identical dimensions, one carrying the scent marks of adult males, the other carrying the scent marks of adult females. A consistent significant preference for either male or female odor was interpreted as a proof that the animals could distinguish between male and female scents.

Procedures:

Experiment I (Epple, 1971). Four adult males and five adult females served as subjects. Nine adult males and 10 adult females were used as donors of scent marks. All monkeys lived in permanent groups and had experienced no direct contact with animals other than their group mates since their arrival in the laboratory. The groups were housed in units of two connecting cages, 2 x 2 1/2 x 3 and 2 x 5 x 3 ft, visually isolated from other groups but in vocal and olfactory contact with most of the marmset colony.

To obtain odor samples, we provided the donors with a fresh perch of white pine (2 ft x 1 1/2 in x 3/4 in), which they were allowed to scent mark for one hour. Since females usually show a higher marking frequency per unit than males (Epple, unpublished

data), we had to eliminate the possibility that the subjects responded to a higher amount of scent on the perch marked by donor
females rather than to the quality of the odors. Therefore the
subjects were given 32 tests in which they were offered a perch
marked by one male and a perch marked by two females, 93 tests
with a perch marked by two males and a perch marked by one female,
63 tests with both stimuli marked by two donors, and 52 tests with
both stimuli marked by one donor. The combinations of donors
were presented in random order.

Both perches were simultaneously introduced into a subject's
home cage for a 15-minute test. They were placed two feet apart,
and their left-right position was counterbalanced to control for
a possible side preference. During the tests, the subjects had
free contact with both stimuli. Each total test period was
divided into intervals of five seconds each; and the subjects
received a score of 1 per interval for contacting, sniffing, and
scent marking either perch. The total time the subjects spent in
contact with either perch was recorded on stopwatches. Two group
mates were always tested together in a counterbalanced combination.
For further details of procedure, see Epple, 1971.

Experiment II (Epple, unpublished). During this experiment, we
used only one male and one female as scent donors. The subjects
were tested singly in one compartment of their double-unit home
cages, while their group mates were confined to the adjacent
compartment. The testing of one subject at a time controlled for
the possibility that during Experiment I the preference of one of
the subjects influenced the choice of the second subject.

Nine adult males and eight adult females, including the
animals tested in Experiment I, served as subjects. The donors
were eight adult males and 10 adult females, all of the subjects
serving as donors for other subjects with whom they did not have
direct contact. The conditions of housing, maintenance, and intergroup contacts were the same as in Experiment I. The collection
of scent marks from the donors and the testing procedures were
identical with those of Experiment I except that marks were collected over a period of 30 minutes only.[1]

Experiment III (Epple and Burlingame, unpublished). In this
experiment, we tested whether, when presented not with individual
scent marks but with samples of pooled scent marks from males and
females, the monkeys would show the same preference as in the
previous tests. The experiments also served to insure that

---

[1]During Experiment I, the fact that the donors did most of their
marking during the first 15 minutes of the one-hour collection
time suggested that a shorter collection time would be sufficient.

chemical manipulation of the scent marks would not destroy the messages they contained or lead the subjects to change the preference shown in previous tests. Ten males and 13 females served as subjects, including one male and four females which had also been used in the previous experiments. Eight adult males and eight adult females served as donors. Collection and testing were arranged in a way that some of the subjects, after having completed their tests, served as donors and some of the donors were subsequently used as subjects. Subjects and donors lived in 4 x 6 x 3 ft double cages in a large colony room and had experienced visual, vocal, and olfactory contact with each other. Some of the subjects had also had direct contact with the donors during other experiments.

Scent marks were collected by providing the donors with a clean, frosted glass plate (2 ft x 3/4 in x 1/16 in). After the donor had marked the plate, the material was washed off with methanol and methylene chloride. The washings, pooled by sex, were concentrated by rotary evaporation at room temperature under nitrogen and adjusted to appropriate volumes. Samples of 0.4 ml of each pool were placed in the center of two precleaned solid aluminum perches (2 ft x 2 in x 1/4 in) and offered to the subjects. Testing procedures were the same as in the previous experiments.

Each subject received one test on the same day the samples were collected or on the following morning. If samples were kept overnight, they were frozen at -60°C. Collection and testing of each subject were repeated four times at intervals of at least one week.

Results:

During the three experiments, we did not manipulate the subjects' motivation to distinguish between the two stimuli nor was it possible to control the reproductive and social state of the donors. Thus in all experiments female donors were used during all phases of their reproductive cycle, including early pregnancy; and dominant and submissive males served as donors in Experiments I and II. Therefore, a high variability of the subjects' responses was expected and did occur.

As a group, the subjects spent about equal time investigating and sniffing both stimuli in all three experiments. However, in Experiment I, two of the four males showed significant time preferences: one male spent more time on the perch marked by males ($p > 0.01$[1]); the other preferred female scent marks ($p > 0.01$[1]) (Epple, 1971). Individual preferences were not analyzed for

[1]$\underline{t}$-test for matched samples.

Experiments II and III.

In all three experiments, however, the subjects discriminated between perches scent marked by males and perches scent marked by females. They showed a significantly higher score of scent marking on the perch carrying male odor (Fig. 1); i.e., as a group, the monkeys applied their own odors more frequently to perches carrying male scent than to perches carrying female scent, although both stimulus odors seemed to be attractive to them. This response was shown regardless of whether the stimulus perches were scent marked by individual donor monkeys or whether odorous material from a pool of donors was applied to them by the investigator. When the scores for male subjects are analyzed separately from those for female subjects (Table 1), it becomes obvious that the trend to prefer male odor is more pronounced in female subjects

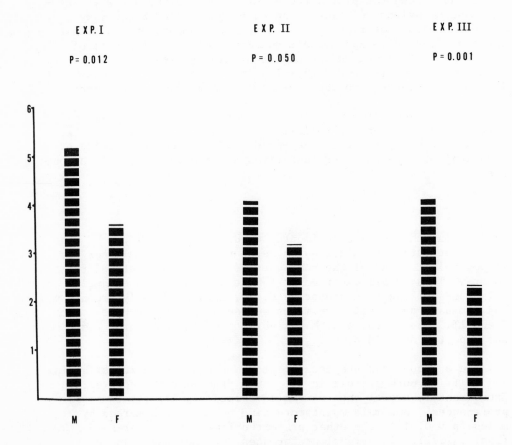

Fig. 1. Choice between scent marks of males (M) and females (F) in Experiments I, II, and III (mean marking scores on stimulus perches; t-test for matched samples).

TABLE 1. *Choice by Male and Female Subjects between Scent Marks of Males and Females in Experiments I, II, and III**

Subjects	Experiment I		Experiment II		Experiment III	
	♂-perch	♀-perch	♂-perch	♀-perch	♂-perch	♀-perch
Mean contact scores during first 5 min of testing						
♂♂	21.719	24.985	18.458	19.064	17.650	15.675
♀♀	16.821	14.849	14.782	12.876($p < 0.05$)	17.154	18.865
Mean scores for sniffing both stimuli						
♂♂	8.130	8.099	7.973	9.013	11.675	9.100($p < 0.05$)
♀♀	5.915	4.873	5.602	4.465($p < 0.02$)	13.519	13.250
Mean scores for marking both stimuli with circumgenital glands						
♂♂	4.876	3.595	3.613	3.115	3.100	1.800 ($0.05 < p < 0.1$)
♀♀	5.542	3.550($p < 0.001$)	4.684	3.392($p = 0.05$)	4.788	2.673($p < 0.01$)

*t-test for matched samples

than in males. In all three experiments, the females scent marked
perches with male odor significantly more frequently than perches
with female odor; this trend is not statistically significant in
the pool of male subjects only. In Experiment II, the females also
sniffed male perches significantly more frequently than female
perches and spent more time in contact with them during the first
five minutes of testing (Table 1). The preferences of the marmosets for the odor of strange males cannot be interpreted as a preference for the stimulus carrying the larger or smaller amount of
scent. The subjects scent marked the stimulus perch carrying male
odor significantly more frequently than that carrying female odor
regardless of whether the male perch was marked by one male and the
female perch by two females or vice versa, or whether both perches
carried the odor of one donor only (Epple, 1971).

    The preference for the intact scent marks of males as well as
for pooled material recovered from male marks showed that the marks
carry information on the sex of the animal that produced them
and that this information is independent of the total amount of
scent present. The question, of course, is which one of the various excretions and secretions (e.g., scent gland secretions,
urine, vaginal discharge) mixed into a scent mark by the performing monkey is the carrier of this information.

    During tests in which we offered the monkeys a choice between
a wooden perch actively marked by a donor male or female and a
second perch carrying a 0.4 ml sample of urine collected from the
same donor animal and applied to the perch by the investigator, the
subjects significantly preferred the perch that carried the natural
scent mark over that carrying urine. They spent more time on the
perches carrying the natural marks and also sniffed and scent
marked them significantly more frequently than the perches impregnated with urine (Fig. 2). These data indicate that intact scent
marks are much more attractive to the monkeys than is urine alone.
Urine, however, which represents one of the constituents of a
natural mark, is also investigated actively. It remains for
further testing to find out whether the monkeys show their preference for male odor over female odor when presented only with urine
samples from males and females and when presented with samples of
scent gland secretions from donors of the opposite sex.

Odors Identifying the Individual

    Behavioral observations suggested that body odor is one of the
means by which marmosets recognize each other. Therefore, we
tested the monkeys' ability to discriminate between the scent marks
of two adults of the same sex (Epple, 1973).

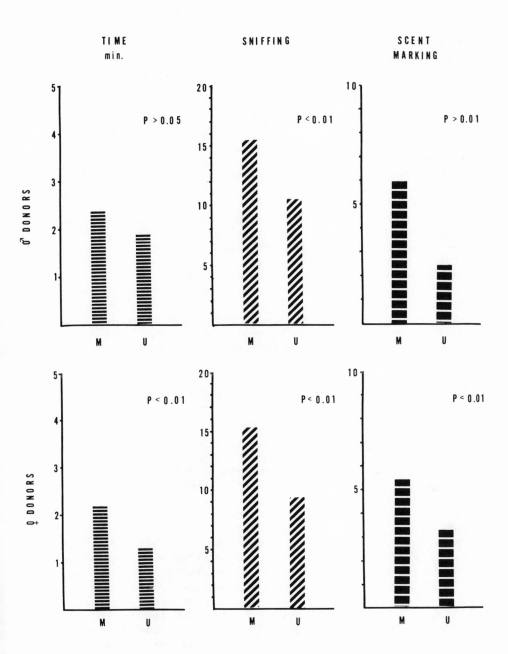

Fig. 2. Choice between scent marks (M) and urine (U) (mean time spent and mean sniffing and marking scores on stimulus perches; t-test for matched samples).

Procedures:

Four adult males and four adult females were used as subjects. They lived as permanently mated male-female pairs under similar conditions as in the previously described experiments.

For individual discrimination to be tested, the subjects had to be motivated to prefer the odor of one individual over that of another. Since the monkeys showed great interest in the scent marks of marmosets they had been fighting with, it was expected that they would prefer the odor of an individual with whom they had recently fought over that of another familiar animal which they had not recently met.

To create "opponents," whose scent marks were presented to the subjects in combination with those of "neutral" donors, a male or a female belonging to a strange group was introduced to the subjects for 10 minutes. This resulted in an aggressive encounter between the subjects and the "opponent." Later, the subjects were given the choice between two wooden perches, one scent marked by an opponent, the second one marked by a neutral donor of the same sex. Six males and four females served as opponents. Ten males and 10 females served as neutral donors. The neutral donors were familiar with the subjects but had not had a recent aggressive encounter with them.

Testing procedures were identical with those used in the previous experiment. For details see Epple, 1973.

Results:

As Figure 3 shows, the monkeys spent significantly more time sniffing and marking the perches carrying the scent of their opponents than those carrying the marks of neutral adults of the same sex. Their preference for the scent of their opponent shows that they distinguished between the scent marks of two adults of the same sex (Epple, 1973). It also shows that each individual marmoset is characterized by a personal body odor. The preference for the odor of an opponent was shown when the stimuli were presented on the day of the encounter between the subjects and the donor. However, it was also shown when the subjects were tested three or more days after the most recent aggressive encounter. This result shows that the monkeys remember the scent of other individuals for several days and also controls for the possibility that the subjects preferred the scent of their recent opponent because it was characteristic of a stressed animal rather than because of its individual quality.

The preference for the odor of recent opponents is obviously motivated by aggression against the opponents. Some of the subjects

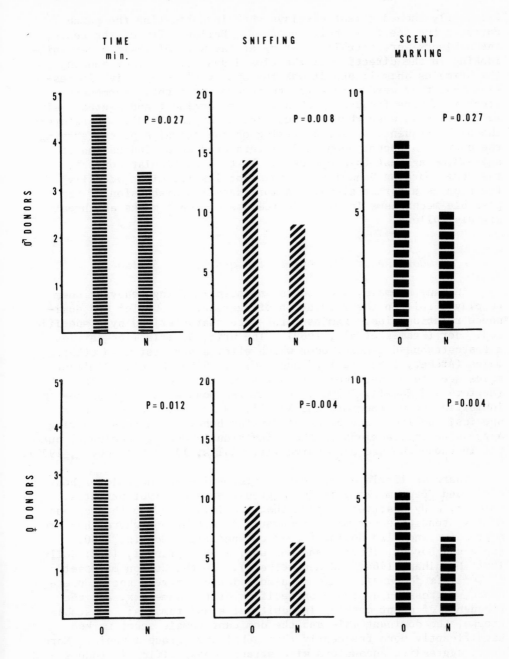

Fig. 3. Choice between scent marks of recent opponents (O) and
neutral donors (N) (mean time spent and mean sniffing and marking
scores on stimulus perches; Walsh-test: Siegel, 1956).

frequently showed threat displays when investigating the perch
carrying the scent of their opponent. During a few of the tests,
the subjects were actually tearing at the wire of their cages while
looking in the direction of the closed door of the room housing
the stimulus animal; and it was the observer's subjective impression that they were very eager to resume the fight. Marmosets
scent mark more frequently when they are dominant and aggressively
motivated than when they are submissive (Epple, 1970). Therefore,
the high frequency of scent marking on the stimulus perch carrying
the odor of a recent oponent is interpreted as an indication of
aggression against this opponent. In this particular context,
the scent-marking behavior, or the odor itself, might actually
function as a threat signal. A more detailed discussion of the
possible mechanisms involved in the use of scent marks as threat
signals follows.

Odors Related to Aggression and Social Status

In many mammals, body odors and scent-marking behavior seem
to play a role in the regulation of aggressive behavior and dominance relationships. Dominant rats are characterized by a specific
body odor (Krames et al., 1969). The urine of male mice carries
a testosterone-dependent odor which elicits aggression in other
males (Archer, 1968; Mugford and Nowell, 1970a, b, c). Dominant
males seem to produce more of this scent than submissive ones
(Mugford and Nowell, 1970c), and females possess an aggression-
inhibiting scent (Mugford and Nowell, 1971a, b). In some other
species, the frequency of scent-marking behavior increases during
aggressive interactions in those individuals who are dominant, but
not in those who are submissive (cf., Ralls, 1971; Mykytowycz, 1970).

There is little experimental information on the role of body
odors and scent marking in the aggressive behavior of primates.
However, many observers report that prosimians and South American
monkeys tend to scent mark frequently in situations where they are
aggressive and also dominant over conspecifics (Bolwig, 1960;
Epple and Lorenz, 1967; Evans and Goy, 1968; Sprankel, 1962; Jolly,
1966; Moynihan, 1964; and many others). In the common marmoset
(*Callithrix jacchus*), we found a correlation between social dominance, aggression against conspecifics of the same sex, and the
frequency of scent marking in adult males and females: in stable
groups, the dominant male and the dominant female scent marked
significantly more frequently than submissive group members. Moreover, aggressive encounters with strange conspecifics introduced
into stable groups strongly stimulated the marking frequency of
dominant group members of the same sex as the strange animal, but
not that of submissive group members (Epple, 1970). These findings
suggest that in the common marmoset a high frequency of scent marking demonstrates and communicates aggressiveness and social domi-

nance. The odor discharged in marking may function as a kind of
olfactory threat signal in intragroup and intergroup encounters,
although this is not its only function.

The pilot study described below was undertaken as a first step
in finding out whether the scent marks of dominant males and submissive males contain information on the social status of the
donors (Epple, 1973).

Procedures:

In this study we offered the subjects, 11 males and 12
females, the choice between a wooden perch carrying the scent
marks of a socially dominant male and a perch carrying the marks
of a submissive male. Five ranked pairs of males, each pair living with a female, were used as donors. The dominant male of each
group was defined as the one who associated most frequently with
the only female of the group (Epple, 1972b). Donors and subjects
had had no previous visual and vocal contact with each other. The
testing procedures were identical with those described for the
previous experiments.

Results:

During the tests, the monkeys preferred the perches carrying
the odor of the dominant males and tended to spend more time in
contact with the marks of dominant males. This tendency, however,
was not statistically significant, except for scores obtained during the first five minutes of testing (Fig. 4). The subjects also
sniffed and scent marked the perches marked by dominant males
significantly more frequently than those marked by submissive
males (Fig. 4). The preference of the monkeys for the odor of
socially dominant males shows that the scent of the dominant male
and/or the scent of the submissive one contained some information
which enabled the subjects to discriminate between them. This
information might be an odor specifically identifying the social
status of the individual. It seems more likely, however, that the
information is based on the quantity of scent present on the stimulus perch rather than on its quality. Dominant marmosets scent
mark more frequently than submissive ones (Epple, 1970); this difference probably resulted in a larger amount of scent on the perch
marked by the dominant donor, which might have been preferred by
the subjects because of its stronger odor. Further tests will be
necessary to show whether perches carrying small amounts of odor
from dominant donors are preferred over perches carrying large
amounts of odor from submissive donors. Such a preference would
suggest odor which might serve as an olfactory threat signal in
aggressive encounters. However, even if the scent marks of marmo-

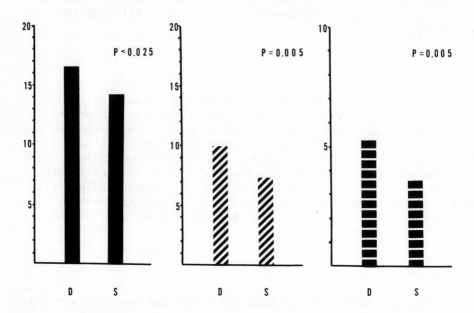

Fig. 4. Choice between scent marks of dominant males (D) and
submissive males (S) (mean scores for contacts during the first 5
min, sniffing, and marking; Wilcoxon matched-pairs signed rank
test: Siegel, 1956).

sets do not contain any odors which characterize social status,
marking might still be involved in the regulation of aggressive
behavior in the following way. A high frequency of scent marking
in an aggressive and dominant animal might be interpreted as an
expression of self-confidence and an act of "self-advertising" by
saturation of the environment with the personal scent of the dominant
individual. Since this scent carries information on the identity
of an individual, submissive animals are exposed to the presence
of a superior throughout much of their home range, even it the
superior is not within sight. Thus, the dominant's odor might
have a stressful effect on animals who have learned, by personal
encounters, that they are inferior to the marking individual. In
this way, the scent of a dominant monkey might be actively involved
in maintaining its social status and in keeping rank relations
within the group stable, while at the same time avoiding the
necessity for contact aggression that could result in wounding the

dominant as well as the submissive animal.

The role of body odor and marking behavior in the control of dominance interactions as suggested above seems not to be limited to the *Callithricidae*. The high scent-marking activity shown by many primates especially during aggressive interactions (Bolwig, 1960; Epple and Lorenz, 1967; Evans and Goy, 1968; Sprankel, 1962; Jolly, 1966; Moynihan, 1964; and others) suggests that odors are involved in the control of intragroup and intergroup aggression in many species. The use of chemical signals in the maintenance of social hierarchies might be especially prominent in species which do not have a large repertoire of visual threat signals to serve as substitutes for contact aggression. Marmoset monkeys and social prosimians would be among those species. An arboreal life and nocturnal activity patterns would also be expected to favor the use of chemical communication in social behavior.

Body odors involved in the regulation of dominance interactions may, to a certain extent, also affect the reproductive success of an individual. In some mammalian species, social dominance increases the chances for reproduction. Many studies show a cessation of reproduction and retardation of maturation in increasing populations of some rodents. The endocrinological mechanisms involved in this regulation of population size and the effects of social stress have recently been reviewed by Brain (1971), Christian (1971), and Leshner and Candland (1972). These investigations show that social stress affects submissive individuals earlier and more severely than dominant ones. In the rabbit *(Oryctolagus cuniculus)*, there is a significant correlation between social dominance, sexual activity, and the weights and secretory activity of the scent glands (cf. Mykytowycz, 1970). Mykytowycz (1970) also points out that intrauterine mortality is much higher in submissive females than in dominant ones. In increasing populations of tree shrews, von Holst (1969) found that submissive individuals suffered permanent regression of the testes and atrophy of maturing follicles, whereas dominant animals remained sexually active. In some species of macaques and baboons, the formation of consort pairs with fully receptive females is, to some degree, correlated with the rank of the males (cf., Rowell, 1972). In marmoset groups maintained under laboratory conditions, only the highest-ranking female breeds, apparently inhibiting reproduction in all other females of the group (Epple, 1974). Among male primates, sexually active dominant rhesus monkeys show a higher level of plasma testosterone than submissive, sexually inactive males (Rose et al., 1971, 1972). In human males, the production rate of testosterone was highly correlated with a measure for aggression derived from a psychological test (Persky et al., 1971).

There is some evidence, then, that even among primates high social rank, aggressiveness, and sexual activity are correlated

with a high level of gonadal steroids in the blood. Gonadal steroids, moreover, control the activity of the scent glands and the
marking behavior of many mammalian species. Gonadectomy results in
atrophy of the scent glands and in a cessation of scent marking in
the males of most species studied so far. The administration of
testosterone to castrated males fully restores the activity of the
glands and marking behavior. The effects of ovariectomy and of
gonadal steroids on the scent glands and on scent marking in
females are less uniform and clearcut. Further details are reported
by Andriamiandra and Rumpler (1968) and Thiessen et al. (1971) and
in the reviews of Ebling (1973) and Mykytowycz (1970).

In summary, the observations reviewed above strongly suggest
that in some species of primates pheromones control to some extent
the chances of individuals to reproduce by affecting their social
status in the group. There seems to be a positive correlation
between high social rank, aggressiveness, high sexual activity,
high levels of gonadal steroids in the blood, active scent glands,
and the production and deposition of chemical signals, some of
which may be involved in the communication and maintenance of
social dominance.

SUMMARY

A very large number of observational reports indicate that
chemical signals play important roles in various areas of the
reproductive and social behavior of primates. However, only three
areas of behavior influenced by chemical signals are discussed in
this paper.

1. Sexual attractants, communicating female receptivity and
arousing male sexual activity, apparently are widely used among
prosimians as well as New World and Old World primates. So far,
however, the chemical structure and hormonal control of sex
attractants have been studied only in the rhesus monkey.

2. Several authors have suggested that pheromones are involved in
the behavioral and physiological correlation of breeding activity
in seasonal primates.

3. Chemical messages identifying the sex of an animal, its species,
its individual identity, and its social status have been demonstrated to exist in a few primate species. Our own studies on the
chemical messages communicated by marmoset monkeys are reviewed and
supplemented with unpublished data. The possible role of these
messages in the control of dominance within the group and their
effect on reproductive success are discussed.

REFERENCES

Andriamiandra, A., and Rumpler, Y., 1968, Rôle de la testostérone sur le déterminisme des glandes brachiales et antébrachiales chez le *Lemur catta*, Compt. Rend. Séan. Soc. Biol. 162:1651-1655.

Archer, J., 1968, The effect of strange male odor on aggressive behavior in male mice, J. Mammal. 49:572.

Baldwin, J.D., 1968, The social behavior of adult male squirrel monkeys (*Saimiri sciureus*) in a seminatural environment, Folia Primat. 9:281-314.

Baldwin, J.D., 1970, Reproductive synchronization in squirrel monkeys (*Saimiri*), Primates 11:317-326.

Bolwig, N., 1960, A comparative study of the behavior of various lemurs, Mem. Inst. Sci. Madagascar, Ser. A. 14:205-217.

Bonsall, R.W., and Michael, R.P., 1971, Volatile constituents of primate vaginal secretions, J. Reprod. Fertil. 27:478-479.

Brain, P.F., 1971, The physiology of population limitation in rodents - a review, Comm. in Behav. Biol. 6:115-123.

Bronson, F.H., 1971, Rodent pheromones, Biol. of Reprod. 4:344-357.

Christian, J.J., 1971, Population density and reproductive efficiency, Biol. of Reprod. 4:248-294.

Curtis, R.F., Ballantine, J.A., Keverne, E.B., Bonsall, R.W., and Michael, R.P., 1971, Identification of primate sexual pheromones and the properties of synthetic attractants, Nature (Lond.) 232:396-398.

Doyle, G.A., Pelletier, A., and Bekker, T., 1967, Courtship, mating and parturition in the lesser bushbaby (*Galago senegalensis moholi*) under semi-natural conditions, Folia Primat. 7:169-197.

Ebling, F.J., 1963, Hormonal control of sebaceous glands in experimental animals, in *The Sebaceous Glands* (Adv. Biol. Skin, Vol. 4; W. Montagna, R.A. Ellis, and A.F. Silver, eds.), pp. 200-219, The MacMillan Co., New York.

Epple, G., 1970, Quantitative studies on scent marking in the marmoset (*Callithrix jacchus*), Folia Primat. 13:48-62.

Epple, G., 1971, Discrimination of the odor of males and females by the marmoset *Saguinus fuscicollis* ssp, in *Proceedings,*

3rd International Congress of Primatology (Zürich, 1970), vol. 3 (H. Kummer, ed.), pp. 166-171, S. Karger, Basel.

Epple, G., 1972a, Social communication by olfactory signals in marmosets, in *International Zoo Yearbook* (J. Lucas and N. Duplaix-Hall, eds.), vol. 12, pp. 36-42, The Zoological Society of London, London.

Epple, G., 1972b, Social behavior in laboratory groups of *Saguinus fuscicollis*, in *Saving the Lion Marmoset; Proceedings of WAPT Golden Lion Marmoset Conference* (D.D. Bridgwater, ed.), pp. 50-58, Wild Animal Propagation Trust, Oglebay Park, Wheeling, W. Va.

Epple, G., 1973, The role of pheromones in the social communication of marmoset monkeys *(Callithricidae)*, J. Reprod. Fertil. (Suppl.), 19:445-452.

Epple, G., 1974, The behavior of marmoset monkeys *(Callithricidae)*, in *Primate Behavior, Developments in Field and Laboratory Research* (L.A. Rosenblum, ed.), vol. 3, Academic Press, New York/London (in press).

Epple, G., and Lorenz, R., 1967, Vorkommen, Morphologie und Funktion der Sternaldrüse bei den Platyrrhini, Folia Primat. 7:98-126.

Evans, C.S., and Goy, R.W., 1968, Social behavior and reproductive cycles in captive ring-tailed lemurs *(Lemur catta)*, J. Zool. Lond. 156:171-197.

Harrington, J., 1974, Olfactory communication in *Lemur fulvus*, in *Proceedings of the Research Seminar on Prosimian Biology, London 1972* (G.A. Doyle, R.D. Martin, and A. Walker, eds.), Duckworth, London (in press).

Hill, W.C.O., 1944, An undescribed feature in the drill *(Mandrillus leucophaeus)*, Nature (Lond.) 153:199.

Hill, W.C.O., 1954, Sternal glands in the genus *Mandrillus*, J. Anat. 88:582.

Holst, D. von, 1969, Sozialer Stress bei Tupajas *(Tupaia belangeri)*, Z. vergl. Physiol. 63:1-58.

Jolly, A., 1966, *Lemur Behavior. A Madagascar Field Study*, University of Chicago Press, Chicago/London.

Klein, L.L., 1971, Observations on copulation and seasonal reproduction of two species of spider monkeys, *Ateles belzebuth* and A. *geoffroyi*, Folia Primat. 15:223-248.

Krames, L., Carr, W.J., and Bergman, B., 1969, A pheromone associated with social dominance among male rats, Psychonom. Sci. 16:11-12.

Latta, J., Hopf, S., and Ploog, D., 1967, Observation on mating behavior and sexual play in the squirrel monkey (*Saimiri sciureus*), Primates 8:229-246.

Leshner, A.T., and Candland, D.K., 1972, Endocrine effects of grouping and dominance rank in squirrel monkeys, Physiol. Behav. 8:441-445.

Lorenz, R., 1972, Management and reproduction of the Goeldi's monkey *Callimico goeldii* (Thomas, 1904) Callimiconidae, Primates, in *Saving the Lion Marmoset, Proceedings of WAPT Golden Lion Marmoset Conference* (D.D. Bridgwater, ed.), pp. 92-109, Wild Animal Propagation Trust, Oglebay Park, Wheeling, W. Va.

McClintock, M., 1971, Menstrual synchrony and suppression, Nature (Lond.) 229:244-245.

Michael, R.P., 1969, The role of pheromones in the communication of primate behaviour, in *Proceedings, 2nd International Congress of Primatology* (Atlanta, 1968), vol. 1 (C.R. Carpenter, ed.), pp. 101-107, S. Karger, Basel.

Michael, R.P., and Keverne, E.B., 1968, Pheromones in the communication of sexual status in primates, Nature (Lond.) 218: 746-749.

Michael, R.P., and Keverne, E.B., 1969, A male sex-attractant pheromone in rhesus monkey vaginal secretions, J. Endocrinol. 46:20-21.

Michael, R.P., and Keverne, E.B., 1970, Primate sex pheromones of vaginal origin, Nature (Lond.) 225:84-85.

Michael, R.P., and Zumpe, D., 1972, Patterns of reproductive behavior, in *Comparative Reproduction of Nonhuman Primates* (E.S.E. Hafez, ed.), pp. 205-242, Charles C. Thomas, Springfield, Ill.

Michael, R.P., Zumpe, D., Keverne, E.B., and Bonsall, R.W., 1972, Neuroendocrine factors in the control of primate behavior, Rec. Prog. Horm. Res. 28:665-706.

Moynihan, M., 1964, Some behavior patterns of platyrrhine monkeys.

I. The night monkey *(Aotus trivirgatus)*, Smithsonian Misc. Coll. 146:1-84.

Mugford, R., and Nowell, N., 1970a, The aggression of male mice against androgenized females, Psychonom. Sci. 20:191-192.

Mugford, R., and Nowell, N., 1970b, The preputial glands as a source of aggression-promoting odors in mice, Physiol. Behav. 6:247-249.

Mugford, R., and Nowell, N., 1970c, Pheromones and their effect on aggression in mice, Nature (Lond.) 226:967-969.

Mugford, R., and Nowell, N., 1971a, Endocrine control over production and activity of the anti-aggression pheromone from female urine, J. Endocrinol. 29:225-232.

Mugford, R., and Nowell, N., 1971b, The relationship between endocrine status of female opponents and aggressive behavior of male mice, Anim. Behav. 19:153-155.

Müller-Schwarze, D., 1974, Olfactory recognition of species, groups and individuals in mammals, in *Pheromones* (M.C. Birch, ed.), Elsevier/Excerpta Medica/North-Holland Publishing Co., Amsterdam (in press).

Mykytowycz, R., 1970, The role of skin glands in mammalian communication, in *Communication by Chemical Signals* (Adv. in Chemoreception, vol. 1; J.W. Johnston, D.G. Moulton, and A. Turk, eds.), pp. 327-360, Appleton-Century-Crofts, New York.

Perkins, E.M., 1966, The skin of the black-collared tamarin *(Tamarinus nigricollis)*,Amer. J. Phys. Anthrop. 25:41-70.

Persky, H., Smith, K.D., and Basu, G.K., 1971, Relation of psychologic measures of aggression and hostility to testosterone production in man, Psychosom. Med. 33:265-277.

Poirier, F.E., 1970, The communication matrix of the nigiri langur *(Presbytis johnii)* of South India, Folia Primat. 13:92-136.

Ralls, K., 1971, Mammalian scent marking, Science 171:443-449.

Rose, R.M., Gordon, T.P., and Bernstein, I.S., 1972, Plasma testosterone levels in the male rhesus: Influence of sexual and social stimuli, Science 178:643-645.

Rose, R.M., Holaday, J.W., and Bernstein, I.S., 1971, Plasma testosterone, dominance rank and aggressive behaviour in male rhesus monkeys, Nature (Lond.) 231:366-368.

Rowell, T.E., 1972, Female reproduction cycles and social behavior in primates, in *Advances in the Study of Behavior* (D.S. Lehrmann, R.A. Hinde, and E. Shaw, eds.), vol. 4, pp. 69-105, Academic Press, New York/London.

Sauer, F., and Sauer, E., 1963, The South-West-African bushbaby of the *Galago senegalensis* group, J. South West Afr. Sci. Soc. Windhoek 16:5-36.

Schaffer, J., 1940, *Die Hautdrüsenorgane der Säugetiere*, Urban und Schwarzenberg, Berlin/Wien.

Seitz, E., 1969, Die Bedeutung geruchlicher Orientierung beim Plumplori, *Nycticebus coucang* Boddaert 1785 *(Prosimii, Lorisidae)*, Z. Tierpsychol. 26:73-103.

Siegel, S., 1956, *Nonparametric Statistics for the Behavioral Sciences*, McGraw-Hill, New York.

Sprankel, H., 1962, Histologie und biologische Bedeutung eines jugosternalen Duftdrüsenfeldes bei *Tupaia glis* (DIARD 1820) in Gefangenschaft, in *Verhandlungen Deutschen Zoologischen Gesellschaft* (Saabrücken, 1961), pp. 198-206.

Thiessen, D.D., Owen, K., and Lindsey, G., 1971, Mechanisms of territorial marking in the male and female Mongolian gerbil *(Meriones unguiculatus)*, J. Comp. Physiol. Psychol. 77:38-47.

Vandenbergh, J.G., 1969, Endocrine coordination in monkeys: Male sexual responses to the female, Physiol. Behav. 4:261-264.

RELATIONSHIPS BETWEEN SCENT MARKING BY MALE MICE AND THE PHEROMONE-INDUCED SECRETION OF THE GONADOTROPIC AND OVARIAN HORMONES THAT ACCOMPANY PUBERTY IN FEMALE MICE [1]

F. H. Bronson and Claude Desjardins

Department of Zoology

The University of Texas at Austin

Many mammals use specific olfactory cues (pheromones) as an integral part of their intrapopulation communication. These compounds can be conveniently categorized according to function as: (a) signalling pheromones, which elicit a more or less immediate change in behavior (if indeed a response does occur), and (b) priming pheromones, which trigger neuroendocrine and endocrine activity. Sex attractants are an example of the former; the regulatory effect of urinary odors on the mouse estrous cycle is an example of the latter. Much of the recent interest in mammalian pheromones stems from the discovery in the 1950's of a series of dramatic primer effects in female mice (e.g., Whitten, 1956; Bruce, 1959). Many contributions have since been made by ethologists, psychologists, and reproductive physiologists; but work by chemists and neurophysiologists is conspicuously absent. Recent reviews include Bronson (1968, 1971, 1974), Bruce (1966, 1967, 1970), Cheal and Sprott (1971), Eisenberg and Kleiman (1974), Gleason and Reynierse (1969), Mykytowycz (1970), Ralls (1971), Schultz and Tapp (1974), Whitten (1966), and Whitten and Bronson (1970).

The present contribution deals with two aspects of pheromone biology in mice: scent marking by males and the (priming) hormonal action of male urinary pheromone in immature females. In regard to the former, the chance observation that urine fluoresces bright blue when viewed under ultraviolet light has led to the discovery of intense urinary scent marking by male mice (Desjardins et al., 1973). A series of studies have now utilized this technique to increase our knowledge about the elicitation of this behavior.

[1]This work was supported by Public Health Grants HD-03803 and HD-07381 from the National Institutes of Health.

Regarding the relationship between pheromones and puberty, the female mouse presents an ideal system for studying priming pheromonal
action, as well as puberty itself, since the final stages of sexual
maturation are easily and profoundly influenced by varying the type
of housing and, hence, the olfactory environment (e.g., Vandenbergh
et al., 1972). As a final objective, we can integrate our present
knowledge of these two aspects of pheromone biology and speculate
about their possible relationships.

SCENT MARKING BY MALE MICE

Any system of communication requires a sender, a receiver, and
a message. Most laboratory experimentation on olfactory communication in mammals has dealt with the behavioral or physiological
responses of the "receiver." Only two pheromones, both signalling
types, have been isolated: a tarsal gland pheromone in the black-
tailed deer (Müller-Schwarze, 1971) and a vaginal sex attractant
in the rhesus monkey (Michael et al., 1971). Our knowledge of the
deposition and transmission of pheromones, i.e., the behavior of
the "sender," is largely naturalistic and anecdotal; two notable
exceptions are a large series of studies on marking with ventral
gland sebum by the gerbil (e.g., Thiessen et al., 1970) and considerable work on chin gland marking by the Australian rabbit
(Mykytowicz, 1970).

Our previous inability to subject the "sender's" behavior to
laboratory experimentation was obviously due to a lack of techniques for visualizing or otherwise detecting pheromone deposition.
Desjardins et al. (1973) have demonstrated the utility of ultraviolet light for visualizing urine and have thereby provided a
most useful tool for studying urinary marking patterns. Studies
using this technique with wild house mice have since demonstrated
that: (a) previously isolated male house mice assiduously mark
filter paper placed on the floors of test chambers at rates as high
as 400 to 500 marks per hour or as much as several thousand marks
in an overnight test; (b) this behavior is elicited primarily by
either biological or physical novelty, i.e., marking quickly habit-
uates during repeated testing, regardless of the presence of other
animals; and (c) marking is largely suppressed by castration and
totally absent in socially subordinated males.

Marking by Previously Isolated Males: the Effect of Subordination

This series of experiments used laboratory-raised descendants
of a wild stock of house mice. All mice were isolated from weaning
until they were tested at 55 to 60 days of age. During the last
two weeks of isolation, they were housed in one side of a 12- by
12-inch test chamber which was divided in half by a wire mesh partition. Figure 1 compares the overnight marking patterns of such

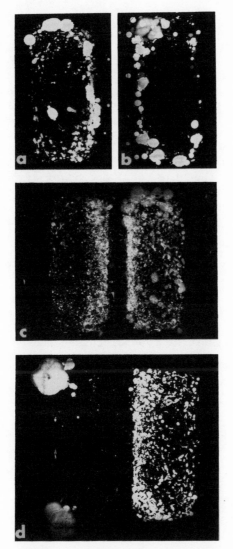

Fig. 1. Overnight urinary marking patterns of two male mice viewed under ultraviolet light; patterns were obtained on 3 consecutive nights. a, b: Two isolated males housed in separate cages; c: the same two previously isolated mice housed in a single cage but separated by a wire mesh partition; and d: the two males after a single 15-min encounter during which dominance was established (dominant male on the right, subordinate on the left).

isolated males with those observed when another male was placed in the other half of the cage. The effect of the second male was obvious. The rate at which urinary marks were deposited increased

dramatically, often to as high as 2,000 to 3,000 marks per 10-hour
test (Desjardins et al., 1973). If, on the other hand, the partition separating two males was raised for 15 minutes and thus allowed
them to fight and establish a social hierarchy, overnight marking
was totally suppressed in the subordinated male; whereas the dominant male continued marking at a relatively high rate (Fig. 1d).
The suppression of marking among subordinates was invariant and
stable across several daily tests, sometimes even when the dominant
male was absent. As would be expected from these findings, the
bladders of subordinate males typically contained at least 20 times
more urine than those of dominants. Studies with tritium-labeled
insulin have verified that the total amount of urine voided is the
same in dominant and subordinate animals, at least as assessed
over an 18-hour period (Desjardins et al., 1973). Indeed, none of
our studies, including those discussed below, indicate differential
rates of urine production. Marks deposited at low rates of marking
thus have the appearance of pools of urine; whereas those deposited
by males marking at a high rate are always in the form seen in
Figure 1b.

Habituation of Marking

After experiments of the type discussed above had been repeated many times, it could be seen that the response pattern of

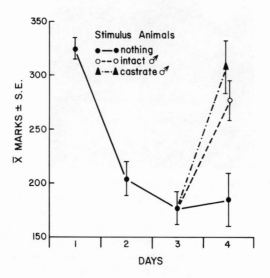

Fig. 2. Mean (± S.E.) numbers of marks deposited by previously isolated males during consecutive, daily 1-hr tests. Males were tested
alone for the first 3 days and alone or in the presence of either
another intact male or a castrated male on the fourth night (see text).

isolated males varied considerably. Much of this variation was traceable to the number of times an animal had experienced the test situation. Figure 2 documents a strong habituation phenomenon associated with repeated exposure to the test chamber (Maruniak et al., 1974). In this experiment, male house mice were reared in isolation in standard mouse cages. At 55 to 60 days of age, one-hour marking tests were run at 8:00 p.m. in the 12 x 12-inch test chambers described above. After the initial test, all males were returned to their home cages until the following morning when they were again placed in the test chambers and re-tested with clean filter paper at 8:00 p.m. that night. The same procedure was followed on the third day. On the fourth night, the animals were again tested, but this time with either another intact male or a castrated male in the other half of the cage. Marked habituation to the test chamber is obvious in Figure 2. Furthermore, marking was returned to the initial high rate on the fourth night of testing by the presence of either an intact or a castrated male. A comparison of Figures 1a and 2 also suggests habituation with increasing test-chamber experience. Figure 1a shows 30 to 80 relatively large "marks" deposited during overnight tests by animals which had previously been housed in the test chambers for two weeks. Figure 2, on the other hand, shows a mean marking rate of over 300 small marks per *hour* during a male's first exposure to the test apparatus.

Two additional experiments confirmed the strong habituation to the test chamber. In one, 0.2 ml of male urine placed in the other half of the test cage was sufficient to significantly elevate marking by test-habituated males; the level of the elevation, however, was only about 50% of that stimulated by the presence of another male. The other experiment showed that marking by test-habituated males was returned to near prehabituation levels by the presence of an ovariectomized female, a PMS-HCG primed female, or a male deermouse. There was little evidence that marking was differentially enhanced by the different types of stimulus animals (Maruniak, et al., 1974).

The foregoing experiments all documented strong habituation to the test chamber. A fourth experiment asked whether habituation would also occur in the presence of another male. Previously isolated adult males were thus exposed to the test apparatus for three days (as described above), but each cage contained two males (same males in all cases) separated by the wire mesh partition. Marked habituation, similar to that shown in Figure 2, occurred even with another male present. Marking was significantly elevated (but not to initial high levels) on the fourth night by the presence of a male deermouse or an estrogen-primed female, but was not altered by switching the test-habituated males so that each was in the presence of a strange male (Maruniak, et al., 1974).

Suppression of Marking by Castration

The effect of castration was tested by exposing both castrated (30 days after the operation) and intact males to the three-day habituation process described above. Mean marking rates for intacts were well over 300 on the first night and had dropped to about 150 by the third night. Mean marking rates for castrates were 55, 51, and 56 for the three nights, respectively.

Conclusions about Urinary Marking by Male House Mice

The general function of scent marking by mammals is still debatable. Indeed, it is probable that there are diverse functions for such behaviors depending upon species, type of social organization, variation in the animal sensing the marks, etc. (see Johnson, 1973). It is obvious, however, that this behavior in combination with different types and/or concentrations of signalling pheromones could serve nicely to transmit information regarding species, sex, age, and possibly an individual "fingerprint". Motivation for the behavior, on the other hand, has been viewed as (a) an intolerance of conspecifics on the part of dominant individuals (Ralls, 1971) or (b) possibly a response to the disruption of an individual's own "odor field" (Eisenberg and Kleiman, 1974). The results of our studies on scent marking argue strongly that there is no need to conceptualize marking, at least for house mouse males, in such socially or physiologically complex frameworks. Our male mice apparently scent marked primarily in response to novelty, either physical (test apparatus) or biological (the initial response to another animal, conspecific or not). Thus the initial response to the test apparatus in all experiments was to mark at a mean rate of 300 to 400 per hour; three days of habituation yielded means between 100 and 175. Habituation occurred whether or not another male was present. Marking was returned to, or at least near, the high (300+) level in test-habituated males by the presence of either an intact or a castrated male. Marking among males habituated to the presence of another male was statistically, but not markedly, elevated by the presence of an estrous female or a deermouse male. Important for determining the ability of a novelty concept to encompass all of these data is the fact that switching stimulus males had no effect on marking by the test male if he had first been habituated to the presence of another male. Thus the high marking rates were elicited only in response to a dramatically new situation; i.e., the first exposure to either the test chamber or the presence of another animal (male or female or even another species). Continual testing in all cases yielded habituation and low marking rates. Intermediate rates of marking were elicited by urine or were observed among male-habituated males that were exposed to a markedly different type of stimulus animal. Novelty as a motivational basis for scent marking, then, better encompasses all of our data on <u>Mus</u> and, in addition,

provides a theoretically more simple conceptual basis for this
behavior than does either of the previously mentioned models, intolerance for conspecifics or odor-field disruption. Marking is,
nevertheless, at least in part androgen dependent and is dramatically and completely absent in socially subordinated males.

Finally, a series of preliminary studies have indicated that
both the technique of UV-visualization and our emerging theories
about the elicitation of this marking behavior among house mice may
have broad application. Females do not mark in the manner shown
in Figures 1 and 2; but they do deposit urine in blue-fluorescing,
medium-size pools which have known olfactory properties (e.g.,
Scott and Pfaff, 1970). Male deermice mark prodigiously. Male ger-
bils do not appear to utilize urine for marking to the same extent
as house mice; but the sebum from their ventral gland, which is
known to be used in scent marking (Thiessen et al., 1970), fluoresces
orange under UV light. The preputial glands of male mice contain
lipid(s) which act as sexual attractants for sexually experienced
females (Caroom and Bronson, 1971); preputial fluid also fluoresces
orange. Desert-evolved species (such as the gerbil), which must
conserve water, have presumably evolved specialized lipid-producing
glands for production of signalling pheromone; most other species
probably use urine as the main vehicle. In both cases, the UV-
visualization technique will be a welcome aid to understanding the
behaviors associated with these modes of pheromone transmission.

PHEROMONES AND PUBERTY

A series of studies by Vandenbergh (1967, 1969; Vandenbergh et
al., 1972) has established that the final, i.e., post-weaning,
stages of sexual maturation in female mice are profoundly affected
by their olfactory environment (see also Castro, 1967; Cowley and
Wise, 1972; and Bronson and Stetson, 1974). Thus maintaining immature females in groups or exposing them to female urine greatly
delays the onset of puberty, but exposure to male urinary odors
markedly accelerates it; social isolation in the absence of male
odor yields an intermediate rate of sexual maturation. The female
mouse thus provides a nearly ideal system for exploring such questions as the physiological action of priming pheromones and the
process of puberty itself. The following experiments dealt with
the hormonal responses to male urinary pheromone in immature female
mice.

Body Weight as a Predictor of Responsiveness to Male Pheromone

Most previous studies of hormonal responses to male urinary
pheromone have been plagued by an inability to detect non- or
slowly responding females. In studies which attempted to elucidate

pheromonal action in recently inseminated females, for example, a certain proportion of such animals did not respond to the male with a blocked implantation. Blood from such females must, nevertheless, be included in the final data analysis since it would not be known until after the fact which females had responded and which had not (Chapman et al., 1970). In a study more pertinent to the present work, Bronson and Stetson (1974) used age of immature females as a criterion for studying their gonadotropic responses to a male's presence. They found that all 25-day-old females eventually responded to the male's pheromone with fertile matings but that ovulation occurred anywhere from two to eight days after the initiation of exposure to the male. There was thus little uniformity in the hormonal responses to the male's presence; hence few solid conclusions about the initial actions of male pheromone could be drawn. Table 1 shows the results of an investigation in which body weight rather than age was used as a predictor of responsiveness among CF-1 females of any age between 26 and 20 days. The data show a marked relationship between body weight and the appearance of ballooned, proestrous-like uteri on the afternoon of the third day after experimental treatment began. The normal adult time sequence of uterine ballooning, ovulation, and fertile matings has previously been established for immature females housed with adult males (Bronson and Stetson, 1974). Thus this study measuring the presence or absence of proestrous uteri provided the basis for a model system in which puberty could be uniformly induced over a three-day period and thereby allowed a relatively undiluted look at hormonal responses

TABLE 1.   *Effects of Social Stimuli and Body Weight on Uterine Growth in Immature Female Mice*

| Weight range (g) | Percentage of Females with Proestrous-like Uteri* | | |
	Grouped	Isolated	Male-exposed
15.5–16.5	0% (0/30)	4% (1/28)	41% (14/34)
16.5–17.5	0% (0/20)	13% (4/30)	62% (20/32)
17.5–18.5	12% (3/25)	10% (3/31)	83% (28/33)

*Immature female mice were housed in groups of 6 per cage from weaning to 26–30 days of age. Animals reaching the indicated body weight ranges were either regrouped (5/cage), isolated, or housed with an adult male at 0900 hr and killed at 1500 hr on the third day.

to the male's presence. Specifically, exposure to the male was initiated at 26 to 30 days of age, but only if body weight was 17.5 to
18.5 g. Individuals failing to respond in a uniform fashion (10 to
20%) could then be detected by their "aberrant" uterine weights and
eliminated from the study before hormone assays were performed
(Bronson and Desjardins, 1974).

Changes in FSH, LH, Estradiol, and Progesterone during Male Exposure

Over 1300 CF-1 females of the appropriate age and weight ranges
were killed at various times over a three-day period after male
exposure was begun at 9:00 a.m. One group of control animals was
killed just before male exposure began; other control females remained in groups until they were killed at the end of the 72-hour
experimental period. Fifteen percent of the females were omitted
from the study on the basis of aberrant uterine weight before hormone assay. Figure 3 shows changes in hormone titers throughout
the three-day experimental period. The obvious initial effect of
male exposure on gonadotropin secretion consisted of a rapid elevation in circulating concentrations of LH (within one hour), elevated
concentrations then being maintained for a period of hours thereafter. In contrast to LH, circulating levels of FSH did not change
significantly during this time. Plasma estradiol concentrations
were already significantly greater at three hours and had increased
15- to 20-fold by 12 hours after male exposure began. Like FSH,
progesterone levels did not change during the first day of male
exposure.

The second day of male exposure was characterized by an almost
total depression of FSH but not LH; a second peak in estradio at 36
hours; and, again, no change in progesterone. Estradiol remained
significantly above the 0- or 72-hour control levels until mid-day
of the third day (52 hours), after which concentrations of this
hormone were in the range of control values. Normal, adult-like
periovulatory changes in FSH, LH, and progesterone commenced during
the third day (Bronson and Desjardins, 1974).

Another study asked whether or not the dramatic initial (1-12
hr) changes in LH and estradiol were in themselves sufficient to
ensure the attainment of a full pubertal cycle in the three-day
experimental period. As little as three or six hours of male exposure followed by isolation in a male-free room resulted in an
increase in uterine weight as assessed on the afternoon of the third
day. However, a significant increase in the number of females
attaining fully ballooned, proestrous-like uteri on the third day
was not obtained with less than 36 hours of male exposure. A period
of 48 hours of cohabitation with a male, on the other hand, was fully
effective in this regard. It is again important to note that previous work on this phenomenon has verified the expected adult-like

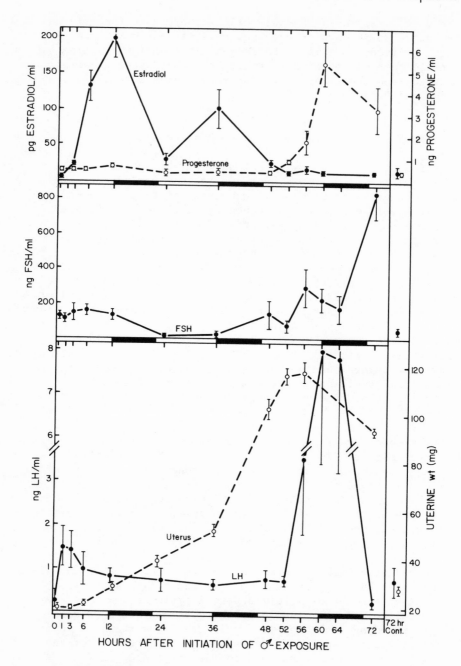

Fig. 3. Changes (mean ± S.E.) in hormone levels of immature females. throughout 3 days of exposure to adult males. LH and FSH values are presented as ng-equivalents of NIH-LH-S1/ml and NIH-FSH-S1/ml.

time sequence of uterine ballooning and ovulation (Bronson and Stetson, 1974).

Pubertal Cycles in the Absence of a Male

We (Stiff et al., 1973) examined gonadotropic, vaginal, and uterine characteristics of CF-1 female mice from before birth until 25 days of age and then of females isolated in the absence of a male from 21 days of age through their first vaginal cycle. The study on the early developing female revealed a massive peak in FSH, probably starting before birth and lasting for more than 15 days (Fig. 4).

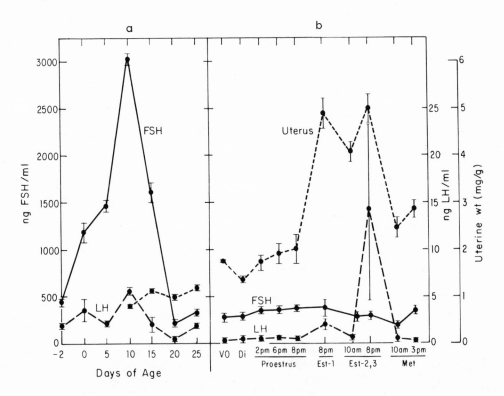

Fig. 4. Uterine weight (mg/g body wt) and plasma levels of FSH and LH a: In immature CF-1 female mice from 17 days of gestation (2 days before birth) to 25 days of age (vagina still closed in all cases); b: dependent upon vaginal stage among isolated females after vaginal opening (VO). Hormone concentrations are expressed as ng-equivalents of NIH-FSH-S1/ml and NIH-LH-S1/ml. Proestrus in this figure represents composite data from females killed after having shown a continuous proestrous smear of any duration from 1 to 12 days (see text).

LH was not markedly variant in the early developing female.

Our study of the "normal" reproductive development in the females isolated at 21 days of age revealed that the first pubertal
cycle is considerably prolonged and often anovulatory if no male
is present. Vaginal opening was accompanied by either diestrous
or proestrous smears or, on occasion, even an estrous smear. Vaginal proestrus often continued up to 12 days; full cornification
usually lasted two days and sometimes as long as four days before
leukocytic invasion, decidedly nonadult-like characteristics. In
addition, changes in the vaginal epithelium were out of phase
with changes in uterine weight, at least in respect to what one
would expect from the adult cycle of this strain. Ballooned uteri
were relatively impossible to predict; when they did occur, they
did so more often in conjunction with a fully cornified smear than
with the expected proestrous smear. The ovulatory flush of LH was
detected in no female showing a proestrous smear, in only three
of 25 females killed at 8:00 p.m. during their first day of a cornified smear, and in six of 25 females killed at 8:00 p.m. on Days
2 or 3 of such a smear. Periovulatory-like, elevated levels of
FSH were detected in only six of 77 females killed between 8:00 p.m.
of the first day of a cornified smear and 10:00 a.m. on the day of
a metestrous smear (Fig. 4).

Oviductal examinations of these females revealed a marked dependence of the presence of eggs on both the stage of the vaginal
cycle and on the age at which a particular stage occurred. Oviducts
of isolated females achieving their first late estrous or metestrous
smear before 36 days of age were typically free of eggs (1/11 of
those showing an estrous smear; 7/19 of those in metestrus). Females attaining their first metestrous smear after 39 days of age
usually had eggs in their oviducts (13/15). Similarly, corpora
lutea were not usually seen in the ovaries except in the case of
females reaching their first metestrus at an advanced age.

In summary, then, females isolated in the absence of a male
from 21 days of age showed prolonged and disorganized vaginal and
uterine cycles; periovulatory changes in gonadotropins were difficult to detect; and these "cycles" were often anovulatory, unless
an animal experienced them for the first time at a relatively
advanced age.

Conclusions about Puberty in Mice

Our recent experiments on pheromones and puberty in female
mice both confirm and extend Vandenbergh's earlier work with this
phenomenon. Both sets of studies show the presence of males to be
of paramount importance to the proper organization of puberty in

immature females. In our stock of mice, ovulatory puberty with
fertile matings and with apparently normal implantation can be induced by a male's presence as early as 22 to 23 days of age. Vaginal opening occurs at about this time in isolated females in the
absence of a male or his odor; the ensuing vaginal and uterine
cycles, however, are greatly prolonged, disorganized, and usually
anovulatory, at least until a relatively advanced age has been
reached (39+ days). First cycles of grouped females are even more
prolonged and disorganized (Stiff et al., 1974). In addition,
unlike the action of male urinary pheromones in adult females,
grouping of immature females is almost totally antagonistic to the
male's pheromonal action (Vandenbergh et al., 1972; Bronson and
Stetson, 1974). Because of this great degree of variability and
because of the ease with which it can be controlled by simply
varying the type of housing and, hence, the olfactory environment,
the immature female mouse provides a possibly unique system for
jointly studying both pheromonal action and the underlying hormonal
bases for puberty itself. It is in this direction that much of our
research has gone.

A comparison of the results shown in Figure 3 with those published in Bronson and Stetson (1974) shows the obvious, crucial
advantage of initiating male exposure at a particular body weight
rather than at a particular age. In the former study, which used
age as the criterion, ovulation occurred but at a quite variable
number of days after male exposure began; attempts to look for
subtle changes in the gonadotropins in response to the male' pheromone were thus unsatisfactory. Use of a selection procedure, based
mainly on body and then uterine weight, yielded a quite uniform
response to the male, and hence, a relatively uniform population
of females for examining hormonal responses associated with this
phenomenon. It should be noted here that our choice of a three-
day time period was largely dictated by the results of early pilot
work. Considering how slowly the pubertal processes progress
in the absence of a male, we undoubtedly could have obtained just as
good a model system by using a greater or lesser body weight in
combination with either a two- or four-day experimental period.

In respect to the action of the male's pheromone, previous
experiments which have combined male exposure with injections of
either PMS or HCG have concluded that the effects of male exposure
can mimic the action of either hormone in the normal ovulatory sequence in immature females; hence, one would suspect that the male
pheromone acts on both FSH and LH (Zarrow et al., 1972). Within
the framework of our three-day model system, however, the initial
effect of the male's presence was only on LH; and the somewhat delayed secondary action was a particularly dramatic surge in estradiol. No concurrent changes were observed in the circulating levels
of either FSH or progesterone. The fact that a male is necessary
for 36 to 48 hours to ensure the completion of the pubertal cycle

shows that the immediate sequential alterations in LH and estradiol
are not in themselves sufficient for completion of the cycle two
days later. Thus hormonal changes occurring on the second day of
male exposure must also be considered as important responses in
pheromone-induced puberty. Estradiol concentrations, which remained significantly above control values until 1300 of the third
day (52 hours after male exposure began; Fig. 3), peaked sharply
for a second time at 36 hours. No immediate explanation for the
second surge in estrogen secretion is suggested by our data on
pituitary gonadotropin release. FSH was totally depressed during
the second day; LH was not. One could hypothesize that there was
an LH basis for the second estrogen peak but that it was missed
because of a 12-hour sampling interval. Thus it would appear that
the male's action rapidly to elevate LH and, hence, estradiol, must
be continued, but at a much lower level, for almost 48 hours if a
full pubertal cycle is to be attained in a three-day period. An
interesting, correlative conclusion is that the male's pheromone
must act to support secretion of LH even in the face of a high titer
of estrogen; FSH secretion, not being affected by the male's pheromone, was totally depressed during the second day, presumably by
the high levels of circulating estrogens.

Generalities about Puberty in Female Mice and Rats

The results of our experiments with hormone assays can be
integrated into a working model which deals with the hormonal bases
of puberty in both the mouse and the rat. A massive peak in FSH
has now been reported in female neonates of both species (Ojeda
and Ramirez, 1972; Stiff et al., 1974). The most obvious correlate of
this neonatal FSH activity is the initiation of the growth of follicles which will be mature at pubescence (Pedersen, 1969). Injections of antibodies to gonadotropin have confirmed the important
role for FSH during neonatal life in the mouse (Eshkol and Lunenfeld, 1972). In addition, the results of studies in which antibodies to estrogen were given to neonatal rat females suggest that
this steroid also plays a role in early follicular development
(Reiter et al., 1972). Whether FSH, estradiol, or both hormones
will eventually be assigned the pivotal role in neonatal ovarian
development remains to be seen. Regardless of this confusion, however, a maximum number of follicles are present in the 21-day-old
mouse (Pedersen, 1969), after which superovulation can be induced
with gonadotropin (Zarrow and Wilson, 1961); the principle is the
same but the timing somewhat different in the rat.

At this relative point in time, an ovulatory, pubertal cycle
can be rapidly induced in the mouse by male pheromone, acting via
LH and estrogen successively; the pubertal changes occur more gradually in the rat. It is also at this relative point in development
when an injection of estrogen induces ovulation during the third

night after injection in both species, its efficiency being enhanced
if progesterone is also given on the morning of the third day (Ying
and Greep, 1971; Bronson, unpublished). Considering both rats and
mice, then, either an acute or a more gradual change in LH could
act on an ovary primed neonatally with FSH so that sufficient estrogen is released to trigger the first periovulatory changes in
gonadotropins. One current view of the relative roles of the two
gonadotropins in the adult cycle attributes the development of the
succeeding crop of follicles to the periovulatory surge in FSH secretion (Daane and Parlow, 1971; Schwartz et al., 1973).

Obvious differences are apparent when the time relations between peak levels of estrogen and LH are compared in the pubertal
mouse and the adult rat. The ovulatory surge of LH actually begins
with or just after the end of a several-hour surge in estrogen
secretion in the adult proestrous rat (cf., Gay et al., 1970;
Shaikh, 1971). The pheromonally induced, pubertal cycle in the
mouse, on the other hand, is characterized by two marked peaks in
peripheral estradiol concentrations, one immediately after male exposure begins and one during the second day, followed by low levels
throughout the third day, i.e., the proestrous equivalent of the
adult cycle. A single injection of estradiol induces an ovulatory
secretion of LH in the immediately prepubertal rat (28 to 30 days
of age), with time latencies (either 1 to 2 or 2 to 3 days) differing in two different reports (cf., Ying et al., 1971; Caligaris et
al., 1972). This estrogen injection is effective at an earlier age
in the rat only if accompanied by a priming injection of estrogen
(Caligaris et al., 1972). This observation is conceptually similar
to our demonstration of two peaks in circulating estradiol in the
immature female mouse. The negative feedback system is apparently
mature well in advance of the positive action of estradiol on LH;
the latter develops rapidly in the immediately prepubertal female
rat and mouse (cf., Ramirez, 1972; Ying and Greep, 1971; Bronson,
unpublished). On the basis of all of the above, it is reasonable
to speculate that: (a) the longer period of time between endogenous
or exogenous estrogen peaks and the ovulatory flush of LH in the
prepubertal animal involves the final maturation of the positive
feedback system and; (b) the male pheromone acts, probably via enhanced secretion of ovarian estrogens, to promote this maturational
stage in the mouse.

URINARY MARKING AND PRIMING PHEROMONE TRANSMISSION

Priming pheromones from male mice are transmitted via their
urine and are either androgen metabolites or the products of tissues maintained by testicular hormones (see review by Bronson,
1971). The discovery of vigorous scent marking by male mice
raises a question about the relationship between this behavior and
the transmission of the physiologically active pheromones. This

was the general subject of an experiment which evaluated the relative efficacy of dominant vs. subordinate males, or of filter papers on which these two social categories had marked, to induce uterine growth in immature females. Adult, previously isolated, CF-1 males were paired for one-half hour a day for seven days during which individual pairs established strong social hierarchies. Some of these pairs were then housed on filter paper in the standard test chambers, separated by the wire mesh partition, but allowed to fight daily for 15 minutes. Immature CF-1 females were isolated at the appropriate age and weight and then exposed in their home cages for 2-1/2 days to (a) a dominant or a subordinate male, or (b) shredded filter papers (cage floors) marked by other dominant or subordinate males. Control females were left in groups of six per cage. Shredded filter paper was placed in the immature female's home cages twice a day; each addition was thus the result of 12 hours of marking on the part of either a dominant or a subordinate male. Figure 5 shows that exposure to marked filter paper had absolutely no effect on uterine weight. The presence of a male, whether dominant or subordinate, had an obviously strong effect on the uterus.

Dominant and subordinate males differ markedly in their reproductive-endocrine characteristics (e.g., Christian and Davis, 1964;

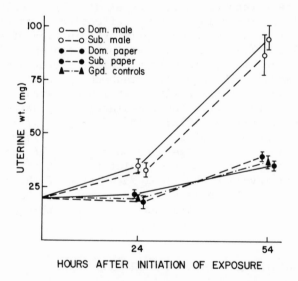

Fig. 5. Uterine weight (± S.E.) of isolated immature females killed before, during, or after a 54-hr period during which they were exposed to the presence of either a dominant or a subordinate male or shredded filter paper (cage floors) marked by either dominant or subordinate males; controls were left in groups of 6 in a male-free room.

Bronson, 1973). They also differ dramatically in their marking behavior, at least when tested in each other's presence (Fig. 1). Whether marking is also suppressed among subordinates housed with females is not yet known. Nevertheless, both dominant and subordinate males both produced and transmitted sufficient pheromone to induce good uterine growth in the females with which they were cohabiting (Fig. 5). Two possible explanations for the lack of effect of the filter paper, even when heavily marked by a dominant male, are that (a) filter paper firmly binds the priming pheromone, or (b) the pheromone is simply too volatile to be effective when accumulated on a single sheet of filter paper for 12 hours. Acceleration of puberty has been induced in immature female mice by male-soiled bedding or by actually placing male urine on the nose (Vandenbergh, 1969; Cowley and Wise, 1972). Adult reproductive effects of this pheromone have also been induced with well-soiled bedding or lab coat material, as well as by both direct and indirect exposure to male urine (e.g., Whitten et al., 1968). Thus there can be little doubt of the presence of a urinary primer as well as its ability to be effectively transferred in soaked bedding material. The most probable explanation for the lack of effectiveness on the part of marked filter paper would thus seem to be that involving high volatility.

A comparison of all of the results discussed in this paper yields some interesting speculations about the relationships between priming and signalling pheromone deposition and efficacy. The energy required for 2000+ marks per night suggests that urinary scent marking by a male mouse is a passive reflex, typically indulged in while travelling around its home range. Sexual maturity, i.e., a sufficient quantity of circulating androgens, and a dominant social position would certainly act as permissive factors in the elicitation of this reflex. Because of its obvious relationship with novel cues, one could certainly speculate that urinary marking would be most common at the periphery of a mouse's home range, particularly during encounter with a strange animal or its odor. Thus this reflexive response combined with the consequent deposition of signalling pheromones could act efficiently to disseminate information regarding species, sex, age, and possibly individual identification.

The real advantage of transmitting information via a chemical-olfactory modality is that it can be accomplished both at a distance and across a time dimension (Bronson, 1971). The latter aspect would argue that signalling pheromones either are of relatively low volatility or are combined with carrier substances. The work summarized here, on the other hand, indicates that the primer pheromones may be quite volatile (at least when collected on filter paper) and, furthermore, that prolonged contact is necessary to induce its full effects, at least in immature females; i.e., 36 to 48 hours of male exposure were necessary to maximize the probability of

attainment of a full pubertal cycle in three days. Thus a pair-
bond or some such close social relationship appears to be a necessary prerequisite for any degree of priming pheromone efficiency.

To extend our speculations to natural populations, puberty is
probably suppressed in an immature female by the presence of her
mother and other female litter mates, regardless of the presence
of an adult male. Since male urinary pheromone (in the absence of
antagonistic female odors) is necessary for the proper organization of puberty, a failure to attain a full ovulatory cycle during
the young female's period of dispersal is also probable. Encoun-
tering male urine during this critical period of migration would
result in information reception, but because of the high volatility
of the primers would probably not induce puberty; this latter process would occur only after a close association with a male in his
home range. Because of the suppression of marking among social
subordinates, a higher probability of pair-bond formation can be
predicted with dominant males. In addition, the preputial glands
that produce lipids with sex attractant functions actually grow
markedly during the attainment of a dominant social position (Bronson and Marsden, 1974). Nevertheless, the current data argue that
puberty-induction would be an invariant correlate of the establishment of a pair-bond (once it had occurred), regardless of the
social rank of the male.

REFERENCES

Bronson, F.H., 1968, Pheromonal influences on mammalian reproduction, Reprod. Sex. Behav. 21:341-361.

Bronson, F.H., 1971, Rodent pheromones, Biol. Reprod. 4:344-357.

Bronson, F.H., 1973, Establishment of social rank among grouped
male mice: relative effects on circulating FSH, LH, and corticosterone, Physiol. Behav. 10:947-951.

Bronson, F.H., 1974, Pheromonal influences on reproductive activities in rodents, in Pheromones (M.C. Birch, ed.), Elsevier/
Excerpta Medica/North-Holland Publishing Co., Amsterdam
(in press).

Bronson, F.H., and Desjardins, C., 1974, Circulating concentrations
of FSH, LH, estradiol, and progesterone associated with the
acute pheromonal induction of puberty in female mice,
Endocrinology (in press).

Bronson, F.H., and Marsden, H.M., 1974, The preputial gland as an
indicator of social dominance in male mice, Behav. Biol.
9:625-628.

Bronson, F.H., and Stetson, M.H., 1974, Gonadotropin release in prepubertal female mice following male exposure: a comparison with the adult cycle, Biol. Reprod. 9:449-459.

Bruce, H.M., 1959, An exteroceptive block to pregnancy in the mouse, Nature (Lond.) 184:105.

Bruce, H.M., 1966, Smell as an exteroceptive factor, J. Anim. Sci. 25:83-89.

Bruce, H.M., 1967, Effects of olfactory stimuli on reproduction in mammals, in *Effects of External Stimuli on Reproduction* (Ciba Found. Study Group No. 26; G.E.W. Wolstenholme and M. O'Conner, eds.), pp.29-42, Little, Brown, Boston.

Bruce, H.M., 1970, Pheromones, Brit. Med. Bull. 26:10-13.

Caligaris, L., Astrada, J.J., and Taleisnik, S., 1972, Influence of age on the release of LH induced by oestrogen and progesterone in immature rats, J. Endocrinol. 55:97-103.

Caroom, D., and Bronson, F.H., 1971, Responsiveness of female mice to preputial attractant: effects of sexual experience and ovarian hormones, Physiol. Behav. 7:659-662.

Castro, B.M., 1967, Age of puberty in female mice: relationship to population density and the presence of adult males, Anais Acad. Bras. Cient. 39:289-291.

Chapman, V.M., Desjardins, C., and Whitten, W.K., 1970, Pregnancy block in mice: changes in pituitary LH and LTH and plasma progestin levels, J. Reprod. Fertil. 21:333-337.

Cheal, M.L., and Sprott, R.L., 1971, Social olfaction: a review of the role of olfaction in a variety of animal behaviors, Psychol. Reports 29:195-243.

Christian, J.J., and Davis, D.E., 1964, Endocrines, behavior, and population, Science 146:1550-1560.

Cowley, J.J., and Wise, D.R., 1972, Some effects of mouse urine on neonatal growth and reproduction, Anim. Behav. 20:499.

Daane, T.A., and Parlow, A.F., 1971, The differential effect of the presence of male and female mice on the growth and development of the young, J. Genet. Psychol. 119:89.

Desjardins, C., Maruniak, J.A., and Bronson, F.H., 1973, Social rank in house mice: differentiation by ultraviolet visualization of urinary marking patterns, Science 182:939-941.

Eisenberg, J.F., and Kleiman, D.G., 1974, Olfactory communication in mammals, Ann. Rev. Ecol. Systematics 3 (in press).

Eshkol, A., and Lunenfeld, B., 1972, Gonadotropic regulation of ovarian development in mice during infancy, in *Gonadotropins* (C.G. Beling and H.M. Gandy, eds.), pp.335-346, Wylie Interscience, New York.

Gay, V.L., Midgley, A.R., Jr., and Niswender, G.D., 1970, Patterns of gonadotropin secretion associated with ovulation, Fed. Proc. 29:1880-1887.

Gleason, K.K., and Reynierse, J.H., 1969, The behavioral significance of pheromones in vertebrates, Psych. Bull. 71:58-73.

Johnson, R.P., 1973, Scent marking in mammals, Anim. Behav. 21: 521-535.

Maruniak, J.A., Owen, K., Bronson, F.H., and Desjardins, C., 1974, Urinary marking in male house mice: responses to novel environmental and social stimuli, Physiol. Behav. (in press).

Michael, R.P., Keverne, E.B., and Bonsall, R.W., 1971, Pheromones: isolation of male sex attractants from a female primate, Science 172:964-966.

Müller-Schwarze, D., 1969, Complexity and relative specificity in a mammalian pheromone, Nature (Lond.) 223:525-526.

Mykytowycz, R., 1970, The role of skin glands in mammalian communication, in *Communication by Chemical Signals* (Adv. Chemoreception, vol. 1; J.W. Johnson, D.G. Moulton, and A. Turk, eds.), pp.327-360, Appleton-Century-Crofts, New York.

Ojeda, S.R., and Ramirez, V.D., 1972, Plasma level of LH and FSH in maturing rats: response to hemigonadectomy, Endocrinology 90:466-472.

Pedersen, T., 1969, Follicle growth in the immature mouse ovary, Acta Endocrin. 62:117.

Ralls, K., 1971, Mammalian scent marking, Science 171:443-449.

Ramirez, V.D., 1972, Maturation of the gonadotropin control system, Acta Endocrin., Suppl. 166:170-176.

Reiter, E.O., Goldenberg, R.L., Vaitukaitis, J.L., and Ross, G.T., 1972, A role for endogenous estrogen in normal ovarian development in the neonatal rat, Endocrinology 91:1537.

Schultz, E.F., and Tapp, J.T., 1974, Olfactory control of behavior in rodents, Psych. Bull. (in press).

Schwartz, N.D., Krone, K., Talley, W.L., and Ely, C.A., 1973, Administration of antiserum to ovine FSH in the female rat: failure to influence immediate events of cycle, Endocrinology 92:1165.

Scott, J.W., and Pfaff, D.W., 1970, Behavioral and electrophysiological responses of female mice to male urine odors, Physiol. Behav. 5:407-411.

Shaikh, Abubakar, A., 1971, Estrone and estradiol levels in the ovarian venous blood from rats during the estrous cycle and pregnancy, Biol. Reprod. 5:297-307.

Stiff, M.E., Bronson, F.H., and Stetson, M.H., 1974, Plasma gonadotropins in prenatal and prepubertal female mice; disorganization of pubertal cycles in the absence of a male, Endocrinology (in press).

Thiessen, D.D., Linzey, G., Blum, S.L., and Wallace, P., 1970, Social interactions and scent marking in the Mongolian gerbil (*Meriones unguiculatus*), Anim. Behav. 19:505-513.

Vandenbergh, J.G., 1967, Effect of the presence of a male on the sexual maturation of female mice, Endocrinology 81:345-359.

Vandenbergh, J.G., 1969, Male odor accelerates female sexual maturation in mice, Endocrinology 84:658-660.

Vandenbergh, J.G., Drickamer, L.C., and Colby, D.R., 1972, Social and dietary factors in the sexual maturation of female mice, J. Reprod. Fertil. 28:397-405.

Whitten, W.K., 1956, Modifications of the oestrous cycle of the mouse by external stimuli associated with the male, J. Endocrinol. 13:399-404.

Whitten, W.K., 1966, Pheromones and mammalian reproduction, in *Advances in Reproductive Physiology* (A. McLaren, ed.), pp.155-177, Academic Press, New York/London.

Whitten, W.K., and Bronson, F.H., 1970, Role of pheromones in mammalian reproduction, in *Communication by Chemical Signals* (Adv. Chemoreception, vol. 1; J.W. Johnson, D.G. Moulton, and A. Turk, eds.), pp.309-325, Appleton-Century-Crofts, New York.

Whitten, W.K., Bronson, F.H., and Greenstein, J.A., 1968, Estrus-

inducing pheromone of male mice:　transport by movement of
air, Science 161:584-585.

Ying, S.-Y., and Greep, R.O., 1971, Effect of age of rat and dose
of a single injection of estradiol benzoate (EB) on ovulation
and the facilitation of ovulation by progesterone (P),
Endocrinology 89:785.

Ying, S.-Y., Fang, V.S., and Greep, R.O., 1971, Estradio benzoate
(EB)-induced changes in serum-luteinizing hormone (LH) and
follicle-stimulating hormone (FSH) in immature female rats,
Fertil. Steril. 22:64-68.

Zarrow, M.X., and Wilson, E.D., 1961, The influence of age on superovulation in the immature rat and mouse, Endocrinology 69:
851-855.

Zarrow, M.X., Christenson, C.M., and Eleftheriou, B.E., 1971,
Strain differences in the ovulatory response of immature mice
to PMS and to the pheromonal facilitation of PMS-induced
ovulation, Biol. Reprod. 4:52-56.

EFFECTS OF PROGESTERONE ON FEMALE REPRODUCTIVE BEHAVIOR IN RATS: POSSIBLE MODES OF ACTION AND ROLE IN BEHAVIORAL SEX DIFFERENCES

Lee-Ming Kow, Charles W. Malsbury, and Donald W. Pfaff

The Rockefeller University

New York, New York

INTRODUCTION

Much recent work has focused on neurophysiological mechanisms that underlie lordosis in female rodents (Pfaff et al., 1972, 1973) and often has been directed at explaining the effects of estrogen on female reproductive behavior. For instance, the neuroanatomy of estradiol-binding cells in the central nervous system of the female rat has been clarified (Pfaff and Keiner, 1973); and the effects of estrogen on peripheral sensory nerve input have been recorded (Kow and Pfaff, 1973/74).

Progesterone, also, has an important role in the control of female rodent mating behavior, since it acts synergistically with estrogen to facilitate this behavior (Beach, 1948; Young, 1961; Powers, 1970; Davidson, 1972). In this paper we review evidence for the hypothesis that progesterone facilitates lordosis by acting on serotonin systems in the female rat brain and elaborate, and add anatomical detail to, the ideas put forward by Meyerson (1964a, 1964b, 1966b, 1968a; cf. Zemlan et al., 1973). Second, in reviewing literature on sex differences in the performance of female rat reproductive behavior, we find evidence that the relative inability of males to perform lordosis reflects an insensitivity to both progesterone (cf. Clemens et al., 1969; Gorski, 1971) and estrogen. Finally, we suggest sites of neural-hormonal interaction where insensitivity to progesterone and estrogen might result in deficient lordosis performance by males.

POSSIBLE MODES OF PROGESTERONE ACTION ON LORDOSIS

The suggestion for a mechanism of progesterone facilitation of lordosis, outlined in Figure 1, is based on previously published notions about the relationship between serotonin and female mating behavior. Since most of the serotonin neurons in the rat brain are found in the midbrain raphe nuclei, the anatomical relations of these nuclei are of special interest. Similarities between the effects of raphe lesions and septal lesions prompted us to formulate the hypothesis outlined below. Briefly, it suggests that progesterone inhibits activity in a raphe-septal loop which includes serotonergic fibers and which inhibits lordosis. Numbered paragraphs review the evidence for the relationships indicated by the corresponding number in Figure 1.

General Effects of Lesions and Reduced Serotonergic Function

1.   Septal lesions heighten irritability (Brady and Nauta, 1953) and lead to a general hypersensitivity at least to somatic and gustatory stimuli (Carey, 1972, 1973).

2.   Lesions of midbrain raphe nuclei heighten aggressiveness (Heller et al., 1962) and lead to general hypersensitivity to somatic stimuli (Lints and Harvey, 1969), hyperactivity (Kostowski et al., 1968), and sleeplessness (Jouvet, 1969; Morgane and Stern, 1972).

3.   Para-chlorophenylalanine (PCPA) injections, which decrease brain levels of serotonin, are followed by heightened sensitivity to external stimuli (Tenen, 1967), hyperactivity (Koe and Weissman, 1966), and sleeplessness (Jouvet, 1968; Morgane and Stern, 1972). Increased sensitivity to electric shock also follows a pharmacological block of serotonergic activity in the septal area; whereas decreased sensitivity follows the application of serotonin there (Persip and Hamilton, 1973).

Conclusion: *Septal lesions, raphe lesions, and PCPA have similar effects on some aspects of overall "excitability."*

Lesions, Serotonin, Progesterone, and Lordosis

4.   In estrogen-primed female rats, septal destruction enhances lordosis (Komisaruk et al., 1972; Nance et al., 1974). Note that preoptic lesions can also enhance lordosis (Law and Meagher, 1958; Powers and Valenstein, 1972a).

5.   In estrogen-primed female rats, PCPA and other agents which antagonize serotonin enhance lordosis (Zemlan et al., 1973;

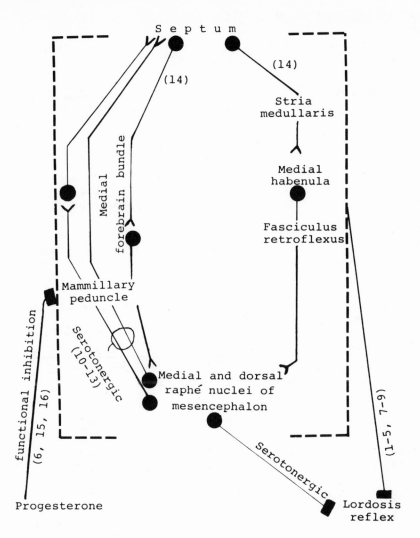

Fig. 1. Model for the possible action of progesterone on lordosis
in female rats. Progesterone is assumed to have a net inhibitory
effect on a system of neurons linking the midbrain raphe nuclei
with the septum. These possibly progesterone-sensitive connections
are indicated within the dashed-line box. All lines on the left
inside the dashed-line box represent projections through the medial
forebrain bundle. Descending connections on reflex loops subserv-
ing lordosis could come via serotonergic fibers from the raphe nuclei themselves or from other parts of the raphe-septal system of
connections. Bracketed numbers in the figure refer to correspond-
ingly numbered paragraphs in the text where the evidence is reviewed.
Symbols: ◁ = excitatory relation; ─■ = inhibitory relation.

Meyerson, 1964a; Ward et al., 1974).

6.   Progesterone enhances lordosis in estrogen-primed ovariectomized female rats (Beach, 1948; Young, 1961).

Conclusion: *Septal lesions, PCPA, and progesterone all can enhance
          lordosis in estrogen-primed female rats.*

Stimulation Effects

7.   In estrogen- and progesterone-treated female hamsters and rats,
septal (or preoptic) electrical stimulation can inhibit lordosis
(Malsbury and Pfaff, 1973; Moss et al., 1973; Zasorin et al., 1974).

8.   Electrical stimulation of raphe nuclei (at low frequencies)
can cause decreased motor activity and decreased responsiveness to
external stimuli (Kostowski et al., 1969; Gumulka et al., 1971;
Liebeskind et al., 1973).

9.   Heightened serotonin function interferes with lordosis in the
female rat (Meyerson, 1964a, 1966b, 1968a).

Conclusion: *Stimulation and serotonin elevation have opposite
          effects from lesions and serotonin depletion on
          lordosis and general "excitability."*

Neurochemical and Neuroanatomical Background

10.   Serotonin-containing cell bodies are found in the dorsal and
medial mesencephalic raphe nuclei (Dahlström and Fuxe, 1965).

11.   Serotonin-containing cells in the raphe nuclei project through
the medial forebrain bundle (Fig. 2) (Conrad et al., 1974; Dahlström
and Fuxe, 1965; Ungerstedt, 1971). Some fibers reach the septum
(Conrad et al., 1974; Ungerstedt, 1971; Kuhar et al., 1972; Guillery, 1957). Midbrain raphe neurons also project through the
fasciculus retroflexus to the habenula (Conrad et al., 1974), and
fibers from the habenular nuclei project forward into the septum
(Smaha and Kaelber, 1973).

12.   Destruction of midbrain raphe nuclei or of medial forebrain
bundle or septum decreases brain serotonin levels (Heller and Moore,
1965).

13.   PCPA depletes brain serotonin (Koe and Weissman, 1966).

14.   Septal output, indirectly at least, reaches the raphe nuclei.
Septal neurons project into the medial forebrain bundle, and some

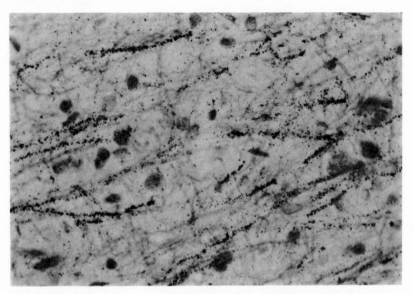

Fig. 2. Autoradiograph of radioactively labelled nerve fibers
coursing through the medial forebrain bundle of the rat. The rat
was sacrificed and its brain sectioned and processed for autoradiog-
raphy 48 hr after injection of ^{3}H-proline into the medial raphe
nucleus. Radioactively labelled proteins formed in raphe neuron
cell bodies and transported by axoplasmic flow make possible the
study of axonal projections. Some raphe projections through the
medial forebrain bundle reach the diagonal bands and septum.
(Sagittal section, counterstained with Luxol Fast Blue; from
Fig. 7B of Conrad et al., 1974.)

medial forebrain bundle fibers reach the midbrain raphe nuclei.
Septal neurons also project through the stria medullaris to the
medial habenula, which in turn projects through the fasciculus
retroflexus to the medial raphe nucleus (Raisman, 1966; Cragg, 1961;
Wolf and Sutin, 1966; Smaha and Kaelber, 1973; Guillery, 1957;
Conrad and Pfaff, 1974).

Conclusion: *Septum and raphe nuclei are reciprocally connected,*
                at least through the medial forebrain bundle and per-
                haps also through the stria medullaris-habenula-
                fasciculus retroflexus system.

Progesterone

15. Radioactive progesterone is taken up throughout the female

rodent brain, with midbrain uptake slightly higher than the rest (Wade and Feder, 1972a; Whalen and Luttge, 1971).

16. Progesterone tends to decrease responses to sensory stimuli by hypothalamic and other neurons. In some neurons, responses to cervical probing may be decreased selectively (Barraclough and Cross, 1963). In other neurons, which are correlated with EEG activity, progesterone may decrease responses to any arousing stimulus (Komisaruk et al., 1967; Ramirez et al., 1967).

Conclusion: *If cells in the mesencephalon and diencephalon affected by progesterone have a net inhibitory effect on lordosis, then the action of progesterone on them should facilitate lordosis.*

Hypothesis

Projections from the midbrain raphe nuclei through serotonergic fibers in the medial forebrain bundle or through fibers in the fasciculus retroflexus-stria medullaris system directly or indirectly reach the septum (paragraphs 10, 11, 12, 13). Projections from the septum through the medial forebrain bundle or through the stria medullaris-fasciculus retroflexus route directly or indirectly reach the midbrain raphe nuclei (paragraph 14). Hypothalamic cells, including progesterone-sensitive ones, could participate in mid-brain-limbic loops at least through the medial forebrain bundle (paragraphs 15, 16). The dynamics of the raphe-septal loop normally maintain a continued excitatory state of its neurons. The net effect of descending projections from the loop, the most important of which may be serotonergic fibers from the raphe, is to inhibit responses to somatosensory input, including inhibition of lordosis (paragraphs 7, 8, 9). Interruption of function in the loop facili-tates responses to somatosensory input, including facilitation of lordosis (paragraphs 1, 2, 3, 4, 5). Therefore, if progesterone decreases the activity and responsiveness of neurons in this loop, it should facilitate lordosis (paragraphs 6, 15, 16). The hypoth-esis is shown pictorially in Figure 1.

Discussion

This hypothesis states that a likely way in which progesterone could facilitate lordosis is by inhibiting activity in a raphe-septal loop which itself inhibits lordosis. Therefore, progester-one could facilitate lordosis by a process of disinhibition. Such a notion can be considered a specific application of the ideas of Beach (1967), who emphasized the net inhibitory effect of forebrain structures on mating responses. A clear prediction from our pre-sent model is that lesions of the midbrain raphe nuclei should facilitate lordosis in estrogen-primed ovariectomized female rats.

Present data suggest that progesterone may act in several brain regions by affecting activity in a raphe-septal loop which influences lordosis. Studies with radioactive progesterone and electrophysiological studies (paragraphs 15, 16) indicate widespread effects. Effects of progesterone on lordosis have been reported after implants of progesterone in the midbrain (Ross et al., 1971) and in the medial hypothalamus (Powers, 1972; Ward et al., 1974).

The present hypothesis does not exclude the possibility that other monoamines and presumed neurotransmitters in the central nervous system affect lordosis. Serotonin is emphasized here because of a coherent body of data which make possible the synthesis of neuroanatomical, neurochemical, and behavioral information. Several experiments from various other laboratories--outside the scope of this review--suggest that other amines also participate in the control of female reproductive behavior.

It is interesting to note an apparent anatomical dissociation between the serotonin systems implicated by the present hypothesis and the topography of estrogen-binding nerve cells in the female rat. For instance, in the midbrain central grey, most estrogen-binding neurons are in the lateral or dorsolateral portions, showing very little overlap with the serotonin cell group B7 (Pfaff and Keiner, 1973; Dahlström and Fuxe, 1965). Likewise, estrogen-binding neurons are not found in the medial raphe nucleus serotonin cell group B8. Finally, even though there are many estrogen-binding neurons in the anterior hypothalamic region (Pfaff and Keiner, 1973), there are relatively few in the suprachiasmatic nucleus, which is thought to contain a high density of serotonergic synaptic endings. What such an anatomical dissociation would imply for the relationship between estrogen and progesterone effects on lordosis is not known at present.

SEX DIFFERENCES IN THE PERFORMANCE OF FEMALE REPRODUCTIVE BEHAVIOR IN THE RAT

Behavioral Bisexuality

The lordosis reflex, although it is regarded as female mating behavior, is not performed only by female rats (or other female rodents), but also can be performed by males. Lordosis by normal adult male rats has long been reported (Stone, 1924; Beach, 1938, 1945), though recently investigators in this field have failed to observe lordosis in the intact adult male rat. However, investigators generally agree that adult male rats can be induced to perform lordosis if castrated and given repeated large doses of estrogen. A question that remains is: how well can the adult male rat perform lordosis? One group of investigators (Whalen and Edwards, 1967;

Meyerson, 1968b; Edwards and Thompson, 1970; Gerall and Kenney, 1970; Pfaff, 1970a; Edwards and Warner, 1972) believe that the ability of castrated male rats to perform lordosis is very slight, even after repeated administration of large doses of estrogen. Another group (Beach, 1945; Davidson, 1969; Davidson and Bloch, 1969; Davidson and Levine, 1969; Aren-Engelbrektsson et al., 1970; Hendricks, 1972; Eriksson and Sodersten, 1973) claim that with repeated estrogen treatments male rats, castrated as adults, can perform good lordosis, in some cases even indistinguishable from that executed by females (Beach, 1945). Generally, it has been concluded that the adult male rat can perform high quality lordosis only when subjected to prolonged daily doses of estrogen much larger than those sufficient to induce lordosis in the ovariectomized female rat (Davidson, 1969; Aren-Engelbrektsson et al., 1970; Hendricks, 1972; Sodersten and Larsson, 1974). Even under this condition, however, the males' performance at best only approximated that of female rats (Davidson, 1969), unless further treatments with pharmacological agents or brain lesions were carried out.

Behavioral bisexuality is not only shown by the male rat, but also by the female. Beach (1942a) reported incomplete masculine mating behavior in prepuberally ovariectomized, virgin female rats. Moreover, most masculine mating behaviors except ejaculation could be observed in intact female rats at all stages of the estrous cycle (Beach and Rasquin, 1942; Sodersten, 1972). Ovariectomy is not necessary for female rats to perform masculine mating behavior. On the contrary, though it does not eliminate the ability of the female rat to perform masculine mating behavior, ovariectomy decreases this capability (Sodersten, 1972). In fact, the ovariectomized female rats' ability can be enhanced by treatment with exogenous estrogen (Sodersten, 1972). As expected, the administration of exogenous androgen to ovariectomized female rats also can enhance performance of masculine mating behavior (Ball, 1940; Beach, 1942a; Pfaff, 1970a; Sodersten, 1972). Thus, unlike male rats, which rarely executed lordosis, female rats could readily perform masculine mating behavior, whether they were ovariectomized, hormone treated, or received no special treatment at all.

From this brief review, it is clear that rats of either sex can perform both masculine and feminine mating behavior but do so with a sexual difference: female rats perform masculine mating behavior readily; whereas male rats perform lordosis only after repeated large doses of estrogen. This difference is illustrated by an investigation conducted by Pfaff (1970a), who subjected gonadectomized rats of both sexes to a series of hormonal treatments and observed and compared masculine and feminine copulatory behaviors in each rat. He reported that, quantitatively, female rats approximated male rats in the performance of masculine mating behavior more closely than male rats approximated feminine behavior. Similar results have been shown by Whalen and Edwards (1967). In part of their experiment, they used rats of both sexes that were sham-gonadectomized at birth. All of these rats (100%) performed their respective homotypical mating behaviors; but although 91% of the female rats performed masculine mating behavior, only 25% of the male rats showed lordosis responses (see Table 1 in Whalen and Edwards, 1967). Thus, although both female and male rats are capable of behavioral bisexuality, the male rats are inferior in this respect because they are less able to perform lordosis.

Neural Bisexuality

Since masculine and feminine mating behaviors involve two entirely different sets of motor patterns, it may be assumed that they are controlled or mediated by two different neural networks or systems. Thus, the behavioral bisexuality observed in male and female rats implies that these two different neural networks, one responsible for masculine, another for feminine, mating behavior, coexist in the nervous system of each individual rat, male or female.

The notion that masculine and feminine mating mechanisms may have different locations in the brain was suggested by the following findings. In postpuberally ovariectomized female rats, Dorner et al. (1968b) reported that implantation of estrogen in the medial central hypothalamus (MCH) induced lordosis; whereas identical implantation in the medial preoptic-anterior hypothalamus (MPOA-AH) induced predominantly masculine behavior. Furthermore, similar results were found when androgen instead of estrogen was implanted; that is, implantation of androgen in the MPOA-AH induced masculine, while implantation in the MCH induced feminine behavior (Dorner et al., 1968a). Similarly, implantation of androgen in the MPOA-AH in postpuberally castrated male rats induced masculine behavior (Davidson, 1966; Lisk, 1967; Dorner et al., 1968a); implantation in the MCH, however, was relatively ineffective in inducing lordosis (Dorner et al., 1968a). The results from intracerebral hormone implantation concur with those obtained by another approach, brain lesion studies. In male rats, lesioning of the MPOA-AH region impaired or abolished masculine mating behavior (Heimer and Larsson, 1966, 1967; Lisk, 1968; Giantonio et al., 1970). Similar lesions in female rats also eradicated the masculine mating behavior but left feminine behavior intact (Singer, 1968). On the other hand, lesions in the hypothalamus posterior to the MPOA in female rats abolished feminine behavior but only partially deterred masculine behavior (Singer, 1968; Law and Meagher, 1958).

Differences between masculine and feminine behavior mechanisms are not confined to the medial hypothalamus and preoptic area. Surgical removal of a large portion (40%) of the cerebral cortex from male rats abolished their masculine behavior, which could not be restored with hormone treatment (Beach, 1940). In female rats, decortication also eliminated masculine behavior (Beach, 1943, 1944), but did not prevent (Larsson, 1962) and even facilitated lordosis whether the decortication was surgical (Beach, 1943, 1944) or functional (Clemens et al., 1967; Ross and Gorski, 1973; Ross et al., 1973). In the male rat, also, functional decortication facilitates lordosis (Clemens, 1971). Similarly, lesions in medial forebrain bundle eradicated or reduced masculine behavior in both male (Hitt et al., 1970; Hitt et al., 1973; Paxinos and Bindra, 1973) and female (Hitt et al., 1970; Modianos et al., 1973) rats, but did not disrupt feminine behavior in female rats (Rodgers and Law, 1967; Hitt et al., 1970; Modianos et al., 1973). On the other hand, habenular lesions caused a reduction in feminine mating behavior in female rats (Rodgers and Law, 1967; Modianos et al., 1974) but had no effect on masculine behavior in males (Modianos et al., 1974). Contrary to the effect of an habenular lesion, destruction of the septum can enhance feminine mating behavior in the female rat (Komisaruk et al., 1972; Nance et al., 1974), as well as in males (Nance et al., 1974). Like septal destruction, complete or partial removal of olfactory bulbs can facilitate the hormonally induced feminine mating behavior in both female (Moss, 1971; Edwards and Warner, 1972) and male rats (Edwards and Warner, 1972).

Both testicular and ovarian sex hormones in the circulatory system can enter the brain and are taken up and retained specifically by neurons in certain brain regions. On the basis of the evidence for neural bisexuality, the pattern and quantity of brain uptake and retention of sex hormones of both types would be expected to be the same, or at least similar, for male and female rats. Both biochemical and autoradiographic studies verify this expectation. Brain uptake and retention of testosterone was similar in both pattern and quantity for rats of both sexes when either biochemical (McEwen et al., 1970a, 1970b) or autoradiographic methods (Pfaff, 1968b) were used. The pattern of estrogen uptake and retention was found to be very similar for rats of both sexes (Pfaff, 1968a, 1968b), though quantitatively the hypothalamus of female rats took up and retained more estrogen than that of male rats (Anderson and Greenwald, 1969; Flerko et al., 1969; McEwen and Pfaff, 1970). For another ovarian hormone, progesterone, the degree of brain uptake specificity is uncertain. In rats progesterone content is higher in the plasma than in the brain, where labelled progesterone was taken up by every region examined with little differences among brain regions in the amount of radioactivity (Seiki et al., 1968; Seiki et al., 1969; Seiki and Hattori, 1971; Whalen and Luttge, 1971; Wade et al., 1973. Whalen and Luttge (1971) found that the pattern of progesterone retention in the brain is similar for both male and female rats, though the level of retention may be higher in females.

In certain aspects of brain anatomy, there may be some detailed,

quantitative sex differences. Measuring nerve cell nuclear and
nucleolar sizes with the light microscope, Pfaff (1966) and Dorner
and Staudt (1968, 1969a, 1969b) found sex differences in hypothalamic and preoptic cell groups. Using the electron microscope,
Raisman and Field (1971) found another sex difference: female rats
have a greater preponderance of spine synapses of non-amygdaloid
fibers in the medial preoptic area. Whether any of these morphological differences are related to the control of reproductive behavior
is not yet clear.

        If a pharmacological agent can facilitate one type of mating
behavior in one sex, could it also do so in the other sex? An
affirmative answer has been given for several agents. With respect
to masculine mating behavior, it is now thought that para-chlorophenylalanine (PCPA) treatment has a facilitatory effect in male rats
(Tagliamonte et al., 1969; Tagliamonte et al., 1970; Malmras and
Myerson, 1971; Ahlenius et al., 1971) and also in female rats
(Sheard, 1969; Singer, 1972). Evidence has also been gathered with
respect to feminine mating behavior. In estrogen-primed, ovariectomized female rats, reserpine (Meyerson, 1964b, 1966a, 1966b),
tetrabenazine (Meyerson, 1964b, 1966a, 1966b; Ahlenius, Engel,
Eriksson, and Sodersten, 1972), or α-MT (α-methyl-p-tyrosine)
(Ahlenius, Engel, Eriksson, Modigh, and Sodersten, 1972; Eriksson
and Sodersten, 1973) can facilitate lordosis, or feminine mating
behavior, as can progesterone. The feminine mating behavior in
male rats castrated as adults and primed with estrogen can also be
facilitated by treatment with reserpine (Meyerson, 1968b), tetrabenazine (Meyerson, 1968b; Larsson and Sodersten, 1971), or α-MT
(Sodersten and Ahlenius, 1972; Eriksson and Sodersten, 1973), even
though progesterone is not effective in facilitating lordosis in
male rats (see discussion below). The effect of PCPA treatment in
ovariectomized,estrogen-primed female rats is still somewhat con-
troversial. Several investigators (Meyerson and Lewander, 1970;
Meyerson, 1972; Ahlenius, Engel, Eriksson, Modigh, and Sodersten,
1972; Zemlan et al. 1973; Eriksson and Sodersten, 1973) have reported that it is facilitatory; others (Segal and Whalen, 1970;
Singer, 1972) have said that PCPA has no affect.

        Other estrogen-sensitive responses are affected in similar ways
in male and female rats. Estrogen treatment is known to increase
the running activity (Stern and Zwick, 1972) and decrease the
electroshock seizure threshold (Woolley and Timiras, 1962) in the
female rat. As one might expect, the running activity of male rats
treated with estrogen increased (Asdell and Sperling, 1963) and
their electroshock seizure threshold decreased (Wooley and Timiras,
1962), although the effective dosage necessary for the latter effect
was higher for male rats.

        According to the available evidence, the nervous system of
each rat, male or female, appears to have two neural mechanisms,

one mediating masculine, the other feminine, mating behavior. To
some extent, they can be facilitated independently by either appropriate intracerebral hormone implantations or pharmacological agents;
or they can be suppressed independently by appropriate brain lesions.
Because of this dual representation, each rat of either sex is
capable to some extent of performing both masculine and feminine
behavioral responses. Why, then, is the female rat better able to
perform lordosis than the male rat?

Estrogen-Sensitivity Hypothesis

     An explanation for this sex difference is that the nervous
system of the male rat is less sensitive to the lordosis-inducing
effect of estrogen than that of the female. This will be designated as the "estrogen-sensitivity hypothesis" (Fig. 3). It is
derived from the previously-mentioned conclusion that although male
rats castrated as adults can be induced by estrogen to perform
lordosis, estrogen administration must be repeated and at a dosage
considerably higher than that sufficient to induce lordosis in female rats (Davidson, 1969; Aren-Engelbrektsson et al., 1970; Hendricks, 1972). Besides this obvious indication, other lines of
evidence also support this hypothesis. As mentioned earlier, the
retention of estrogen by the hypothalamus is quantitatively higher
in female than in male rats (Flerko et al., 1969; Anderson and
Greenwald, 1969; McEwen and Pfaff, 1970). This inferior capacity
to retain estrogen in the hypothalamus may cause the male rat to be
less sensitive to estrogen, and thus to require more than females
to achieve a certain level of hypothalamic-bound estrogen. Moreover,
investigators using scintillation counting (McEwen and Pfaff, 1970;
McEwen et al., 1970a, 1970b) and autoradiographic techniques (Pfaff,
1968a, 1968b) have shown that the uptake patterns for radioactive
testosterone and estrogen are very similar in male and female rat
brains. These findings suggest a functional overlap of these two
sex hormones in inducing lordosis, which was in fact found to exist.
In female rats, testosterone, either implanted into the appropriate
brain region (Dorner et al., 1968a; Pfaff, 1970b), or administered
systemically (Pfaff, 1970a; Whalen and Hardy, 1970; Pfaff and Zigmond, 1971), can substitute for estradiol in inducing the display
of lordosis. But the dosage of testosterone has to be much higher
(more than 20 times) than that of estradiol. In male rats, the
display of lordosis cannot be induced by testosterone in the same
dosages that were effective in females, with either intracerebral
implantation (Dorner et al., 1968a) or systemic administration
(Pfaff, 1970a; Pfaff and Zigmond, 1971). This sex difference in
rats in their response to the lordosis-inducing effect of a "weak
estrogen," testosterone, also could be a basis for the inference
that the nervous system of the male rat is less responsive to estrogens than that of the female. Finally, although administration
of estrogen can lower the electroshock seizure threshold in both

female and male rats, more estrogen is required to achieve reduction
of the threshold for the male (Woolley and Timiras, 1962). This
observation again indicates that brain sensitivity to estrogen is
less in the male.

Strong support for the estrogen-sensitivity hypothesis comes
from studies on the effects of neonatal hormone treatments. It is
well known that the mating behavior of the adult rat is greatly
affected by neonatal hormonal condition. The capability of male
rats for performing lordosis in response to ovarian hormone treatment is greatly enhanced if they were castrated shortly (within
one week) after birth (Feder and Whalen, 1965; Grady et al., 1965;
Feder et al., 1966; Whalen and Edwards, 1967; Pfaff and Zigmond,
1971; Whalen et al., 1971; Hendricks, 1972). On the other hand,
treating the female rat neonatally with androgen, or even with estrogen in large doses ("androgenization" or "estrogenization,"
respectively) greatly depresses or abolishes the lordosis response
of the animal as an adult to ovarian hormone treatment (Harris and
Levine, 1965; Feder et al., 1966; Whalen and Edwards, 1967; Meyerson, 1968b; Mullins and Levine, 1968; Clemens et al., 1969; Clemens
et al., 1970; Edwards and Thompson, 1970; Gerall and Kenney, 1970;
Pfaff and Zigmond, 1971; Whalen et al., 1971). Similarly, if neonatally castrated male rats were treated with androgen or estrogen
soon after castration, no enhancement of lordosis response could be
observed later in the adults (Feder and Whalen, 1965; Whalen and
Edwards, 1967; Hendricks, 1972). Thus, the presence of androgen
(or large doses of estrogen) in neonatal rats of either genetic sex
greatly reduces their capacity to perform as adults in response to
ovarian hormones.

It is understood that in female rats gonadectomized as adults,
treatment with estrogen alone can induce the display of lordosis
(Davidson et al., 1968; Pfaff, 1970a; Hendricks, 1972). The behavioral responses or the brain sensitivity to estrogen can therefore be evaluated by subjecting gonadectomized adult rats to treatment with estrogen alone. Thus neonatally androgenized (Edwards
and Thompson, 1970; Pfaff and Zigmond, 1971; Whalen et al., 1971)
or estrogenized (Edwards and Thompson, 1970) female rats showed
less and/or weaker lordosis responses to estrogen alone than did
control females. Similarly, neonatally castrated and androgenized
male rats responded with lordosis to estrogen alone less than did
males castrated neonatally without androgenization (Feder et al.,
1966; Hendricks, 1972). Conversely, male rats castrated neonatally
showed better lordosis response than males castrated postpuberally,
when treated with estrogen alone as adults (Pfaff and Zigmond, 1971;
Whalen et al., 1971; Hendricks, 1972). Furthermore, the decrease
in estrogen sensitivity caused by androgenization was inversely
proportional to the dosage of androgen administered neonatally
(Gerall and Kenney, 1970). Finally, the decrease in estrogen responsiveness is not confined to feminine mating behavior; for

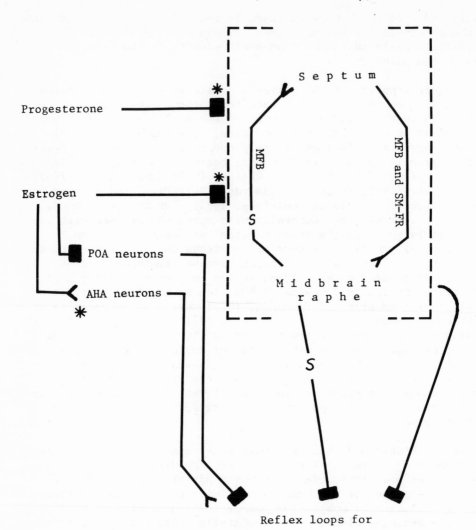

Reflex loops for

feminine mating behavior

Fig. 3. Description of ways in which sex differences in the neural response to hormones could explain sex differences in the perform- ance of lordosis.

The box described by the dashed line encloses the circuits described in more detail in Fig. 1; these are raphe-septum connec- tions which have a net inhibitory effect upon reflex loops control- ling lordosis. Such reflex loops presumably are located in the lower brainstem and spinal cord. Descending connections from the circuits described within the dashed lines and in Fig. 1 could come from the raphe nuclei themselves or from other nuclei in the

running activity in response to estrogen also decreased in female
rats given androgen neonatally, and the more androgen given the
greater the reduction (Gerall et al., 1972).

In addition to its influence on behavior, the presence of neonatal androgen can also affect the estrogen-uptake capability of
the brain. Estrogen uptake by the hypothalamus has been found to
be higher in normal female rats than in females treated neonatally
with androgen (McEwen and Pfaff, 1970). This finding, together
with the one we mentioned earlier that the retention of estrogen by
the hypothalamus is higher in female than in male rats, suggests
that in the male it is the presence of neonatal androgen that causes
the reduction of estrogen concentration in the hypothalamus. Such
a reduction may, in turn, be one of the mechanisms that make male
rats less responsive behaviorally to estrogen than females.

In sum, the evidence discussed above indicates that the response to estrogen is quantitatively less in male than in female
rats. Moreover, the reduction in the estrogen-uptake capability
of the male brain, probably because of the presence of neonatal androgen, may be one of the mechanisms underlying this sex difference
in estrogen-sensitivity.

Progesterone-Sensitivity Hypothesis

Progesterone is known to facilitate the display of lordosis in

---

connections described in Fig. 1.
Progesterone is thought to inhibit activity in this raphe-
septal loop, as described in Fig. 1. Evidence reviewed in the text
suggests that this effect of progesterone is absent or reduced in
male rats.
Lesion evidence described in the text suggests that preoptic
neurons may inhibit lordosis while anterior hypothalamic neurons
facilitate it. Estrogen is thought to have a net inhibitory effect
on preoptic neurons and a net excitatory effect on anterior hypothalamic neurons (Bueno and Pfaff, 1974). As described in the text,
the effect of estrogen in the anterior hypothalamus may be reduced
in male rats. Estrogen may also have a progesterone-like action on
the raphe-septal loop summarized in the dashed-line box, since under
some conditions estrogen can substitute for progesterone in facilitating lordosis in previously estrogen-primed female rats (Kow and
Pfaff, 1974).
Abbreviations: MFB, medial forebrain bundle; SM, stria medullaris; FR, fasciculus retroflexus; POA, preoptic area; AHA anterior
hypothalamic area. Symbols: ⤙ = excitatory relation; ⊸■ =
inhibitory relation; -s- = serotonergic fibers; ✱ = functions in
which males (and androgenized females) are deficient.

both the intact (Powers, 1970) and the ovariectomized, estrogen-primed female rat (Boling and Blandau, 1939; Beach, 1942b; Lisk, 1960, 1969b; Edwards et al., 1968; Green et al., 1970; Nadler, 1970; Whalen and Gorzalka, 1972; Eriksson and Sodersten, 1973). Furthermore, in the normal cycling female rat (Powers, 1970), or in the ovariectomized rat primed with a small dosage of estrogen (Powers and Valenstein, 1972b), progesterone not only can facilitate but also is essential for inducing lordosis.

This lordosis-facilitating effect of progesterone in females cannot be observed in the male rat castrated as an adult and primed with estrogen (Davidson, 1969; Davidson and Bloch, 1969; Davidson and Levine, 1969; Meyerson, 1968b; Aren-Engelbrektsson et al., 1970; Clemens et al., 1970; Edwards and Thompson, 1970; Whalen et al., 1971), even with a dosage ten times higher than the effective dosage for females (Davidson and Levine, 1969). According to Davidson and Levine (1969), the sexual difference in estrogen sensitivity is quantitative, but in progesterone sensitivity the difference is qualitative. The sexual difference in response to progesterone is not confined to feminine mating behavior, but appears in other types of response as well. With progesterone treatment, the electroshock seizure threshold in female rats was rapidly and significantly increased, but not in males (Wooley and Timiras, 1962). Estrogen can stimulate the activity of L-cystine-aminopeptidase in the hypothalamus and paleopallium of both male and female rats; whereas progesterone exerts this stimulating effect only in the female (Bickel et al., 1972). In addition, there is also an interesting parallel sexual difference in response to progesterone in such soliciting behaviors as hopping, darting, and ear-wiggling, which are displayed by normal cycling female rats during estrus. In ovariectomized, estrogen-primed females, the display of soliciting behaviors can be induced with subsequent progesterone administration (Hardy and DeBold, 1971). In the male rat, castrated as an adult and primed with estrogen, no such behaviors can be observed after the administration of progesterone or other lordosis-facilitating agents (Meyerson, 1968b; Aren-Engelbrektsson et al., 1970; Larsson and Sodersten, 1971). Of the many possible explanations for this sexual difference, the most likely is that male rats, unlike females, do not respond to progesterone.

As we have discussed previously (see Fig. 1), there are indications that certain structures in the brain have inhibitory influences on the neural substrate mediating the lordosis reflex and that in the female rat progesterone can retard such inhibitory influences and hence facilitate lordosis behavior by disinhibition. In the male rat, the inhibitory neural structures probably do not respond to progesterone (Fig. 3); thus the result is an absence of progesterone facilitation of lordosis.

This lack of responsiveness to progesterone, like the reduction

in responsiveness to estrogen, seems to be due to the neonatal presence of androgen. Davidson and Levine (1969) reported that whereas
male rats castrated as adults did not respond to progesterone, males
castrated neonatally responded with facilitation of lordosis. Similarly, Lisk (1969a) reported that male rats castrated within five
days after birth showed the facilitatory effect of progesterone.
Conversely, female rats can be made unresponsive to the lordosis-
facilitating effect of progesterone by neonatal treatment with androgen (Clemens et al., 1969; Edwards and Thompson, 1970; Whalen
et al., 1971) or estrogen (Edwards and Thompson, 1970). Thus, high
levels of androgen in the neonatal period, whether from testicular
secretion or experimental intervention, drastically lower sensitivity
to progesterone in adulthood.

Progesterone-like Action of Estrogen

        In discussing the differences in hormonal sensitivities under-
lying behavioral sex differences, one should note that the distinction between the actions of estrogen and progesterone may not be
absolute; for there are indications that under some conditions estrogen can act like progesterone. Wade and Feder (1972a) found two
estrogen-uptake systems in the brain of guinea pigs. One, located
in the hypothalamus, can be saturated with estrogen; the other,
which is widely spread over almost the entire brain, cannot. These
data are reminiscent of earlier results with rats (Eisenfeld and
Axelrod, 1965, 1966; Kato and Villee, 1967a, b; Pfaff, 1968a,
1968b). Wade and Feder further discovered that if a guinea pig is
given a small amount of unlabelled estrogen to saturate the hypothalamic sites and then given radioactively labelled estrogen, the
pattern of uptake of this labelled estrogen is the same as that of
labelled progesterone. These findings suggest that under some conditions estrogen can act on brain structures responsive to progesterone. In another brain uptake study, Wade and Feder (1972b) found
that the uptake patterns for labelled progesterone, 20α-hydroxy-
pregn-α-en-3-one (20α-OHP), estradiol, and corticosterone are similar
in pattern but different quantitatively. In every brain region
examined the uptake is highest for progesterone, followed by 20α-OHP,
estradiol, and corticosterone, in order. Both 20α-OHP and corticosterone as well as progesterone can facilitate lordosis, the order
of effectiveness being progesterone, 20α-OHP, then corticosterone
(see discussion by Wade and Feder, 1972b). From such correlations,
it is reasonable to predict that estradiol would also share the
lordosis-facilitating effect with progesterone and that it would be
more effective than corticosterone but less than 20α-OHP. In fact,
in an intracerebral hormone implantation study it was found that
implantation of progesterone in the mesencephalic reticular formation
of ovariectomized, estrogen-primed rats can facilitate lordosis
(Ross et al., 1971) and that similar implantation of estrogen can
achieve the same effect, but with longer latency. Moreover, the

effect of progesterone in inducing LH release in estrogen-primed
rats can be duplicated by substituting additional estrogen for the
progesterone (Caligaris et al., 1971); and both progesterone and
estrogen, when injected separately, can block the LH release in female rats caused by cerebral cortex stimulation (Taleisnik et al.,
1962). From these observations, we can hypothesize that estrogen
can act in two ways. First, a small amount of estrogen would prime
the estrogen-sensitive sites in the hypothalamus, preoptic area, and
limbic system to induce the display of lordosis. Then, if a larger
amount were given subsequently, the hypothalamic sites would be
saturated; and surplus estrogen would act like progesterone to suppress inhibitory influences from other brain structures (Fig. 1, 3).

According to this hypothesis, an additional, or second, dose
of estrogen could, like progesterone, induce lordosis in estrogen-
primed, ovariectomized female rats; and this effect should occur
within a few hours of its administration. These predictions were
indeed confirmed in an experiment conducted in our laboratory (Kow
and Pfaff, 1974). In short, we found that all of the ovariectomized
female rats that had been primed with a first, small dose of estrogen (3 µgEB/rat) two days earlier could be induced to perform lordosis within four hours by a subcutaneous injection of a second, large
dose of estrogen (400 µgEB/rat). The small first dose induced prac-
tically no lordosis. Further when the first priming dose of estrogen was replaced by oil, then the "second estrogen" treatment did
not induce lordosis within 24 hours of its administration, but would
do so eventually. The second estrogen treatment appeared to "substitute" for progesterone in facilitating lordosis. But, as expected,
it is less effective than progesterone. The average peak lordosis
quotient of the group of rats treated with double estrogen doses was
32.6; that of the group of rats primed first with 3 µgEB/rat and
then given 400 µgP/rat was 60.8. Lower doses in the second estrogen
treatment were also effective.

Previously, we explained the sex difference in the performance
of lordosis with two hypotheses, estrogen-sensitivity and proges-
terone-sensitivity, and treated the two as entirely distinct.
However, now that we have found that estrogen can also act like progesterone, it seems possible that male rats, unresponsive to progesterone, are also unresponsive to this progesterone-like effect of
estrogen. Thus part of the data encompassed by the estrogen-
sensitivity hypothesis may be explained by progesterone-like effects
of estrogen.

SUMMARY

1.   Several aspects of progesterone action on female reproductive
behavior in rodents can be explained by assuming that progesterone
inhibits the activity of serotonergic neurons in midbrain raphe nuclei.

These neurons are thought to participate in midbrain-limbic circuits whose overall effect is to inhibit lordosis.

2.   Sex differences in female reproductive behaviors of rats probably reflect both a reduction of estrogen responsiveness and an absence of progesterone responsiveness in male rats. In particular, male rats may be deficient in certain interactions between gonadal hormones and raphe-septal circuits which inhibit lordosis.

ACKNOWLEDGMENTS

This work was supported by NIH grant HD-05751 and by a grant from the Rockefeller Foundation for the study of reproductive biology and was assisted by the bibliographic services of the Brain Information Service. We thank G. Zummer for help in preparing the manuscript and the figures.

REFERENCES

Ahlenius, S., Engel, J., Eriksson, H., Modigh, K., and Sodersten, P., 1972, Importance of central catecholamines in the mediation of lordosis behaviour in ovariectomized rats treated with estrogen and inhibitors of monoamine synthesis, J. Neural Transmission 33:247-255.

Ahlenius, S., Engel, J., Eriksson, H., and Sodersten, P., 1972, Effects of tetrabenazine on lordosis behaviour and on brain monoamines in the female rat, J. Neural Transmission 33: 155-162.

Ahlenius, S., Eriksson, H., Larsson, K., Modigh, K., and Sodersten, P., 1971, Mating behavior in the male rat treated with p-chlorophenylalanine methylester alone and in combination with pargyline, Psychopharmacologia 20:383-388.

Anderson, C.H., and Greenwald, G.S., 1969, Autoradiographic analysis of estradiol uptake in the brain and pituitary of the female rat, Endocrinology 85:1160-1165.

Aren-Engelbrektsson, B., Larsson, K., Sodersten, P., and Wilhelmsson, M., 1970, The female lordosis pattern induced in male rats by estrogen, Horm. Behav. 1:181-188.

Asdell, S.A., and Sperling, G.A., 1963, Sex steroid hormones and voluntary exercise in rats: a correction, J. Reprod. Fertil. 5:123-124.

Ball, J., 1940, The effect of testosterone on the sex behavior of female rats, J. Comp. Psychol. 29:151-165.

Barraclough, C.A., and Cross, B.A., 1963, Unit activity in the hy-
    pothalamus of the cyclic female rat: effect of genital stimuli
    and progesterone, J. Endocrinol. 26:339-359.

Beach, F.A., 1938, Sex reversals in the mating pattern of the rat,
    J. Genet. Psychol. 53:329-334.

Beach, F.A., 1940, Effects of cortical lesions upon the copulatory
    behavior of male rats, J. Comp. Psychol. 29:193-244.

Beach, F.A., 1942a, Male and female mating behavior in prepuberally
    castrated female rats treated with androgens, Endocrinology
    31:673-678.

Beach, F.A., 1942b, Importance of progesterone to induction of
    sexual receptivity in spayed female rats, Proc. Soc. Exptl.
    Biol. 51:367-371.

Beach, F.A., 1943, Effects of injury to the cerebral cortex upon
    the display of masculine and feminine mating behavior by female
    rats. J. Comp. Psychol. 36:169-198.

Beach, F.A., 1944, Effects of injury to the cerebral cortex upon
    sexually-receptive behavior in the female rat, Psychosom. Med.
    6:40-55.

Beach, F.A., 1945, Bisexual mating behavior in the male rat: effect
    of castration and hormone administration, Physiol. Zool. 18:
    390-402.

Beach, F.A., 1948, *Hormones and Behavior*, Cooper Square Pub., New
    York.

Beach, F.A., 1967, Cerebral and hormonal control of reflexive mechan-
    isms involved in copulatory behavior, Physiol. Rev. 47:289-316.

Beach, F.A., and Rasquin, P., 1942, Masculine copulatory behavior in
    infant and castrated female rats, Endocrinology 31:393-409.

Bickel, M., Kuhl, H., Tan, J.S.E., and Taubert, H.-D., 1972, Evi-
    dence of a sex-specific effect of testosterone and progesterone
    upon L-cystine-aminopeptidase activity in the hypothalamus and
    paleopallium of the rat, Neuroendocrin. 9:321-331.

Boling, J.L., and Blandau, R.J., 1939, The estrogen-progesterone
    induction of mating responses in the spayed female rat,
    Endocrinology 25:359-364.

Brady, J.F., and Nauta, W.J.H., 1953, Subcortical mechanisms in emo-
    tional behavior: affective changes following septal forebrain le-

sions in the albino rat, J. Comp. Physiol. Psychol. 46:339-346.

Bueno, J., and Pfaff, D.W., 1974, Effects of estradiol on resting
    discharge and responsivity of preoptic and hypothalamic neurons
    in female rats (in preparation).

Caligaris, L., Astrada, J.J., and Taleisnik, S., 1971, Release of
    luteinizing hormone induced by estrogen injection into ovari-
    ectomized rats, Endocrinology 88:810-815.

Carey, R.J., 1972, A neuroanatomical investigation of enhanced
    cutaneous and gustatory responsivity associated with septal
    forebrain injury, J. Comp. Physiol. Psychol. 80:449-457.

Carey, R.J., 1973, A neuroanatomical investigation of disturbances
    in thirst-related behavior associated with injury to the sep-
    tum (presented to the Eastern Psychological Association, April
    1973, unpublished).

Clemens, L.G., 1971, Perinatal hormones and the modification of
    adult behavior, in: *Steroid Hormones and Brain Function* (C.H.
    Sawyer and R.A. Gorski, eds.), pp. 203-209, University of
    California Press, Los Angeles.

Clemens, L.G., Hiroi, M., and Gorski, R.A., 1969, Induction and
    facilitation of female mating behavior in rats treated neo-
    natally with low doses of testosterone propionate, Endocrin-
    ology 84:1430-1438.

Clemens, L.G., Shryne, J., and Gorski, R.A., 1970, Androgen and
    development of progesterone responsiveness in male and female
    rats, Physiol. Behav. 5:673-678.

Clemens, L.G., Wallen, K., and Gorski, R.A., 1967, Mating behavior:
    facilitation in the female rat after cortical application of
    potassium chloride, Science 157:1208-1209.

Conrad, L., and Pfaff, D.W., 1974, Projections of preoptic and
    septal neurons in rats studied with proline-H[3] autoradiography,
    Program and Abstracts, Society for Neuroscience (4th Annual
    Meeting), abstr. (in press).

Conrad, L.C.A., Leonard, C.M., and Pfaff, D.W., 1974, Connections
    of the median and dorsal raphe nuclei in the rat: an auto-
    radiographic and degeneration study, J. Comp. Neurol. (in
    press).

Cragg, B.G., 1961, The connections of the habenula in the rabbit,
    Exptl. Neurol. 3:388-409.

Dahlström, A., and Fuxe, K., 1965, Evidence for the existence of monoamine-containing neurons in the central nervous system. I. Demonstration of monoamines in the cell bodies of brain stem neurons, Acta Physiol. Scand. 62, Suppl. 232:1-55.

Davidson, J.M., 1966, Activation of the male rat's sexual behavior by intracerebral implantation of androgen, Endocrinology 79: 783-794.

Davidson, J.M., 1969, Effect of estrogen on the sexual behavior of male rats, Endocrinology 84:1365-1372.

Davidson, J.M., 1972, Hormones and reproductive behavior, in: Reproductive Biology (H. Balin and S. Glasser, eds.), pp. 877-918, Excerpta Medica, Amsterdam.

Davidson, J.M., and Bloch, G.J., 1969, Neuroendocrine aspects of male reproduction, Biol. Reprod. 1:67-92.

Davidson, J.M., and Levine, S., 1969, Progesterone and heterotypical behavior in male rats, J. Endocrinol. 44:129-130.

Davidson, J.M., Rodgers, C.H., Smith, E.R., and Bloch, G.J., 1968, Stimulation of female sex behavior in adrenalectomized rats with estrogen alone, Endocrinology 82:193-195.

Dorner, G., and Staudt, J., 1968, Structural changes in the preoptic anterior hypothalamic area of the male rat, following neonatal castration and androgen substitution, Neuroendocrin. 3:136-140.

Dorner, G., and Staudt, J., 1969a, Structural changes in the hypothalamic ventromedial nucleus of the male rat, following neonatal castration and androgen treatment, Neuroendocrin. 4: 278-281.

Dorner, G., and Staudt, J., 1969b, Perinatal structural sex differentiation of the hypothalamus in rats, Neuroendocrin. 5: 103-106.

Dorner, G., Docke, F., and Moustafa, S., 1968a, Homosexuality in female rats following testosterone implantation in the anterior hypothalamus, J. Reprod. Fertil. 17:173-175.

Dorner, G., Docke, F., and Moustafa, S., 1968b, Differential localization of a male and a female hypothalamic mating centre, J. Reprod. Fertil. 17:583-586.

Edwards, D., and Thompson, M., 1970, Neonatal androgenization and estrogenization and the hormonal induction of sexual receptivity in rats, Physiol. Behav. 5:1115-1119.

Edwards, D.A., and Warner, P., 1972, Olfactory bulb removal facili-
    tates the hormonal induction of sexual receptivity in the fe-
    male rat, Horm. Behav. 3:321-332.

Edwards, D.A., Whalen, R.E., and Nadler, R.D., 1968, Induction of
    estrus: estrogen-progesterone interactions, Physiol. Behav.
    3:29-33.

Eisenfeld, A.J., and Axelrod, J., 1965, Selectivity of estrogen
    distribution in tissues, J. Pharm. Exptl. Therapeut. 150:
    469-475.

Eisenfeld, A.J., and Axelrod, J., 1966, Effect of steroid hormones,
    ovariectomy, estrogen pretreatment, sex and immaturity on the
    distribution of ^{3}H-estradiol, Endocrinology 79:38-42.

Eriksson, H., and Sodersten, P., 1973, A failure to facilitate
    lordosis behavior in adrenalectimized and gonadectomized
    estrogen-primed rats with monoamine synthesis inhibitors,
    Horm. Behav. 4:89-97.

Feder, H.H., and Whalen, R.E., 1965, Feminine behavior in neonatally
    castrated and estrogen-treated male rats, Science 147:306-307.

Feder, H.H., Phoenix, C.H., and Young, W.C., 1966, Suppression of
    feminine behavior by administration of testosterone propionate
    to neonatal rats, J. Endocrinol. 34:131-132.

Flerko, B., Mess, B., and Illei-Donhoffer, A., 1969, On the mechan-
    ism of androgen sterilization, Neuroendocrin. 4:164-169.

Gerall, A.A., and Kenney, A. McM., 1970, Neonatally androgenized
    females' responsiveness to estrogen and progesterone, Endo-
    crinology 87:560-566.

Gerall, A.A., Stone, L.S., and Hitt, J.C., 1972, Neonatal androgen
    suppresses female responsiveness to estrogen, Physiol. Behav.
    8:17-20.

Giantonio, G.W., Lund, N.L., and Gerall, A.A., 1970, Effect of dien-
    cephalic and rhincephalic lesions on the male rat's sexual
    behavior, J. Comp. Physiol. Psychol. 73:38-46.

Gorski, R.A., 1971, Gonadal hormones and the perinatal development
    of neuroendocrine function, in: Frontiers in Neuroendocrinology,
    1971 (L. Martini and W.F. Ganong, eds.), pp. 237-290, Oxford
    University Press, New York.

Grady, K.L., Phoenix, C.H., and Young, W.C., 1965, Role of the
    developing testis in differentiation of the neural tissues
    mediating behavior, J. Comp. Physiol. Psychol. 59:176-182.

Green, R., Luttge, W.G., and Whalen, R.E., 1970, Induction of receptivity in ovariectomized female rats by a single intravenous injection of estradiol-17β, Physiol. Behav. 5:137-141.

Guillery, R.W., 1957, Degeneration in the hypothalamic connexions of the albino rat, J. Anat. 91:91-115.

Gumulka, W., Samanin, R., Valzelli, L., and Consolo, S., 1971, Behavioural and biochemical effects following the stimulation of the nucleus *Raphis dorsalis* in rats, J. Neurochem. 18: 533-535.

Hardy, D.F., and DeBold, J.H., 1971, The relationship between levels of exogenous hormones and the display of lordosis by the female rat, Horm. Behav. 2:287-297.

Harris, G.W., and Levine, S., 1965, Sexual differentiation of the brain and its experimental control, J. Physiol. 181:379-400.

Heimer, L., and Larsson, K., 1966/67, Impairment of mating behavior in male rats following lesions in the preoptic-anterior hypothalamic continuum, Brain Res. 3:248-263.

Heller, A., and Moore, R.Y., 1965, Effect of central nervous system lesions on brain monoamines in the rat, J. Pharmacol. Exptl. Therapeut. 150:1-9.

Heller, A., Harvey, J.A, and Moore, R.Y., 1962, A demonstration of a fall in brain serotonin following central nervous system lesions in the rat, Biochem. Pharmacol. 11:859-866.

Hendricks, S.G., 1972, Androgen modification of behavioral responsiveness to estrogen in the male rat, Horm. Behav. 3:47-54.

Hitt, J.C., Bryon, D.M., and Modianos, D.T., 1973, Effects of rostral medial forebrain bundle and olfactory tubercle lesions upon sexual behavior of male rats, J. Comp. Physiol. Psychol. 82: 30-36.

Hitt, J.C., Hendricks, S.G., Ginsberg, S.I., and Lewis, J.H., 1970, Disruption of male, but not female, sexual behavior in rats by medial forebrain bundle lesions, J. Comp. Physiol. Psychol. 73:377-384.

Jouvet, M., 1969, Biogenic amines and the states of sleep, Science 163:32-41.

Kato, J., and Villee, C.A., 1967a, Preferential uptake of estradiol by the anterior hypothalamus of the rat, Endocrinology 80: 567-575.

Kato, J., and Villee, C.A., 1967b, Factors affecting uptake of estradiol-6,7-^{3}H by the hypophysis and hypothalamus, Endocrinology 80:1133-1138.

Koe, B.K., and Weissman, A., 1966, p-chlorophenylalanine: a specific depletor of brain serotonin, J. Pharmacol. Exptl. Therapeut. 154:499-516.

Komisaruk, B.R., Larsson, K., and Cooper, R., 1972, Intense lordosis in the absence of ovarian hormones after septal ablation in rats, Program and Abstracts, Society for Neuroscience (2nd Annual Meeting), p. 230, abstr. #51.10.

Komisaruk, B.R., McDonald, P.G., Whitmoyer, D.I., and Sawyer, C.H., 1967, Effects of progesterone and sensory stimulation on EEG and neuronal activity in the rat, Exptl. Neurol. 19:494-507.

Kostowski, W., Giacalone, E., Garattini, S., and Valzelli, L., 1968, Studies on behavioural and biochemical changes in rats after lesion of midbrain raphe, Eur. J. Pharmacol. 4:371-376.

Kostowski, W., Giacalone, E., Garattini, S., and Valzelli, L., 1969, Electrical stimulation of midbrain raphe: biochemical, behavioral and bioelectrical effects, Eur. J. Pharmacol. 7:170-175.

Kow, L.-M., and Pfaff, D.W., 1973/74, Effects of estrogen treatment on the size of receptive field and response threshold of pudendal nerve in the female rat, Neuroendocrin. 13:299-313.

Kow, L.-M., and Pfaff, D.W., 1974, Facilitation of lordosis in estrogen-primed, ovariectomized rats with a second estrogen treatment (in preparation).

Kuhar, M.J., Aghajanian, G.K., and Roth, R.H., 1972, Tryptophan hydroxylase activity and synaptosomal uptake of serotonin in discrete brain regions after midbrain raphe lesions: correlations with serotonin levels and histochemical fluorescence, Brain Res. 44:165-176.

Larsson, K., 1962, Spreading cortical depression and the mating behavior in male and female rats, Z. Tierpsychol. 9:321-331.

Larsson, K., and Sodersten, P., 1971, Lordosis behavior in male rats treated with estrogen in combination with tetrabenazine and nialamide, Psychopharmacologia 21:13-16.

Law, T., and Meagher, W., 1958, Hypothalamic lesions and sexual behavior in the female rat, Science 128:1626-1627.

Liebeskind, J.C., Guilbaud, G., Besson, J.-M., and Oliveras, J.-L.,

1973, Analgesia from electrical stimulation of the periaque-
ductal gray matter in the cat: behavioral observations and
inhibitory effects on spinal cord interneurons, Brain Res.
50:441-446.

Lints, C.E., and Harvey, J.A., 1969, Altered sensitivity to foot-
shock and decreased brain content of serotonin following brain
lesions in the rat, J. Comp. Physiol. Psychol. 67:23-31.

Lisk, R.D., 1960, A comparison of the effectiveness of intravenous,
as opposed to, subcutaneous injection of progesterone for the
induction of estrous behavior in the rat, Can. J. Biochem.
38:1381-1383.

Lisk, R.D., 1967, Neural localization for androgen activation of
copulatory behavior in the male rat, Endocrinology 80:754-761.

Lisk, R.D., 1968, Copulatory activity of the male rat following
placement of preoptic-anterior hypothalamic lesions, Exptl.
Brain Res. 5:306-313.

Lisk, R.D., 1969a, Progesterone: biophasic effects on the lordosis
response in adult or neonatally gonadectomized rats, Neuroendocrin. 5:149-160.

Lisk, R.D., 1969b, Progesterone: role in limitation of ovulation
and sex behavior in mammals, Trans. N.Y. Acad. Sci. 31:593-601.

Malmras, C.O., and Meyerson, B.J., 1971, p-chlorophenylalanine and
copulatory behavior in the male rat, Nature (Lond.) 232:398-400.

Malsbury, C., and Pfaff, D.W., 1973, Suppression of sexual receptivity in the hormone-primed female hamster by electrical
stimulation of the medial preoptic area, Program and Abstracts,
Society for Neuroscience (3rd Annual Meeting), p. 122, (abstr.
#5.6).

McEwen, B.S., and Pfaff, D.W., 1970, Factors influencing sex hormone uptake by rat brain regions. I. Effects of neonatal
treatment, hypophysectomy, and competing steroids on estradiol
uptake, Brain Res. 21:1-16.

McEwen, B.S., Pfaff, D.W., and Zigmond, R.E., 1970a, Factors influencing sex hormone uptake by rat brain regions. II. Effects
of neonatal treatment and hypophysectomy on testosterone uptake, Brain Res. 21:17-28.

McEwen, B.S., Pfaff, D.W., and Zigmond, R.E., 1970b, Factors influencing sex hormone uptake by rat brain regions. III. Effects
of competing steroids on testosterone uptake, Brain Res. 21:
29-38.

Meyerson, B.J., 1964a, Central nervous monoamines and hormone induced estrus behaviour in the spayed rat, Acta Physiol. Scand. 63, Suppl. 248:5-32.

Meyerson, B.J., 1964b, Estrus behaviour in spayed rats after estrogen or progesterone treatment in combination with reserpine or tetrabenazine, Psychopharmacologia 6:210-218.

Meyerson, B.J., 1966a, Oestrous behaviour in oestrogen treated ovariectomized rats after chlorpromazine alone or in combination with progesterone, tetrabenazine or reserpine, Acta Pharmacol. Toxicol. 24:363-376.

Meyerson, B.J., 1966b, The effect of imipramine and related antidepressive drugs on estrous behaviour in ovariectomized rats activated by progesterone, reserpine or tetrabenazine in combination with estrogen, Acta Physiol. Scand. 67:411-422.

Meyerson, B.J., 1968a, Amphetamine and 5-hydroxytryptamine inhibition of copulatory behaviour in the female rat, Annales Medicinae Experimentales et Biologiae Fennae 46:394-398.

Meyerson, B.J., 1968b, Female copulatory behavior in male and androgenized female rats after estrogen/amine depletor treatment, Nature (Lond.) 217:683-684.

Meyerson, B.J., 1972, Monoamines and female sexual behavior, Psychopharmacologia 26 (Suppl.):132.

Meyerson, B.J., and Lewander, T., 1970, Serotonin synthesis inhibition and estrous behavior in female rats, Life Sci. 9:661-671.

Modianos, D.T., Flexman, J.E., and Hitt, J.C., 1973, Rostral medial forebrain bundle lesions produce decrements in masculine, but not feminine, sexual behavior in spayed female rats, Behav. Biol. 8:629-636.

Modianos, D.T., Hitt, J.C., and Flexman, J.E., 1974, Habenular lesions produce decrements in feminine, but not masculine, sexual behavior in rats, Behav. Biol. 10:75-87.

Morgane, P.J., and Stern, W.C., 1972, Relationship of sleep to neuroanatomical circuits, biochemistry, and behavior, Ann. N.Y. Acad. Sci. 193:95-111.

Moss, R.L., 1971, Modification of copulatory behavior in the female rat following olfactory bulb removal, J. Comp. Physiol. Psychol. 74:374-382.

Moss, R.L., Paloutzian, R.F., and Law, O.T., 1974, Electrical stimulation of forebrain structures and its effect on copulatory as well as stimulus-bound behavior in ovariectomized hormone-primed rats, Phys. Behav. (in press).

Mullins, R.F., Jr., and Levine, S., 1968, Hormonal determinants during infancy of adult sexual behavior in the female rat, Physiol. Behav. 3:333-338.

Nadler, R.D., 1970, A biphasic influence of progesterone on sexual receptivity of spayed female rats, Physiol. Behav. 5:95-97.

Nance, D.W., Shryne, J., and Gorski, R.A., 1974, Septal lesions: effects on lordosis behavior and pattern of gonadotropin release, Horm. Behav. 5:73-81.

Paxinos, G., and Bindra, D., 1973, Hypothalamic and midbrain neural pathways involved in eating, drinking, irritability, aggression, and copulation in rats, J. Comp. Physiol. Psychol. 82:1-14.

Persip, G.L., and Hamilton, L.W., 1973, Behavior effects of serotonin or a blocking agent applied to the septum of the rat, Pharmacol. Biochem. Behav. 1:139-147.

Pfaff, D.W., 1966, Morphological changes in the brains of adult male rats after neonatal castration, J. Endocrinol. 36: 415-416.

Pfaff, D.W., 1968a, Uptake of ^{3}H-estradiol by the female rat brain. An autoradiographic study, Endocrinology 82:1149-1155.

Pfaff, D.W., 1968b, Autoradiographic localization of radioactivity in rat brain after injection of tritiated sex hormones, Science 161:1355-1356.

Pfaff, D.W., 1970a, Nature of sex hormone effect on rat sex behavior: specificity of effects and individual patterns of response, J. Comp. Physiol. Psychol. 73:349-358.

Pfaff, D.W. 1970b, Synergistic and antagonistic effects of sex hormones on rat sex behavior, Amer. Zool. 10:478 (abst. #23).

Pfaff, D., and Keiner, M., 1973, Atlas of estradiol-concentrating cells in the central nervous system of the female rat, J. Comp. Neurol. 151:121-158.

Pfaff, D.W., and Zigmond, R.E., 1971, Neonatal androgen effects on sexual and non-sexual behavior of adult rats tested under various hormone regimes, Neuroendocrin. 7:129-145.

Pfaff, D.W., Diakow, C., Zigmond, R.E., and Kow, L.-M., 1973,
    Neural and hormonal determinants of female mating behavior in
    rats, in: *The Neurosciences*, (F.O. Schmitt et al., eds.),
    vol. III, pp. 621-646, M.I.T. Press, Boston.

Pfaff, D.W., Lewis, C., Diakow, C., and Keiner, M., 1972, Neuro-
    physiological analysis of mating behavior responses as hormone-
    sensitive reflexes, in: *Progress in Physiological Psychology*,
    (E. Stellar and J.M. Sprague, eds.), vol. 5, pp. 253-297,
    Academic Press, New York.

Powers, J.B., 1970, Hormonal control of sexual receptivity during
    the estrous cycle of the rat, Physiol. Behav. 5:831-835.

Powers, J.B., 1972, Facilitation of lordosis in ovariectomized rats
    by intracerebral progesterone implants, Brain Res. 48:311-325.

Powers, B., and Valenstein, E.S., 1972a, Sexual receptivity:
    facilitation by medial preoptic lesions in female rats,
    Science 175:1003-1005.

Powers, J.B., and Valenstein, E.S., 1972b, Individual differences
    in sexual responsiveness to estrogen and progesterone in
    ovariectomized rats, Physiol. Behav. 8:673-676.

Raisman, G., 1966, The connexions of the septum, Brain 89:317-348.

Raisman, G., and Field, P.M., 1971, Sexual dimorphism in the pre-
    optic area of the rat, Science 173:731-733.

Ramirez, V.D., Komisaruk, B.R., Whitmoyer, D.I., and Sawyer, C.H.,
    1967, Effects of hormones and vaginal stimulation on the EEG
    and hypothalamic units in rats, Amer. J. Physiol. 212:1376-
    1384.

Rodgers, C.H., and Law, O.T., 1967, The effects of habenular and
    medial forebrain bundle lesions on sexual behavior, Psychonom.
    Sci. 8:1-2.

Ross, J.W., and Gorski, R.A., 1973, Effects of potassium chloride
    on sexual behavior and the cortical EEG in the ovariectomized
    rat, Physiol. Behav. 10:643-646.

Ross, J., Claybaugh, C., Clemens, L.G., and Gorski, R.A., 1971,
    Short latency induction of estrous behavior with intracerebral
    gonadal hormones in ovariectomized rats, Endocrinology 89:
    32-38.

Ross, J.W., Gorski, R.A., and Sawyer, C.H., 1973, Effects of cor-
    tical stimulation on estrous behavior in estrogen-primed

ovariectomized rats, Endocrinology 93:20-25.

Segal, D.S., and Whalen, R.E., 1970, Effect of chronic administration of p-chlorophenylalanine on sexual receptivity of the female rat, Psychopharmacologia 16:434-438.

Seiki, K., and Hattori, M., 1971, A more extensive study on the uptake of labelled progesterone by the hypothalamus and pituitary gland of rats, J. Endocrinol. 51:793-794.

Seiki, K., Higashida, M., Imanishi, Y., Miyamoto, M., Kitagawa, T., and Kotani, M., 1968, Radioactivity in the rat hypothalamus and pituitary after injection of labelled progesterone, J. Endocrinol. 41:109-110.

Seiki, K., Miyamoto, M., Yamashita, A., and Kotani, M., 1969, Further studies on the uptake of labelled progesterone by the hypothalamus and pituitary of rats, J. Endocrinol. 43:129-130.

Sheard, M., 1969, The effect of p-chlorophenylalanine on behavior in rats; relation to brain serotonin and 5-hydroxy-indoleacetic acid, Brain Res. 15:524-528.

Singer, J.J., 1968, Hypothalamic control of male and female sexual behavior in female rats, J. Comp. Physiol. Psychol. 66: 738-742.

Singer, J.J., 1972, Effect of p-chlorophenylalanine on the male and female sexual behavior of female rats, Psychol. Rep. 30:891-893.

Smaha, L.A., and Kaelber, W.W., 1973, Efferent fiber projections of the habenula and the interpeduncular nucleus. An experimental study in the opossum and cat, Exptl. Brain Res. 16: 291-308.

Sodersten, P., 1972, Mounting behavior in the female rat during the estrous cycle, after ovariectomy, and after estrogen or testosterone administration, Horm. Behav. 3:307-320.

Sodersten, P., and Ahlenius, S., 1972, Female lordosis behavior in estrogen-primed male rats treated with p-chlorophenylalanine or alpha-methyl-p-tyrosine, Horm. Behav. 3:181-189.

Sodersten, P., and Larsson, K., 1974, Lordosis behavior in castrated male rats treated with estradiol benzoate or testosterone propionate in combination with an estrogen antagonist, MER-25, and in intact male rats, Horm. Behav. 5:13-18.

Stern, J.J., and Zwick, G., 1972, Hormonal control of spontaneous activity during the estrous cycle of the rat, Psychol. Rep. 30:983-988.

Stone, C.P., 1924, A note on "feminine" behavior in adult male rats, Amer. J. Physiol. 68:39-41.

Tagliamonte, A., Tagliamonte, P., Gessa, G.L., and Brodie, B.B., 1969, Compulsive sexual activity induced by p-chlorophenyl-alanine in normal and pinealectomized male rats, Science 166:1433-1435.

Tagliamonte, P., Tagliamonte, A., Stein, S., and Gessa, G., 1970, Inhibition of sexual behavior in male rats by monoamine oxi-dase inhibitor (MOAI): reversal of this effect of p-chloro-phenylalanine (PCPA), Clin. Res. 18:671.

Taleisnik, S., Caligaris, L., and De Olmos, J., 1962, Luteinizing hormone release by cerebral cortex stimulation, Amer. J. Physiol. 203:1109-1112.

Tenen, S.S., 1967, The effects of p-chlorophenylalanine, a sero-tonin depletor, on avoidance acquisition, pain sensitivity and related behavior in the rat, Psychopharmacologia 10: 204-219.

Ungerstedt, U., 1971, Stereotaxic mapping of the monoamine pathways in the rat brain, Acta Physiol. Scand., Suppl. 367.

Wade, G.N. and Feder, H.H., 1972a [1,2-^{3}H] progesterone uptake by guinea pig brain and uterus: differential localization, time-course of uptake and metabolism, and effects of age, sex, estro-gen-priming and competing steroids, Brain Res. 45:525-543.

Wade, G.N., and Feder, H.H., 1972b, Uptake of [1, 2-^{3}H] 20α-hydroxypregn-4-en-3-one, [1,2-^{3}H] corticosterone, and [6,7-^{3}H] estradiol-17β by guinea pig brain and uterus: comparison with uptake of [1,2,-^{3}H] progesterone, Brain Res. 45:545-554.

Wade, G.M., Harding, C.G., and Feder, H.H., 1973, Neural uptake of [1,2-^{3}H] progesterone in ovariectomized rats, guinea pigs, and hamsters: correlation with species differences in be-havioral responsiveness, Brain Res. 61:357-367.

Ward, I.L., Crowley, W.R., and Zemlan, F.P., 1974, Monoaminergic mediation of female sexual behaviors, J. Comp. Physiol. Psychol. (in press).

Whalen, R.G., and Edwards, D.A., 1967, Hormonal determinants of the development of masculine and feminine behavior in male and female rats, Anat. Rec. 157:173-180.

Whalen, R.E., and Gorzalka, B.B., 1972, The effects of progesterone and its metabolites on the induction of sexual receptivity in rats, Horm. Behav. 3:221-226.

Whalen, R.E., and Hardy, D.F., 1970, Induction of receptivity in female rats and cats with estrogen and testosterone, Physiol. Behav. 5:529-533.

Whalen, R.E., and Luttge, W.G., 1971, Differential localization of progesterone uptake in brain. Role of sex, estrogen pretreatment and adrenalectomy, Brain Res. 33:147-155.

Whalen, R.E., Luttge, W.G., and Gorzalka, B.B., 1971, Neonatal androgenization and the development of estrogen responsivity in male and female rats, Horm. Behav. 2:83-90.

Wolf, G., and Sutin, J., 1966, Fiber degeneration after lateral hypothalamic lesions in the rat, J. Comp. Neurol. 127:137-156.

Wooley, D.E., and Timiras, P.S., 1962, The gonad-brain relationship: effects of female sex hormones on electroshock convulsions in the rat, Endocrinology 70:196-209.

Young, W.C., 1961, The hormones and mating behavior, in: *Sex and Internal Secretions* (W.C. Young, ed.), Vol. 2, pp. 1173-1239. Williams & Wilkins, Baltimore.

Zasorin, N., Malsbury, C., and Pfaff, D.W., 1974, Disruption of lordosis by septal stimulation in female hamsters (in preparation).

Zemlan, F.P., Ward, I.L., Crowley, W.R., and Margules, D.L., 1973, Activation of lordotic responding in female rats by suppression of serotonergic activity, Science 179:1010-1011.

THE RELATIONSHIP BETWEEN FETAL HORMONES AND THE DIFFEREN- TIATION OF THE CENTRAL NERVOUS SYSTEM IN PRIMATES[1]

John A. Resko

Oregon Regional Primate Research Center, Beaverton

As many research reports attest over the years, sex steroids such as testosterone affect the differentiation of the central nervous system (CNS) (reviewed by Harris, 1964). The effects of prenatal or perinatal hormones on the development of the CNS can usually be divided into two categories: 1) effects on how gonado- tropins are released in adulthood; and 2) effects on sexual be- havior.

In rodents with short periods of gestation, such as the rat, testosterone given shortly after birth androgenizes the CNS of spayed females so that gonadotropins are produced acyclically and masculine behaviors are elicited from genetic females in adult- hood. In rodents with longer periods of gestation, such as the guinea pig, testosterone given before birth produces a similar effect (Phoenix, et al., 1959). However, no one has shown whether testosterone produces the same effects in primates on the release of gonadotropins. Female rhesus monkeys whose mothers had been treated with testosterone showed a preovulatory surge of estradiol, ovulated, and developed a functional corpus luteum (Goy and Resko, 1972), an indication that gonadotropins are released cyclically in the androgenized rhesus monkey. Animals thus treated, however, showed masculine tendencies in the display of infant behaviors (rough and tumble play, play initiation, and threat) during the first two years of life (Phoenix et al., 1968). Male rhesus monkeys, on the other hand, that are castrated as adults can, like

[1]This article is publication No. 666 of the Oregon Regional Primate Research Center, supported by NIH Grants RR-00163 and HD-05969-01.

females, release LH in response to pulsatile injections of estrogen (Karsch et al., 1973); intact males cannot.

These studies seem to indicate that primates differ from rodents in the way testosterone affects the differentiation of those areas of the brain that mediate gonadotropin release but not behavior. But the relationship of androgens to CNS development in primates is uncertain since 1) the level of the hormone circulating in the fetus has not been rigidly controlled experimentally, as is the case for the pseudohermaphroditic female rhesus monkeys that were androgenized with testosterone in our laboratory; and 2) the meaning of androgenization at the cellular level is not understood, as is the case when LH release was induced by estrogen in castrated male rhesus monkeys.

The fact that exogenous testosterone produces an effect is no proof that testosterone endogenous to the pregnant condition is the molecular substance responsible for sexual differentiation. To be able to state more clearly both the qualitative and quanitative nature of the hormonal environment in which the fetal rhesus monkey develops, we needed to characterize the quantities of several classes of steroid hormones in the umbilical circulation at various times during gestation. Preliminary studies, in which testosterone in fetal plasma was measured by gas-liquid chromatography with electron-capture detection, indicated that the quantities of testosterone found in pools of umbilical artery plasma differed markedly in male and female fetuses at various times during gestation (Resko, 1970).

Having developed a radioimmunoassay for testosterone, we went back recently and compared the quantities of this hormone in individual male and female fetuses from 59 to 163 days of gestation (Resko et al., 1973). The results are shown in Figure 1. On the average, the quantities of testosterone in plasma from the fetal male greatly exceeded those from the fetal female. However, in the latter, testosterone differed only in concentration and variation. The concentrations of testosterone in cord blood from the male fetus varied greatly from fetus to fetus. In Table 1, the concentrations of testosterone in the umbilical artery have been classified as high, medium, or low. Although one male fetus had as little as 300 pg/ml of testosterone, a value that falls within the range for females, most of the other males had over 1000 pg/ml in their plasma. Most of the female fetuses had less than 500 pg/ml except one which had an intermediate concentration of 775 pg/ml testosterone; none fell into the high category. These data seem to indicate that the fetal nervous systems of some males (a small percentage) are exposed to smaller amounts of androgen than other males and that some fetal females (a small percentage) are exposed to endogenous levels of testosterone similar to those of males. These data are complicated by the fact that very little

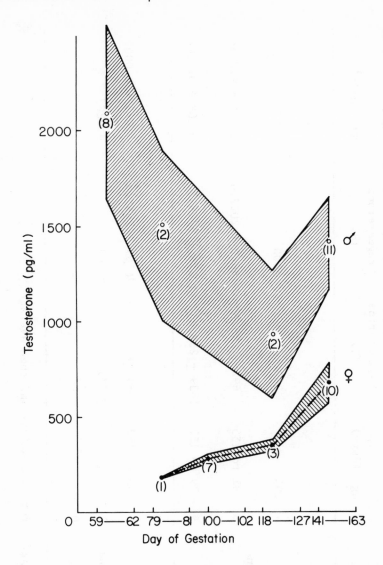

Fig. 1. Concentrations of testosterone in the umbilical artery plasma of the rhesus monkey; data are presented as means (circles) + S.E. (shaded areas); upper curve is males; lower curve, females (data taken from Resko, et al., 1973).

is known about how various animals differ from one gestational period to another. What, then, is the basal level of testosterone in the fetus and how can we best determine this level?

TABLE 1. *Classification of Fetuses according to the Level of Testosterone in the Umbilical Circulation*

Classification	N	Age (days)	mean pg/ml ± S.E.		
			Testosterone (T)	Progesterone (P)	T/P Ratio[a]
Males					
Low testosterone (< 500 pg/ml)	1	100	300	5000	0.06
Intermediate testosterone (500 to 1000 pg/ml)	3	81 to 155	692 ± 93.6	6660 ± 2027.6	0.146 ± 0.0738
High testosterone (> 1000 pg/ml)	9	79 to 157	1839 ± 288.8	12276 ± 3806.5	0.308 ± 0.100
Females					
Low testosterone	6	100 to 163	348 ± 22.9	26167 ± 7955.8	0.023 ± 0.0009
Intermediate testosterone	1	163	775	27000	0.0287
High testosterone	0	--	--	--	--

[a]T/P ratio computed in pg/ml for each animal and then averaged

*TABLE 2. A Comparison of the Concentrations of Testosterone in
Plasma from the Umbilical Vessels of Male, Female, and
Castrated Fetal Rhesus Monkeys*[a]

Gestational Age (range)	N[b]	Fetal Sex	Testosterone (mean pg/ml ± S.E.)	
			Artery	Vein
(100 to 157)	8	♂	1385 + 249	698 + 166
(100 to 153)	10	♀	442 + 73	361 + 89
(150 to 156)	3	⚥[c]	309 + 64	-----

[a]Data taken from Resko et al., 1973
Statistics: Male, testosterone in artery vs vein: p < 0.05.
Female, testosterone in artery vs vein: p > 0.05. Artery, male
vs female: p < 0.01. Vein, male vs female: p > 0.05. Artery castrated male vs intact: p < 0.02 (comparisons were made with 7 intact
males, 150 to 157 days of gestation)

[b]Number of determinations

[c]Animals were castrated on Day 100 of gestation

The source of the sex difference in the quantities of testosterone in the umbilical artery can be traced to the fetal testis
(Table 2). On Day 150 of gestation, the plasma from three male
fetuses that had been castrated on Day 100 showed quantities of
testosterone similar to those found in females. These data are
consistent with in vitro data which showed that the fetal testis
but not the ovary synthesizes testosterone and androstenedione
from pregnenolone (Resko, 1970). When the data for testosterone
in the umbilical artery and vein of both males and females (Table
2) are compared, the findings are similar to those in the in vitro
work. In the fetal male, almost twice as much testosterone was
found in the umbilical artery as in the umbilical vein; but in
the fetal female, the levels in the artery and vein were nearly
equal. Castration of the male reduced the concentration of testosterone in the umbilical artery to levels that approached those
of the vein. Thus, the testis is a major contributor to the
testosterone pool in the male fetus; but testosterone is produced
from such other sources as the placenta and the adrenal gland or
from peripheral conversions from precursor substances.

The relationship between the amount of endogenous androgen

circulating in the primate fetus and the degree of differentiation
has yet to be determined. The answer is complicated by the difficulties of 1) distinguishing between the effects of different
amounts of testosterone and the effects of different thresholds of
sensitivity to androgen by the developing anlagen of the CNS; and
2) determining the length of time that these tissues are exposed
to androgen action. For example, 30 µg of testosterone propionate
administered on postnatal Day 5 to a neonatal rat produced an androgenizing effect on the development of the central nervous system
within 12 hr after administration (Arai and Gorski, 1968). Smaller
doses of androgen (0.5 to 0.05 µg) given twice daily (Sheridan et
al., 1973) produced similar effects but needed 10 days of exposure.
These latter data seem more physiological since the neonatal rat
had only about 270 pg/ml of testosterone in its circulatory system
on Day 1 after birth (Resko et al., 1968), a level that remained
relatively constant until Day 5 when it dropped to 210 pg/ml. By
postnatal Day 10, the level of testosterone had dropped to 90
pg/ml. There is evidence to suggest (Napoli and Gerall, 1970) that
differentiation is not an all-or-none process but that different
amounts of testosterone produce different levels of androgenization in the female. Exogenous testosterone does not further androgenize the CNS of the neonatal male rat since an increase in male
behavior in adulthood does not occur (Whalen, 1968). Neonatal
treatment of intact male mice (Campbell and McGill, 1970) and rats
(Baum, 1972) with testosterone, however, produced precocious mating.

The quantitative effects of testosterone on brain differentiation can best be illustrated in the hamster. Swanson (1971)
reported that males castrated in adulthood showed feminine behaviors when injected with estrogen and progesterone and masculine
behaviors when injected with testosterone. However, when castrated
in adulthood, intact males treated with testosterone neonatally
lost their ability to display feminine behavior in response to
estrogen and progesterone and displayed only masculine behavior.
It is as if neonatal treatment with testosterone had induced greater
androgenization. Evidence for graded effects of testosterone on
primate brain differentiation is scarce, but the levels of endogenous testosterone (Table 1) suggest that such a mechanism operates
in the development of the CNS of primates.

In the rhesus monkey, larger amounts of progesterone are found
in the circulatory system of the fetus than of the mother (Hagemenas
and Kittinger, 1972) and in the fetal circulation of the female
than of the male (Hagemenas and Kittinger, 1972; Macdonald et al.,
1973). The sex difference in the quantities of progesterone after
100 days of gestation has been confirmed in our laboratory. The
results are shown in Figure 2. In both males and females, more
progesterone is quantified in the umbilical vein than in the
umbilical artery.

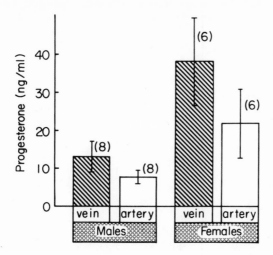

Fig. 2. Concentrations of progesterone in plasma from the umbilical vessels of the fetal rhesus monkey during the last trimester of gestation; data are presented as means (bars) ± S.E. (closed lines).

The fact that more progesterone was found in both the umbilical vein and artery of females suggests that the female placenta produces more progesterone than the male placenta.

The quantities of testosterone and progesterone and the ratios of the two hormones in the umbilical artery of the same fetus during the last trimester of pregnancy are shown in Table 3. Ten male fetuses showed an average concentration of 1217 ± 245 (S.E.) pg/ml of testosterone in plasma. The mean concentration of progesterone in the same plasma was 7.3 ± 1.5 ng/ml and the testosterone to progesterone ratio (T/P) was 0.289 ± 0.092. In females, however, the average concentration of testosterone was much lower, 409 ± 64 pg/ml (N = 8); and the mean concentration of progesterone was much higher (26.3 ± 6.7 ng/ml with a T/P ratio of 0.023 ± 0.009). These sex differences in testosterone and progesterone values stimulate speculation about interaction of these two hormones in the natural process of sexual differentiation. It is the absolute quantity of testosterone that is important for psychosexual development or is it the ratio of androgen (testosterone) to antiandrogen (progesterone) that affects development?

There is much evidence to suggest that progesterone antagonizes the action of androgens. For example, Kincl and Magueo (1965) and Cagnoni et al. (1965) found that in the neonatal rat progesterone prevented the androgenizing action of testosterone on

TABLE 3.   A Comparison of the Ratio of Androgen (Testosterone) to Antiandrogen (Progesterone) in Umbilical Artery Plasma of the Fetal Rhesus Monkey

Gestational Age (range)	N^a	Fetal Sex	Testosterone (T) (mean pg/ml $\pm$ S.E.)	Progesterone (P) (mean ng/ml $\pm$ S.E.)	Ratio $T/P^b \pm$ S.E.
(100 to 157)	10	♂	1217 $\pm$ 245	7.3 $\pm$ 1.5^c	0.289 $\pm$ 0.092
(100 to 163)	7	♀	409 $\pm$ 64	26.3 $\pm$ 6.7^c	0.023 $\pm$ 0.009

[a]Number of animals

[b]T/P ratio computed in pg/ml for each animal, then averaged

[c]Concentrations of progesterone, male vs female, differ significantly: $\underline{t}$ = 2.75, 15 d.f., $p < 0.05$

reproductive function if it was administered at the same time as testosterone. Progesterone antagonized androgen-induced aggressive behavior in the mouse (Erpino and Chappelle, 1971) and sexual behavior in the guinea pig (Diamond, 1966). Similar effects of progesterone were reported in birds by Erickson et al. (1967), Murton et al. (1969), and Erpino (1969).

The theory that progesterone limits the action of androgen on the central nervous system of the adult mammal is not new, but how exogenous progesterone effects this process is debatable. Some say it does so by inhibiting the secretion of gonadotropins by the pituitary gland and thereby lowering testicular androgen production (Murton et al., 1969). Others supply evidence that it directly inhibits androgen at a neural site (Erickson et al., 1967; Erpino, 1969; Diamond, 1966).

In addition to the sites of action mentioned above, progesterone acts directly on the biosynthetic pathways in the gonad by inhibiting key enzymes necessary for androgen secretion. Mahajan and Samuels (1962) reported that progesterone inhibited 17-demolase activity in the rat testis in vitro and that the consequent buildup of progesterone prevented the biosynthesis of androgens. Similarly, when progesterone was administered in silastic capsules to female rhesus monkeys during the late follicular phase of the cycle, the preovulatory surges of estradiol and testosterone in the systemic circulation were inhibited, but not the basal levels of LH (Hess and Resko, 1972). In general, however, the function of progesterone in the development of the nervous system is even less clear than its function in the adult animal. If progesterone can be antiandrogenic in the fetus, then the sex differences seen in fetal plasma must have some physiological effect on the regulation of androgen action or androgen biosynthesis.

The high levels of progesterone in the female rhesus fetus would seem to negate the action of the testosterone found in the female fetus. The lower levels in the male would allow testosterone to act on the neural anlagen responsible for the release of gonadotropin and for the display of reproductive behaviors in adulthood. High levels of progesterone in the male would probably negate the action of testosterone and produce some degree of androgenization of the male nervous system.

Figure 3 depicts how testosterone and progesterone probably interact to effect differentiation of the CNS. The plasma of the female fetus possesses about 400 pg/ml of testosterone, which represents a pool composed of secretions of the fetal ovary and adrenal gland and peripheral conversions from androstenedione and other precursors. Under peripheral conversions, I have also included placental contributions to the fetal testosterone pool. From whatever origin, this testosterone makes its way to the

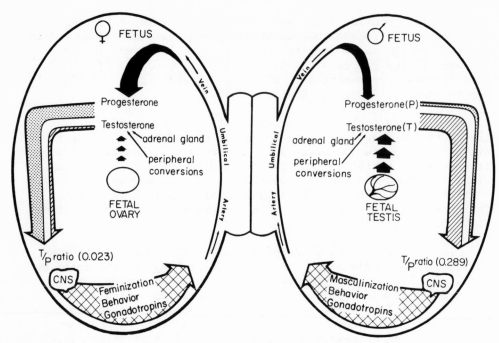

Fig. 3. Possible interaction of steroid hormones on the differentiation of the central nervous system in the fetal rhesus monkey. The widths of the shaded sections of the arrows represent the quantities of progesterone (stippled areas) and testosterone (cross-hatched areas) found in the fetal circulation.

developing nervous system and would probably androgenize it, were it not for the large quantities of progesterone found in the female. In the male, however, the fetal testis secretes testosterone and thus elevates the quantities of this male hormone. Superimposed upon the absolute quantity of testosterone is the ratio of testosterone to progesterone. High levels of progesterone in the female fetus cancel out the action of the testosterone produced from various sources. On the other hand, lower levels of progesterone in the male allow testosterone to produce its effect but in different degrees.

REFERENCES

Arai, Y., and Gorski, R.A., 1968, Critical exposure time for androgenization of the developing hypothalamus in the female rat, Endocrinology 82:1010–1014.

Baum, M.J., 1972, Precocious mating in male rats following treatment with androgen or estrogen, J. Comp. Physiol. Psychol.

78:356-367.

Cagnoni, M., Fantini, F., Morace, G., and Ghetti, A., 1965, Failure
    of testosterone propionate to induce the "early androgen" syn-
    drome in rats previously injected with progesterone, J. Endo-
    crinol. 33:527-528.

Campbell, A.B., and McGill, T.E., 1970, Neonatal hormone treatment
    and sexual behavior in male mice, Horm. Behav. 1:145-150.

Diamond, M., 1966, Progestogen inhibition of normal sexual behavior
    in the male guinea pig, Nature (Lond.) 209:1322-1324.

Erickson, C.J., Bruder, R.H., Komisaruk, B.R., and Lehrman, D.S.,
    1967, Selective inhibition by progesterone of androgen-induced
    behavior in male ring doves (Streptopelia risoria), Endocrin-
    ology 81:39-44.

Erpino, M.J., 1969, Hormonal control of courtship behaviour in the
    pigeon (Columbia livia), Anim. Behav. 17:401-405.

Erpino, M.J., and Chappelle, T.C., 1971, Interactions between andro-
    gens and progesterone in mediation of aggression in the mouse,
    Horm. Behav. 2:265-272.

Goy, R.W., and Resko, J.A., 1972, Gonadal hormones and behavior of
    normal and pseudohermaphrodite non-human female primates,
    Recent Prog. Horm. Res. 28:707-733.

Hagemenas, F.C., and Kittinger, G.W., 1972, The influence of fetal
    sex on plasma progesterone levels, Endocrinology 91:253-256.

Harris, G.W., 1964, Sex hormones, brain development and brain
    function, Endocrinology 75:627-648.

Hess, D.L., and Resko, J.A., 1973, The effects of progesterone on
    the patterns of testosterone and estradiol concentrations in
    the systemic plasma of the female rhesus monkey during the
    intermenstrual period, Endocrinology 92:446-453.

Karsch, F.J., Dierschke, D.J., and Knobil, E., 1973, Sexual differ-
    entiation of pituitary function: apparent difference between
    primates and rodents, Science 179:484-486.

Kincl, F.A., and Magueo, M., 1965, Prevention by progesterone of
    steroid-induced sterility in neonatal male and female rats,
    Endocrinology 77:859-862.

Macdonald, G.J., Yoshinaga, K., and Greep, R.O., 1973, Serum pro-
    gesterone values in Macaca mulatta near term, Amer. J. Phys.

Anthropol. 38:201-206.

Mahajan, D.K., and Samuels, L.T., 1962, Inhibition of steroid 17-
desmolase by progesterone, Fed. Proc., 21:209Abs.

Murton, R.K., Thearle, R.J.P., and Lofts, B., 1969, The endocrine
basis of breeding behavior in the feral pigeon (*Columbia
livia*). I. Effects of exogenous hormones on the pre-incubation
behavior of intact males, Anim. Behav. 17:286-306.

Napoli, A.M., and Gerall, A.A., 1970, Effect of estrogen and antiestrogen on reproductive function in neonatally androgenized
female rats, Endocrinology 87:1330-1337.

Phoenix, C.H., Goy, R.W., Gerall, A.A., and Young, W.C., 1959,
Organizing action of prenatally administered testosterone
propionate on the tissues mediating mating behavior in the
female guinea pig, Endocrinology 65:369-382.

Phoenix, C.H., Goy, R.W., and Resko, J.A., 1968, Psychosexual differentiation as a function of androgenic stimulation, in
Perspectives in Reproduction and Sexual Behavior (M. Diamond,
ed.), p. 33-49, Indiana University Press, Bloomington.

Resko, J.A., 1970, Androgen secretion by the fetal and neonatal
rhesus monkey, Endocrinology 87:680-687.

Resko, J.A., Feder, H.H., and Goy, R.W. 1968, Androgen concentrations in plasma and testis of developing rats, J. Endocrinol.
40:485-491.

Resko, J.A., Malley, A., Begley, D., and Hess, D.L., 1973, Radioimmunoassay of testosterone during fetal development of the
rhesus monkey, Endocrinology 93:156-161.

Sheridan, P.J., Zarrow, M.X., and Denenberg, V.H., 1973, Androgenization of the neonatal female rat with very low doses of
androgen, J. Endocrinol. 57:33-45.

Swanson, H.H., 1971, Determinations of the sex role in hamsters by
the action of sex hormones in infancy, in *The Influence of
Hormones on the Nervous System* (D.H. Ford, ed.), p. 424-440,
S. Karger, Basel.

Whalen, R.E., 1968, Differentiation of the neural mechanisms which
control gonadotropin secretion and sexual behavior, in
Perspectives in Reproduction and Sexual Behavior (M. Diamond,
ed.), pp. 303-340, Indiana University Press, Bloomington.

SOCIAL FACTORS AFFECTING THE DEVELOPMENT OF

MOUNTING BEHAVIOR IN MALE RHESUS MONKEYS [1]

Robert W. Goy, Kim Wallen, and David A. Goldfoot

Department of Psychology, University of Wisconsin, and

Wisconsin Regional Primate Research Center, Madison

INTRODUCTION

In primates as well as in other mammalian species, the development of normal sexual behavior in the male depends upon adequate social experience (Nissen, 1954; Valenstein et al., 1955, Mason, 1960; Gerall, 1963; Harlow et al., 1966; Evans, 1967; Beach, 1968; Gruendel and Arnold, 1969; Davenport and Rogers, 1970; Riesen, 1971). In such nonprimate mammals as rats, guinea pigs, and dogs, the effects of early social deprivation on the later expression of sexual behavior seem to be less profound and less permanent than in primates (Kagan and Beach, 1953; Valenstein and Goy, 1957; Zimbardo, 1958; Beach, 1968). These effects are often transitory and can be eliminated with repeated opportunities for sexual and social experience at advanced ages. In male rhesus monkeys, by contrast, deprivation of early social experience has led to long-lasting and probably permanent inadequacies in sexual behavior (Senko, 1966; Missakian, 1969).

One factor that may account for the severe and relatively permanent effects of early social deprivation on sexual behavior in primates is the development of various pervasive affectional disorders, which have been extensively described (Harlow and Harlow, 1962; Mason, 1963; Harlow et al., 1965; Harlow and Harlow, 1966). These disorders may impair or prevent the development of adequate sexual responses even when opportunities for social and

[1]This article is Publication No. 13-031 of the Wisconsin Regional Primate Research Center.

sexual experience are provided later. If corresponding affective disorders occur in rats or guinea pigs, they have not been reported. However, Gerall (1965) suggested aberrant types of social interactions which occur more frequently in isolate guinea pigs than in socially reared animals and may reflect disturbances in affect.

The techniques used in these studies to restrict early social experience were designed specifically to keep the development of emotional disturbances at a minimum. To some extent our efforts were successful. Monkeys separated at three months of age from their mothers, then reared with one-half hour of daily contact with peers did not develop any appreciable bizarre or autistic patterns of self-clasping, self-biting, and social withdrawal, which are characteristic of socially isolated infant monkeys (Harlow and Suomi, 1971). Of the 31 males in the present study that were reared with limited peer contact, only eight displayed self-clasping in social situations, and then only for brief periods during the first two years of life. None of these subjects reared with restricted opportunities for social experience displayed the social withdrawal syndromes characteristic of isolate-reared monkeys. Nevertheless, the conditions imposed were restrictive enough to prevent the normal development of sexual behavior. In rhesus males, the development of adequate sexual behavior in adulthood is related to the development of a specific type of mounting behavior which involves clasping the partner's legs with the feet, the foot-clasp mount (Goy and Goldfoot, 1973). Accordingly, in this study we concentrated on the factors that influence the rate or age at which this type of mounting behavior expresses itself.

Our results suggest that when the affectional relations among peers are less severely disrupted than they are during complete social deprivation, late socialization has very positive effects on the development of sexual behavior in males.

EXPERIMENT I: STUDIES OF THE DEVELOPMENT OF SEXUAL AND SOCIAL BEHAVIOR DURING THE FIRST YEAR OF LIFE

Methods

Subjects

Forty-five male and 36 female rhesus monkeys born in the laboratories of the Oregon and Wisconsin Regional Primate Research Centers from 1965 to 1972 were used in this series of experiments. The type of rearing condition, including the ages of separation from mothers and the amount of time allowed for peer interaction each day, was an inherent part of the design. Two distinct rearing conditions were used: 1) peer rearing, which involved maternal separation at 90 days and no peer contact except for 30-minute daily observation periods for the first year of life; and 2) mother-infant rearing, which involved continuous contact with mothers and peers throughout the first year of life. Each of these conditions is described in detail below.

Rearing Conditions

Peer-reared groups. Infants composing the peer-reared groups were raised by their mothers in separate cages for the first 90 days of life and then removed and housed individually. Immediately after the separation from their mothers, these infants (five or six per group) were transported daily to a playroom (Fig. 1) where they were allowed to interact for one-half hour per day, five days a week, for 20 successive weeks. Their sexual and social inter-actions were recorded during this 100-day run. The infants were returned to their individual cages each day immediately after the observation period. Each observation group had the same member composition throughout the study. The age range within a group was as small as possible, the largest difference in birthdates being two months and the median difference in ages being one month. Sex ratios varied with the groups formed but had no determinable influence on the measures reported. Both males and females were present in every group. Eleven peer groups were used for this experiment.

Mother-infant rearing. Pregnant females housed in separate cages were allowed to give birth and raise their infants for the first one to two months. A mother-infant group was formed whenever five or six mothers that had given birth within a month of one another were available for study. When the infants were about six weeks to two months of age, they and their mothers were transferred to a large indoor pen (Fig. 2) where they remained during the first year of study. Each of the five groups in this experiment (three with six mother-infant pairs and two with the five mother-infant pairs) contained male and female infants.

During the first month after the establishment of the group, mothers within a group were observed for evidence of excessive aggression. If the group was not stable, either the most aggres-sive mother or the one most often attacked was removed with her infant. One mother-infant pair was removed from two of the five groups before formal testing began.

Daily one-half hour observations of social and sexual behaviors were started when the infants reached a mean age of 90 days and were continued five days a week for 20 consecutive weeks. The first 100 days of observation were used as the data base for this paper. Results are reported only for the male infants in each group, even though all groups studied were composed of both male and female infants.

Fig. 1. One of three similar observation rooms used to study the behavior of peer groups. Infants were allowed to interact in this room for one-half hour per day during the course of the study.

Observation Techniques

In collecting the data for both peer and mother-infant groups, we used similar definitions of behaviors and noted on checklist score sheets the actual frequencies of social and sexual inter-actions, preserving the identity of the initiator and recipient of the behavior. All sexual behaviors were scored for all animals

Fig. 2. One of seven identical pens at the Wisconsin Regional
Primate Research Center used to house and study groups of infants
living with their mothers.

throughout the entire testing period. Play and agonistic behaviors of infants were scored by means of a "shifting focal animal"
technique which recorded the behaviors initiated by one predeter-
mined animal and any recipient of the behavior. Every five minutes
a different member of the group would become the focal animal,
until each infant was followed in this manner for five minutes.
The order of focus was changed daily in a counterbalanced fashion.
Generally, teams of two trained observers participated in data

collection. For the mother-infant groups and for certain peer
groups, 25-minute data are available for agonistic behaviors; but
only five-minute data are used here for analysis of the first year
of life.

Definitions of the behaviors used in this article are:

1. <u>Abortive mount</u>. Any improperly oriented mount accompanied by
pelvic thrusting, including stand to head, stand to side, and ventral lie-on. For this study, seated ventral-ventral with pelvic
thrusting was included as an abortive mount form.

2. <u>No-foot-clasp mount</u>. A mount in which the partner is approached
from the rear. Hands are placed on the back or perineum of the
partner, with both feet on the floor (Fig. 3) and thrusting directed toward the perineum. This behavior, as well as the foot-
clasp mount defined below, was scored whether intromission occurs
or not, since it has been rarely observed under these conditions
and with animals of these ages.

Fig. 3. The no-foot-clasp mount (see text for description).

3. <u>Foot-clasp mount</u>. A mount similar to the no-foot-clasp mount in orientation, except that the mounter grasps the ankles or hocks of the partner with one or both of his rear feet so that the partner must accommodate the entire weight of the mounter (Fig. 4). This is the stereotyped adult form of the mount.

4. <u>Aggression</u>. A vigorous bite with or without headshaking given to another animal.

5. <u>Woofing threat</u>. A behavior in which the threatening animal orients toward his partner, retracts his ears, and often with frowning face vocalizes with a bark. He often appears ready to lunge at his partner.

6. <u>Fear grimace</u>. A stereotyped submissive behavior in which the lips are fully retracted to expose the teeth.

7. <u>Rough-and-tumble play</u>. A vigorous form of wrestling play with whole body involvement.

8. <u>Grooming</u>. Stereotyped spreading and picking of a partner's fur.

Fig. 4. The foot-clasp mount (see text for description).

Analysis of the Data

Since mother-infant groups lived together throughout the entire period of study in the first year, the one-half hour per day observation of these groups constituted a time-sampling technique of data collection and represented only a portion of the total behaviors displayed by infants. In contrast, the one-half hour per day observation of peer groups in their social context constituted an entirely different experimental approach; and data so collected represented 100 per cent of the possible occurrences of behaviors measured. The two types of data collection necessarily produce frequencies of behavior which might be assumed to have differing rate functions, even if all other possible differences between the groups were equivalent. The absolute frequencies of behaviors generated in each of these situations are therefore not directly comparable, despite the fact that observations were performed one-half hour per day for each of the two types of rearing conditions. Formal statistical comparisons used as the basis for theoretical hypotheses or for interpretations of results were thus limited to comparisons of the proportion of animals in each rearing condition displaying a given behavior. Nonetheless, frequency data with limited statistical tests are presented in tables and in the text to give a general impression of the levels of behaviors in the different rearing conditions; but the reader is reminded that the nature of the metrical bias necessarily involved in this study makes frequency comparison difficult to interpret.

Results and Discussion

Mounting Behavior

The most striking differences between the males observed when only peers were present and the males studied in the presence of mothers and other infants were in the performance of mounting behavior. The study of mounting behavior involved recording the frequency of occurrence for essentially three types of mounts: 1) abortive mounts, 2) no-foot-clasp mounts (Fig. 3), and 3) foot-clasp mounts (Fig. 4). Males observed under the two rearing conditions differed greatly in the distribution of mounting over these three categories (Table 1). For males in the mother-infant groups, the foot-clasp mount predominated; and abortive as well as no-foot-clasp mounts were infrequent. For males in the peer-reared groups, a nearly opposite distribution in types of mount was found. For these males, abortive mounts were on the average the most frequent and foot-clasp mounts the least frequent. Differences between males reared under the two conditions are reflected both in the medians reported in Table 1 and in the percentage of the total

TABLE 1. Influence of Rearing Condition on Performance of Mounting Behavior by Rhesus Males in 100 Days of Observation during the First Year of Life

Rearing Condition	N	Abortive Mounts[†]		No-foot-clasp Mounts[†]		Foot-clasp Mounts[†]	
		Median (QR)*	% of total	Median (QR)	% of Total	Median (QR)	% of Total
Peer-reared (3–10 mo. old)	19	36.0(11–78)	50.5	18.0(6–53)	46.9	1.0(0–3.5)	2.9
Mother-Infant (3–10 mo. old)	14	8.0(1–15)	21.0	3.0(0.5–7)	12.3	25.0(8–39)	66.7

*Interquartile Range, Q_1 to Q_3

[†]Data on all three types of mounts were obtained for only 19 of the 31 peer-reared males during the first year of life.

mount frequencies represented by each type of mount.

The 19 peer-reared males displayed a total of 2386 mounts of all types during the 100 days of observation; the 14 mother-infant males showed a total of only 705. The percentages of each mount type listed in Table 1 are presumed to represent the nature of the differences between these two rearing conditions. In addition, for foot-clasp mounts, the difference between rearing conditions cannot be adequately explained by the amount of time available to show the behavior. In this instance, mother-infant-reared males showed a median frequency of 25 foot-clasp mounts compared with a median frequency of only one foot-clasp mount for peer-reared males. Thus the difference in frequency of this type of mount between rearing conditions was opposite from the difference in total mounting frequency. In this sense, the absolute frequency of foot-clasp mounting reflects a very real difference between the two rearing conditions despite the fact that temporal conditions of social interaction and opportunities to display the behavior were not identical. Moreover, this difference in frequency is of special importance if we recall that development of the foot-clasp mount is predictive of copulatory success in adulthood (Goy and Goldfoot, 1973).

During the 100-days of observation during the first year of life, foot-clasp mounting showed a progressive development in mother-infant males and little or no development in peer-reared males (Fig. 5). By the fortieth day of the 100-day run, 57 per cent of the mother-infant males but only 21 per cent of the peer-reared males were displaying the foot-clasp mount. At the end of 100 days, 86 per cent of mother-infant males but only 68 per cent of peer-reared males had exhibited the behavior at least once. These differences between the two rearing conditions were not marked; and what they do not indicate is that among mother-infant reared males, the foot-clasp mount, once acquired, occurred consistently across blocks of daily observations. In contrast, peer-reared males displayed this type of mount inconsistently and did not stop using the less mature mount behaviors in the 100-day run. Of the 13 peer-reared males displaying the foot-clasp mount, six displayed the behavior only during a single 10-day block during the 100-day observation period. In contrast, all of the 12 mother-infant males that showed foot-clasp mounts displayed the behavior during more than one 10-day block. Mother-infant males displayed the foot-clasp mount during 6.3 ± 0.61 of the 10 blocks on the average; whereas peer-reared males averaged only 2.5 ± 0.52 of the 10 blocks. This difference between the two was highly significant ($t = 4.68$, df = 23, $p < 0.01$).

Social Behaviors

Males from the two rearing conditions also differed markedly

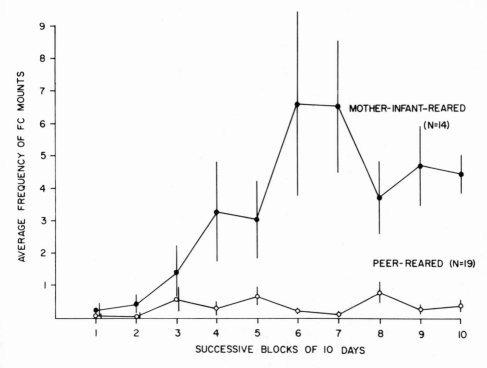

Fig. 5. Mean frequencies per male per 10-day block of foot-clasp (FC) mounting during the first year of life for peer-reared and mother-infant-reared male rhesus (vertical lines represent standard errors).

in their display of agonistic behaviors. Fear grimacing, a predominant response among peer-reared males was seldom observed in mother-infant-reared males (Table 2). Similarly, displays of woofing threat and aggression were frequent in peer-reared groups but almost completely absent in mother-infant groups (Table 2). The males from the two rearing conditions differed not only in the frequency of fear grimacing but also in the proportion of animals showing it. Among mother-infant males, nine of 14 never showed the fear grimace during observations over the 100 days. Only 10 of 31 peer-reared males had a similar absence of fear-grimacing behavior. The difference between these proportions was statistically significant ($X^2 = 4.05$, df = 1, $p < 0.05$). Comparable differences in the proportions displaying the behaviors prevailed for aggressive interactions and woofing threat (Table 2). Thus even though the lack of identical testing conditions cautions against direct comparisons of the absolute frequencies shown, the males from the two rearing conditions differed extensively in the proportion of individuals ever displaying these agonistic behaviors.

TABLE 2. Average Frequency per Male Rhesus of Agonistic Behaviors Shown during 100 Days of Observation in the First Year of Life*

Rearing Condition	N	Fear Grimace		Aggression		Woofing Threat†	
		Mean ± S.E.	% Displaying	Mean ± S.E.	% Displaying	Mean ± S.E.	% Displaying
Peer-reared	31	16.4 ± 4.2	68	12.8 ± 2.4	97	12.5 ± 4.3	89
Mother-Infant	14	0.3 ± 0.15	36	0.06± 0.04	14	0.40± 0.2	21

*Each male was scored for 5 min of the daily 30-min observation period.

†Data on this behavior were obtained for only 19 of the peer-reared males during the first year of life.

Mother-infant and peer-reared males did not differ significantly in terms of the proportions of individuals displaying rough-
and-tumble play: all of the males in each of the rearing conditions displayed this behavior. The quantitative difference in
frequency between the two, however (50.3 $\pm$ 7.3 per male per 20-day
block for peer-reared males compared with 16.4 $\pm$ 1.8 for mother-
infant males),was statistically significant ($\underline{t}$ = 3.11, df = 43,
p < 0.01). The most relevant observation is that the peer-reared
males were capable of engaging in play and did so at levels sur-
passing, or at least comparable to, those of mother-infant males.

Summary

In summary, observations during the first year of life on
males reared under these two conditions indicate that the only
differences in the development of play behavior were quantitative.
The two rearing conditions did differ markedly in levels of agonistic interaction and in development of foot-clasp mounting.
Peer-reared males showed high levels of agonism and poor development of foot-clasp mounting; whereas mother-infant-reared males
showed very low levels of agonism and high frequencies of foot-
clasp mounting.

EXPERIMENT II: STUDIES OF THE DEVELOPMENT OF SEXUAL AND SOCIAL BEHAVIOR DURING THE SECOND YEAR OF LIFE

The deficiency in foot-clasp mounting exhibited by peer-reared
males cannot be accounted for by the social apathy or withdrawal
syndrome so characteristic of monkeys reared in social isolation
(Harlow and Harlow, 1966). Peer-reared males obviously did not
suffer from this syndrome; the data indicate high levels of social
play, and most of them (23 of 31) never showed any sign of the
autistic patterns of self-clasping and self-biting. Thus, we concluded that peer-reared males showed less severe affectional disturbances than monkeys reared under harsher forms of social deprivation. Lacking the severe manifestations of total social isolation, these infants were therefore suitable for testing the
hypothesis that extensive socialization after a socially restrictive
rearing period in the first year permits the development of more
normal sexual behavior. Accordingly, studies of behavior in the
second year of life were carried out for 28 of the 31 peer-reared
males after differential amounts of social experience intervened
between the first and second period of observation.

Methods

Conditions of housing for some of the peer-reared males were

altered before the commencement of behavioral studies during the
second year of life. To study the effects of extensive social
experience at the end of the first 100-day period of observation,
we allowed four of the original peer-reared groups to live continuously for four to six months in large communal cages. Each of
these four was housed separately so that social experience was
limited to members of that peer group. This opportunity for socialization was provided for 12 males and their female peers; they are
henceforth referred to as the socialized peers. Records of behavior were not kept during this period of group housing. At
the end of the socialization period, these monkeys were again
placed in individual cages; and a second period of observation of
their interactions in daily half-hour test periods commenced. The
remaining 16 of the peer-raised males available for continued
studies (three of the original 31 were assigned to another project)
were left in individual cages, having only visual and auditory
contact with other infants; they are henceforth termed the nonsocialized peers.

Housing conditions were also changed for all mother-infant-
reared males. At one year of age, they were separated from their
mothers and placed in individual cages so that their behavior
could be studied in daily half-hour periods of social interaction
in a manner exactly comparable to the socialized and nonsocialized
peers. Thus the problems of metrical bias in comparing data during
the first year of life were circumvented in studies during the
second year of life, and uniform conditions prevailed for all
subjects.

At the beginning of this second observation period, all males
were approximately 13 to 14 months old and were living in individual cages. This observation period consisted of daily 30-
minute sessions of social interaction under the same conditions of
testing previously described for groups during the first year of
life. Observations were carried out for 10 successive weeks, five
days per week. During this time, records were kept on each individual's frequency of performance of fear grimacing, aggression,
woofing threat, rough-and-tumble play, grooming, and the three
types of mounting behavior previously described. Since conditions
of testing were now identical for all three types of observation
groups in this experiment, statistical analyses on frequencies as
well as proportions were appropriate and were used in the analyses
of the data for the second year of life.

Results and Discussion

Mounting Behavior

Table 3 shows that most of the peer-reared males provided with

TABLE 3. *Mounting Behavior by Rhesus Males in 50 Days of Observation during the Second Year of Life*

Subjects	N	Abortive Mounts		No-foot-clasp Mounts		Foot-clasp Mounts		
		Median (QR)*	% of Total Mounts	Median (QR)	% of Total Mounts	Median (QR)	% of Total Mounts	% of Males Displaying Foot-clasp Mounts
Socialized Peers	12	3.5(1.5-7.5)	11.9	2.0(0-3)	6.5	19(1-62.5)	81.6	75.0
Nonsocialized Peers	16	5.5(1-29)	44.9	3.5(0-11)	53.2	0(0-1)	1.9	12.5
Mother-Infant	14	1.5(0-6)	2.6	0(0-1)	0.1	184(118-122)	97.3	92.8

*Interquartile Range, Q_1 to Q_3

continuous group housing for four to six months between the first
and second periods of observation developed foot-clasp mounting.
These socialized peers showed 19 foot-clasp mounts on the average
during the 50 days of observation. Nonsocialized peers, in contrast, failed to develop this type of mounting behavior. Both
peer groups, however, were clearly deficient in foot-clasp mounting
when contrasted with the mother-infant males, which displayed a
median of 184 foot-clasp mounts on testing in their second year
under exactly the same conditions as the two types of peer groups.
An analysis of variance for all three groups of males was significant (F = 29.62, df = 2,39, p < 0.001. Mother-infant males showed
significantly more foot-clasp mounts than the socialized peers ($\underline{t}$ =
4.43, df = 24, p < 0.0001), and the socialized peers showed significantly more foot-clasp mounts than the nonsocialized peers ($\underline{t}$ =
2.73, df = 26, p < 0.02). These findings suggest that providing
extensive socialization experience with peers in the presence of
mothers during the first year of life results in higher frequencies
of foot-clasp mounting than rearing systems offering extensive
social experience at the end of the first year of life with peers
but without mothers. The variables of age at which socialization
was provided and at which mothers were present are confounded;
hence we are unable to estimate from these experiments which of
these variables is more important to the development of foot-clasp
mounting.

When the males from our three categories were compared for
proportions of mounts within each mount type, the following profiles emerged. For nonsocialized peers, the abortive mount remained predominant for the second year of study, as it was for the
first year in these same animals. For mother-infant males, the
change from continuous group living to housing in individual cages
did not bring about any change in the distribution of mounting;
for them the foot-clasp mount continued to predominate. Socialized
peers, however, showed a dramatic shift in distribution from the
first to the second year. Whereas abortive mounts had predominated during the first year, the foot-clasp mount became the predominant mount in the second year of life after extended socialization.

Despite changes in the average frequency of foot-clasp mounting for socialized peers as well as for mother-infant males, these
two types differed in their responses to the repeated daily trials
of social interaction (Fig. 6). Mother-infant males showed
increases in the frequency of mounting throughout the 50 days of
observation. Socialized peers did not show any corresponding
increases, and the average frequency of foot-clasp mounts across
blocks remained relatively constant. For nonsocialized peers, the
frequency of foot-clasp mounting remained close to zero throughout
the 50-day period.

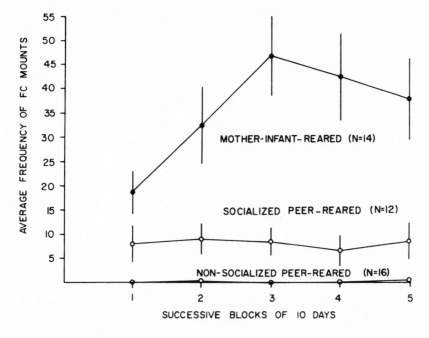

Fig. 6. Mean frequencies per 10-day block of foot-clasp (FC) mounting in the second year of life for male rhesus previously reared under three different conditions (vertical lines represent standard errors).

Social Behaviors

During the second year of life, males from the three types of groups did not differ significantly in frequencies of fear grimacing or woofing threat. Aggression, however, was less frequently displayed by mother-infant males than by peer-reared males ($\underline{t}$ = 2.39, df = 39, p < 0.05; see Table 4). The socialization of peer-reared males did not measurably affect the frequency of display of these agonistic behaviors, and peer-reared males receiving no intervening socialization showed a decline in agonistic behaviors from the first to the second year which was entirely comparable to that of socialized peers.

Rough-and-tumble play was shown at nearly equal frequencies by the two types of peer groups during the second year; whereas the mother-infant-reared males showed significantly lower frequencies than the peer groups over the 50-day period of observation

TABLE 4.   Average Frequency per Male Rhesus of Agonistic Behaviors
            Shown during 50 Days of Observation in the Second Year
            of Life*

Subjects	N	Fear Grimace Mean ± S.E.	Aggression Mean ± S.E.	Woofing Threat Mean ± S.E.
Socialized Peers	12	2.28 ± 0.63	3.73 ± 1.1	9.0 ± 3.9
Nonsocialized Peers	16	2.61 ± 0.81	3.57 ± 1.2	3.9 ± 2.7
Mother-Infant	14	3.02 ± 1.3	0.81 ± 0.34	6.2 ± 2.3

*Each male was scored for 25 min of the daily 30-min observation
session of the frequency of these behaviors. However, the values
in the table are adjusted to the estimated frequency for 5 min of
scoring for each male for easier comparison with the frequencies
presented in Table 2.

($F = 7.03$, $p < 0.01$). However, as Table 5 indicates, by the
fifth 10-day block of testing, the mother-infant males reached
levels of rough play comparable to those of the peer-reared males.
These data suggest that the males in the mother-infant rearing
condition were suffering from the depressive effects of weaning,
which occurred immediately before the second observation period.
A maternal separation syndrome has been described repeatedly by
other laboratories (Hinde and Spencer-Booth, 1971), and it has
as a major characteristic a decrease in play activity. Our results
suggest that rough play was severely affected by the trauma of
maternal separation but that other categories of social behavior,
including sexual and agonistic behaviors, were less obviously
affected. Except for the temporary depression of rough-and-tumble
play in mother-infant males in the early weeks of the second observation period, the profiles of rough play behaviors for males from
all groups were remarkably similar.

There were differences among the groups in the frequency of
grooming behavior. This behavior was not analyzed for frequency
of performance by males, since it was more often shown by female
members of the group. Therefore, the total frequency of grooming

TABLE 5. *Average Frequency per Rhesus Male of Rough-and-Tumble Play during the Second Year of Life**

Subjects	N	Successive 10-Day Blocks (Mean ± S.E.)					Average (± S.E.) of 5 Blocks
		1	2	3	4	5	
Socialized Peers	12	17.3 ± 3.3	21.1 ± 4.4	23.6 ± 3.9	17.2 ± 2.9	27.8 ± 4.2	23.6 ± 3.8
Nonsocialized Peers	16	19.8 ± 4.6	20.2 ± 3.5	22.2 ± 4.7	21.9 ± 4.4	21.6 ± 3.7	21.6 ± 3.5
Mother-Infant	14	0.6 ± 0.25	2.6 ± 1.2	6.6 ± 3.1	11.3 ± 2.4	18.5 ± 3.4	7.9 ± 1.4

*Each male was scored for 5 min of the daily 30-min observation session.

shown in each group, regardless of which individuals performed the response, was used as the basis for comparing the different rearing conditions. For four groups of socialized peers, the average frequency per 10-day block was 6.65 ± 2.5 bouts of grooming. For non-socialized peers, the frequency was 0.6 ± 0.2, and the difference between these two categories was statistically significant ($t = 2.82$, $p < 0.05$). Mother-infant reared groups groomed 17.8 ± 2.9 times per 10-day block; but intergroup variability was large, and the difference between this rearing condition and nonsocialized peers was of marginal statistical significance ($t = 2.18$, $0.05 < p < 0.10$).

DISCUSSION

Three different rearing conditions were investigated in these experiments. Each allowed for the development of play and mounting behavior among infants, and none produced animals demonstrating the bizarre deficits of severe social isolation (Harlow and Harlow, 1966). Males reared in these three conditions did not differ conspicuously in the overall frequency of mounting behavior or in the proportion of males that showed mounting behavior. It was the pattern of types of mount that distinguished males from the differing conditions and that further distinguished socialized and non-socialized peers.

Deficiency in a specific type of mounting has been reported several times in studies dealing with social experience and the development of sexual behavior in mammals. Thus early isolation in guinea pigs produces a specific deficiency in the expression of posterior mounts with clasps in adulthood, but not in the overall frequency of mounting (Valenstein et al., 1955; Gerall, 1963). Rats reared in social isolation showed deficiencies in posterior mounts with thrusts (Gruendel and Arnold, 1969). Similarly, early isolation in dogs caused marked deficiencies in the performance of mounts with a posterior orientation but not in the overall frequency of all types of mounts (Beach, 1968). Our results are, of course, limited to the juvenile and prepuberal manifestations of these behaviors in social groups; whereas the other investigators were dealing with the performance of pairs of animals in adulthood. Nevertheless, the parallel between these studies of adult nonprimate mammals and our prepuberal monkeys is striking; in each species studied, the specific deficiency in mounting behavior is in that type which is most characteristic of the normal adult.

Beach (1968) postulated that in males reared with limited social experience the deficient expression of the normal adult mount is related to the failure to develop some other less specific type of behavior that constitutes a basic response pattern out of which mounting would be differentiated. For rhesus monkeys, the play

behavior repertoire has been suggested as the basic response pattern
that performs this function as well as indicates the development of
appropriate affectional relations among age-mates (Harlow, 1965).

Our results are not entirely consistent with this view of the
role of play in the development of foot-clasp mounting postures in
rhesus males. The two rearing conditions studied during the first
year of life appeared equally effective in aiding the development
of play behavior in 100 percent of the males. Moreover, during the
second year of life socialized peers, nonsocialized peers, and
mother-infant-reared males did not differ in the proportion of
males showing this behavior. Thus the expression of play per se
cannot account for the differential expression of foot-clasp mounting by these three experientially diverse groups.

From our results, the missing behavior that seems to be important to both the development and expression of foot-clasp mounting
is not a deficiency in a generalized prototypical pattern, but rather
the virtual absence of aggression in the social environment at early
ages. The single characteristic which differentiated mother-infant
males from peer-group males in the successful expression of foot-
clasp mounting was the nearly complete absence of aggression in the
first year of life and the significantly lower levels of aggression
in the second year. When the level of aggression between age-mates
was high, as it was in the peer-rearing system in the first year of
life, foot-clasp mounting was infrequent; and abortive mounts predominated. When the frequency of aggression between age-mates was
extremely low, as it was in the mother-infant rearing condition,
foot-clasp mounting predominated and was expressed by nearly all
males at this early age. Even when foot-clasp mounting was displayed by socialized peers, as it was during the second year of life,
high levels of aggression appeared to inhibit its expression. When
socialized peers and mother-infant males were compared, there was
a strong inverse relationship between the frequency of aggression
toward age-mates and frequency of foot-clasp mounting. We believe
that the virtual absence of aggression among age-mates reflects
the development of affectional relationships which must be present
before foot-clasp mounting can be expressed. Moreover, for the
rearing conditions which we have studied, deficiencies in emotional
relationships between age-mates rather than deficiencies in presumed fundamental or prototypical motor response patterns seem to
account for the failure to develop foot-clasp mounting.

Our hypothesis about the necessity of developing a positive
social environment is consistent with the observed sexual inadequacies of isolate-reared male rhesus studied by other investigators.
Male rhesus reared under these conditions are not subjected to
aggression by age-mates during early development, since none are
present; but their sexual inadequacies are, if anything, more profound than those of males reared in the peer condition. Therefore,

probably neither the presence nor the absence of aggression causes
the inhibition of foot-clasp mounting, but rather the lack of
development of a basic positive social environment in which the
animals can interact with one another without excessive fear. The
development of foot-clasp mounting and the incidence of grooming
may equally reflect the development of the necessary positive social
relationships, just as the expression of aggression at very low
frequencies may reflect the same condition.

This point of view does not rule out the possibility that
deficiencies in foot-clasp mounting reflect deficiencies in the
emotional relationship between age-mates which underlies a partner's
willingness to tolerate a foot-clasp mount. The partner's unwill-
ingness to hold the sexual presentation response and to support
the weight of the male while he is performing the foot-clasp mount
may be part of this deficiency. Regardless of the contribution of
this factor to the failure of peer-reared males to show foot-clasp
mounts, the developmental failure of an underlying emotional relationship with age-mates is, in our view, the fundamental cause of
the sexual deficit.

The experiments discussed here show that if opportunities to
interact continuously with peers are provided as late as one year
of age, marked improvement in mounting behavior can result. We
have unpublished data showing that continuous social experience
with peers can still be effective when it is provided at the end of
the second year of life. The majority of males reared with late
socialization procedures were found to be successful copulators in
adulthood (see Goy and Goldfoot, 1973); those animals were not described as socialized peers in the earlier article, but rather as
males showing the development of foot-clasp mounting early in life.
These results stand in contrast to previous unsuccessful attempts
to improve sexual behavior by socialization of isolate-reared
males at relatively advanced ages (Senko, 1966; Missakian, 1972).

Of the many reasons for the differences obtained, certainly
one could be that individuals with less profound and pervasive
emotional disorders more easily develop the necessary positive
social relationships with age-mates. Or perhaps, opportunities for
social experience were provided the isolate-reared males in the
previous unsuccessful efforts to improve their sexual behavior at
too late an age. Thus the often-stated hypothesis of critical
(Scott, 1962) or sensitive periods (Hinde, 1966) of development,
during which experience can produce stronger effects than at later
ages. There is a suggestion in our own experiments that such an
hypothesis might be appropriate. Socialized peers never achieved
levels of performance of the foot-clasp mount equal to those of
mother-infant males, and one of the relevant differences between
these groups was the age at which continuous social experience was
allowed. This difference in age, however, was confounded with

differences in the character of the individuals participating in the socilization process, a factor which has recently been shown to be significant in the emotional rehabilitation of isolate-reared monkeys (Harlow and Suomi, 1971). This latter discovery suggests that effects of socialization may depend as much upon the type of partner involved as upon the age at which it occurs.

What practical applications are suggested from these findings? For juvenile rhesus monkeys, a nonhostile social climate is obviously necessary for the development of behavioral and emotional response systems that will ensure future reproductive success. Peers afforded only limited daily contact with one another cannot structure such a nonhostile environment, and monkeys reared in this manner have little chance of becoming effective copulators in adulthood (Goy and Goldfoot, 1973). Peers reared in this same manner initially but provided continuous contact with age-mates at about one year of age have a greater probability of copulating as adults, but are less capable of effective adult copulatory behavior than feral males (unpublished observations). Therefore, the only rearing program known to us that can successfully produce breeding rhesus males is one that incorporates long periods of social contact with age-mates and low levels of aggression between them. The mother-infant rearing system through the first year of life has these characteristics; we have noted that, despite the usual aggressive encounters between the mothers, aggression between infants is rare. Because of this circumstance, we are advancing the hypothesis that in the social environment in which the infants are developing, aggression between infants is more adverse to sexual outcomes than aggression between adults. Whether the males reared in this system can display adequate and regular sexual behavior in adulthood will be determined in the next two to four years.

ACKNOWLEDGMENTS

This work was supported by grants RR-00163 and RR-00167 from the National Institutes of Health to the Oregon and Wisconsin Regional Primate Research Centers, respectively, as well as by grants MH-08634 and MH-21312 from the National Institute of Mental Health. Special thanks are due Jens Jensen, Warren Schmidt, Mary Collins, and Jane Cords for their assistance in the collection of these data and Nancy Arnold for her help in preparing the manuscript.

REFERENCES

Beach, F.A., 1968, Coital behavior in dogs: III. Effects of early isolation on mating in males, Behaviour 30:217-238.

Davenport, R.K., and Rogers, C.M., 1970, Differential rearing of the chimpanzee: A project survey, The Chimpanzee 3:337-360.

Evans, C.S., 1967, Methods of rearing and social interaction in
    Macaca nemestrina, Anim. Behav. 15:263-266.

Gerall, A.A., 1963, An exploratory study of the effect of social
    isolation variables on the sexual behaviour of male guinea
    pigs, Anim. Behav. 11:274-282.

Gerall, A.A., 1965, Effects of social isolation and physical con-
    finement on motor and sexual behavior of guinea pigs, J. Person-
    ality Soc. Psychol. 2:460-464.

Goy, R.W., and Goldfoot, D.A., 1973, Experiential and hormonal fac-
    tors influencing development of sexual behavior in the male
    rhesus monkey, in *The Neurosciences: Third Study Program*
    (F.O. Schmitt and F.G. Worden, eds.), vol. 3, Chapter 50,
    pp.571-581, MIT Press, Cambridge, Mass.

Gruendel, A.D., and Arnold, W.J., 1969, Effects of early social
    deprivation on reproductive behavior of male rats, J. Comp.
    Physiol. Psychol. 67:123-128.

Harlow, H.F., 1965, Sexual behavior in the rhesus monkey, in *Sex
    and Behavior*, (F.A. Beach, ed.) pp.234-265, John Wiley and
    Sons, New York.

Harlow, H.F., and Harlow, M.K., 1962, The effect of rearing condi-
    tions on behavior, Bull. Menninger Clinic 26:213-224.

Harlow, H.F., and Harlow, M.K., 1966, Learning to love, Amer. Scien-
    tist 54:244-272.

Harlow, H.F., and Suomi, S.F., 1971, Social recovery by isolation-
    reared monkeys, Proc. Nat. Acad. Sci. 68:1534-1538.

Harlow, H.F., Dodsworth, R.O., and Harlow, M.K., 1965, Total social
    isolation in monkeys, Proc. Nat. Acad. Sci. 54:90-96.

Harlow, H.F., Joslyn, W.D., Senko, M.G., and Dopp, A., 1966, Behav-
    ioral aspects of reproduction in primates, J. Anim. Sci.
    25:49-67.

Hinde, R.A., 1966, *Animal Behaviour*, McGraw-Hill, New York.

Hinde, R.A., and Spencer-Booth, Y., 1971, Effects of brief separa-
    tion from mother on rhesus monkeys, Science 173:111-118.

Kagan, J., and Beach, F.A., 1953, Effects of early experience on
    mating behavior in male rats, J. Comp. Physiol. Psychol.

46:204-208.

Mason, W.A., 1960, The effects of social restriction on the behavior of rhesus monkeys: I. Free social behavior, J. Comp. Physiol. Psychol. 53:582-589.

Mason, W.A., 1963, Social development of rhesus monkeys with restricted social experience, Perceptual Motor Skills 16:263-270.

Missakian, E.A., 1969, Reproductive behavior of socially deprived adult male rhesus monkeys (*Macaca mulatta*), J. Comp. Physiol. Psychol. 69:403-407.

Missakian, E.A., 1972, Effects of adult social experience on patterns of reproductive activity of socially deprived male rhesus monkeys (*Macaca mulatta*), J. Personality Soc. Psychol. 21:131-134.

Nissen, H.W., 1954, Development of sexual behavior in chimpanzees, in *Genetic, Psychological and Hormonal Factors in the Establishment of Sexual Behavior in Mammals* (unpub. sym. proc., Lawrence, Kan., 1954; copy in Univ. of Kan. Libraries, Lawrence).

Riesen, A.H., 1971, Nissen's observations on the development of sexual behavior in captive-born, nursery-reared chimpanzees, The Chimpanzee 4:1-18.

Scott, J.P., 1962, Critical periods in behavioral development, Science 138:949-958.

Senko, M.G., 1966, The effects of early, intermediate, and late experiences upon adult macaque sexual behavior (unpub. M.S. thesis, Univ. of Wis., Madison).

Valenstein, E.S., and Goy, R.W., 1957, Further studies of the organization and display of sexual behavior in male guinea pigs, J. Comp. Physiol. Psychol. 50:115-119.

Valenstein, E.S., Riss, W., and Young, W.C., 1955, Experiential and genetic factors in the organization of sexual behavior in male guinea pigs, J. Comp. Physiol. Psychol. 48:397-403.

Zimbardo, P.G., 1958, The effects of early avoidance training and rearing conditions upon the sexual behavior of the male rat, J. Comp. Physiol. Psychol. 51:764-769.

THE ROLE OF ANDROGENS IN THE SEXUAL BEHAVIOR OF ADULT MALE RHESUS MONKEYS[1]

Charles H. Phoenix

Oregon Regional Primate Research Center, Beaverton

Many factors determine the kind and amount of sexual behavior displayed by adult male rhesus monkeys, but our focus here is on the role of androgens in the control of this behavior.

When individually housed adult male rhesus monkeys were paired for 10 minutes with females around the time of ovulation, mating with ejaculation occurred in about 80% of the cases (Phoenix, et al., 1968). Likewise, when ovariectomized females were treated with 10 μg of estradiol benzoate for 12 or 13 days and paired, mating frequencies were approximately the same as those obtained at the time of ovulation (Phoenix et al., 1973; Phoenix, unpublished observation).

In the present study, all of the males and females were wild born, adult, sexually sophisticated, and adapted to laboratory conditions. In addition, the females had served as subjects in tests of sexual behavior before and after ovariectomy. All of the animals had been in a laboratory environment for several years; that is, they were caged individually, housed in an air-conditioned room, and given tests of sexual behavior in an adjoining room.

Our research animals are kept in what some would describe as unnatural conditions. In many respects, laboratory housing resembles the conditions in some of our more enlightened prisons.

[1]This article is publication No. 668 of the Oregon Regional Primate Research Center, supported in part by grants RR-00163 and HD-05969 from the National Institutes of Health, United States Public Health Service.

Therefore, our description of the sexual behavior of the male rhesus may no more resemble his sexual behavior in nature than would a description of the sexual behavior of an inmate on a weekend pass resemble the sexual behavior of the average human male in nature. However, it occurs to me that the rhesus monkey, living one step ahead of the bulldozer, eating the DDT-sprayed crops of irate villagers, or just managing to survive in tourist-ridden preserves, is himself living under less-than-natural conditions. The real point is that just as environments differ, so do behaviors in different environments. The overly simple distinction between "natural" and "unnatural" conditions carries too much surplus meaning and is not, therefore, useful. What is more important is to show how and to what extent environmental factors modify behavior, if indeed they do. Sexual behavior in the forest canopy may be totally different from the same behavior in a 10-foot, stainless steel cube; but the difference may be irrelevant to the problem being studied. In any event, in reporting his findings, the experimenter must be mindful of the conditions in which his subjects live.

In our stainless steel and plexiglass world, therefore, we gave 20 males 10 weekly 10-minute tests of sexual behavior and then castrated half the group (Phoenix et al., 1973). A sex test consisted simply of pairing a male with an estrogen-treated female and recording the frequency of occurrence of various components of sexual and sex-related behaviors. The moment an item of behavior first occurred after the onset of the test was also noted and is referred to here as the "latency of response."

After the 10 weekly tests and just before castration or sham castration (control group), blood samples were taken from each male; and the concentration of testosterone in the systemic circulation was determined by means of gas liquid chromatography with electron capture (Resko and Phoenix, 1972). The mean amount of testosterone for the group was 4.59 ng/ml ($\pm$ 0.87 S.E.). Four samples were lost in processing; therefore, the mean was based on samples from 16 animals. There was no statistically significant correlation (p > 0.05) between testosterone levels and ejaculation frequency (ρ = 0.03), latency to ejaculation (ρ = 0.33), intromission frequency (ρ = 0.04), and mounting frequency (ρ = 0.10). Such behaviors as aggression and threat occurred so infrequently as not to warrant being tested for statistical significance.

At one-half hour, two hours, and five hours, and at 13 weeks after castration, mean blood levels of testosterone dropped to 2.0, 1.1, 0.7, and 0.03 ng/ml respectively. At 13 weeks when 5 ml of plasma were analyzed, only one animal showed detectable amounts of testosterone.

One week after operation, weekly testing was resumed for both castrated and sham-castrated males and continued for 25 consecutive weeks. The animals were again tested during weeks 36 to 40 and

51 to 55. The frequency of ejaculation showed a relatively slow
but constant decline so that during the last block of five tests a
year after castration, ejaculation occurred during only 8% of the
tests. Three of the 10 males ejaculated during these five weekly
tests, and an additional two males achieved intromission but did
not ejaculate. The percentage of tests in which ejaculation,
intromission, and mounting occurred for controls and experimental
animals is shown in Figures 1, 2, and 3.

One year after castration, when mean ejaculation frequency
was at 8%, blood samples were again taken from the castrates and
testosterone concentrations were determined by radioimmunoassay
(Resko and Phoenix, 1972). The mean concentration of testosterone
for the group was 298.7 pg/ml (+ 42.1 S.E.). No significant correlations were found between testosterone concentrations and
ejaculation frequency, ejaculation latency, intromission frequency,
or mounting frequency. Therefore, there is no correlation between
testosterone levels and sexual performance before or after castration; but when there is a significant reduction in testosterone
concentrations like that after castration, levels of sexual
performance drop significantly. We conclude, therefore, that
testosterone does play a crucial role in maintaining the levels of
sexual performance characteristic of adult rhesus males. Since we
know that castration does not eliminate all testosterone from the
systemic circulation, we question whether the small remaining
amount of testosterone accounts for the level of sexual behavior

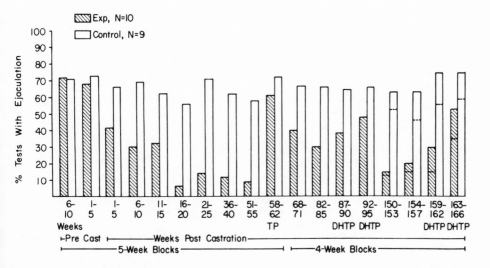

Fig. 1. Percentage of tests with ejaculation per 4- or 5-wk block
before and after castration. Testosterone propionate (TP) and
dihydrotestosterone propionate (DHTP) were injected IM (1 mg/kg).

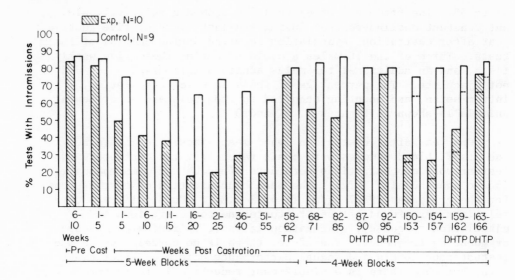

Fig. 2. Percentage of tests with intromissions per 4- or 5-wk block before and after castration. Testosterone propionate (TP) and dihydrotestosterone propionate (DHTP) were injected IM (1 mg/kg).

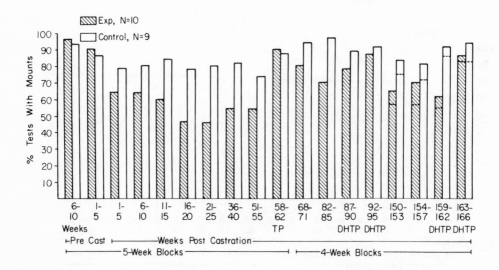

Fig. 3. Percentage of tests with mounts per 4- or 5-wk block before and after castration. Testosterone propionate (TP) and dihydrotestosterone propionate (DHTP) were injected IM (1 mg/kg).

that persists for at least a year after castration. In male rats,
for example, adrenal hormones did not account for the persistence
of sexual behavior after castration (Bloch and Davidson, 1968).
Studies with other species, including dogs (Schwartz and Beach,
1954), hamsters (Warren and Aronson, 1956), and cats (Cooper and
Aronson, 1958), reached similar conclusions. In view of these
results, it is unlikely that adrenal hormones are responsible for
the persistence of sexual behavior in the rhesus monkey after castration; but the possibility should be checked.

Daily injections (IM) of testosterone propionate (TP) were
begun 57 weeks after castration and continued for five weeks. The
first sexual behavior test was given one week after the first
injection and continued weekly for five weeks. Briefly, sexual
behavior was restored to its precastration level, and control and
experimental males did not differ significantly in any item of
behavior. After four weeks of injections, there was little change
in performance; therefore, in subsequent studies, tests were conducted and analyzed in four-week blocks.

I had assumed that six weeks after the termination of treatment with TP, sexual behavior would have returned to prehormone
treatment levels. I was mistaken. In four weekly tests starting
six weeks after the last TP injection, ejaculation occurred during
40% of the tests, and 70% of the males ejaculated at least once;
intromission was achieved by nine of the 10 males. In four weekly
tests beginning 20 weeks after the end of treatment with TP, 50%
of the males ejaculated; and 80% achieved intromission. Since the
average weight of the males at the time of TP treatment was about
10 kg, each male had received approximately 350 mg of TP over the
five-week period. The behavior level of the males 20 weeks after
this treatment was somewhat higher than it had been 20 weeks after
castration (Fig. 1, 2, 3).

Several reports indicate that testosterone affects genital
tract tissues by being converted to the active metabolite, dihydrotestosterone (DHT), in the target tissues, but that the hormone
fails to maintain sexual behavior in the castrated male rat
(McDonald et al., 1970; Feder, 1971; Whalen and Luttge, 1971).
To test whether the ineffectiveness of systemically injected DHT
was due to its failure to reach the relevant brain areas in sufficent
quantities, Johnston and Davidson (1972) implanted dihydrotestosterone propionate (DHTP) directly into the hypothalamus. But here
again, DHTP was found to be relatively ineffective in restoring
sexual behavior in the castrated male rat.

Unlike the rat, castrated rhesus monkeys showed an increase in
sexual behavior after systemic IM injections of DHTP. A preliminary study with DHTP was carried out between postcastration weeks
87 and 95 (25 weeks after TP treatment) (Phoenix, 1973) and was

repeated with slight modification in procedure a year later. The effects of DHTP (1 mg/kg body weight) on the percentage of tests with ejaculation, intromission, and mounting in the preliminary study are shown in Figures 1, 2, and 3.

Three years after castration and one year after the last injection of DHTP, the 19 males were again tested (one control male had become ill and was dropped from the study, and a substitute was found for a second control male that was dropped because his aggressive attacks endangered the life of female partners). Test duration was extended to 30 minutes or until the male ejaculated. Eight tests at weekly intervals were given between postcastration weeks 150 and 157 (Fig. 1, 2, 3). Figure 1 shows the percentage of tests with ejaculation in 30-minute tests for weeks 150 to 153 (15% for experimental, 64% for controls) and the percentage of tests with ejaculation within the first 10 minutes of the tests (12.5% for experimental, 53% for controls). During weeks 154 to 157, ejaculation occurred in 20% of the tests among castrates and in 64% among intact males. During the first 10 minutes of the tests, ejaculation occurred in 15% and 56% of the tests for castrated and intact males, respectively. Three of the castrated males ejaculated, and an additional three males achieved intromission but failed to ejaculate.

A followup of preliminary testing of sexual performance during treatment with DHTP was undertaken during weeks 159 to 166 after castration, two years after TP treatment and one year after treatment with DHTP. The study confirmed and extended the findings of the preliminary work (Fig. 1, 2, 3). All but one castrated male achieved intromission and ejaculated at least once during the eight weekly tests. Although DHTP unquestionably activated sexual behavior in long-term castrated males, the same dosage took twice as long as TP to activate a level of sexual performance that did not differ statistically from the performance level of the control group (Phoenix, unpublished data).

Recent work with DHTP suggests that except for the rat, DHTP is effective in maintaining sexual behavior in castrated males of other rodent species (Goy, personal communication; Whalen, personal communication; Luttge and Hall, 1973) as well as in rhesus monkey. What appear to be discrepancies in the literature on the effects of castration in man may simply indicate broad individual variability similar to that in castrated rhesus monkeys. The greater uniformity found in response to castration among laboratory animals is probably related to the fact that most studies were carried out on inbred strains of mice, rats, or guinea pigs.

The behavior of two or our rhesus males, #3531 and #1071, provides an example of the extreme variability in performance after castration and treatment with TP and DHTP (Table 1). After castration with or without TP or DHTP treatment, #1071 did not ejaculate

TABLE 1. *Percentage of Tests with Ejaculation (per Four- or Five-week Block) by Three Males before and after Castration and under Hormone Treatment*[1]

Male No.	5-Week Blocks							4-Week Blocks								
	Precastration		No Rx					Postcastration								
								TP	No Rx		DHTP		No Rx		DHTP	
	1	2	3	4	5	6	7	8	9	10	11	12	13	14	15	16
3531	100	100	80	80	40	80	40	100	100	100	100	100	100 (100)	50 (75)	100 (100)	75 (75)
3536	80	60	0	40	0	20	20	100	100	50	100	100	0 (0)	0 (0)	25 (50)	100 (100)
1071	40	20	0	0	0	0	0	0	0	0	0	0	0 (0)	0 (0)	0 (0)	0 (50)

[1]Percentages in parentheses based on 30-min tests; others on 10-min tests.

*TABLE 2.   The Percentage of Tests in Which Two Intact Male Rhesus
Monkeys Ejaculated when Paired on Eight Occasions with
Each of Two Females*

Females

		20	14	$\overline{X}$
M				
a	91	100.0	12.5	56.25
l				
e	50	0	87.5	43.75
s				
	$\overline{X}$	50.0	50.0	

[1]Mean percentages for each animal are indicated in lower and right-
hand table margins.

in 10-minute tests; but after five weeks of DHTP injections, he
ejaculated during two of four 30-minute tests.   Male #3531, on the
other hand, ejaculated in less than 10 minutes during 75% of eight
tests administered three years after castration and in the absence
of exogenous hormone.   The behavior of a third male, #3536, more
clearly resembles the values that represent mean performance of the
castrated group over the three years (Table 1).

After castration, even the most vigorous males show an obvious
decline in the frequency of ejaculation.   In such cases, at least,
androgens are crucial for optimum sexual performance (maximum frequency and minimum latency of ejaculation); for some individuals,
they are crucial for even a minimum display of sexual behavior.
However essential testosterone may be for optimum sexual performance, hormones alone are not enough.   The effects of a nonhormonal variable on sexual performance are clearly depicted in
Table 2; it shows the percentage of eight tests during which two
intact males ejaculated when tested with each of two females.   Male
#50 never ejaculated with female #20 but ejaculated in seven of
eight tests with female #14.   If female #20 was an undesirable
female, male #91 was unaware of the fact since he ejaculated in
100% of the tests with her.   That individual preferences exist is
clear enough, but the explanation remains obscure.   Adult testosterone levels are probably not the answer.

REFERENCES

Bloch, G.J., and Davidson, J.M., 1968, Effects of adrenalectomy and experience on postcastration sex behavior in the male rat, Physiol. Behav. 3:461-445.

Cooper, M., and Aronson, L.R., 1958, The effect of adrenalectomy on the sexual behavior of castrated male cats, Anat. Rec. 131: 544 (abstr.).

Feder, H.H., 1971, The comparative actions of testosterone propionate and 5α-androstan-17β-ol-3-one propionate on the reproductive behaviour, physiology and morphology of male rats, J. Endocrinol. 51:241-252.

Johnston, P., and Davidson, J.M., 1972, Intracerebral androgens and sexual behavior in the male rat, Horm. Behav. 3:345-357.

Luttge, W.G., and Hall, N.R., 1973, Differential effectiveness of testosterone and its metabolites in the induction of male sexual behavior in two strains of albino mice, Horm. Behav. 4: 31-43.

McDonald, P., Beyer, C., Newton, F., Brien, B., Baker, R., Tan, H. S., Sampson, C., Kitching, P., Greenhill, R., and Pritchard, D., 1970, Failure of 5α-dihydrotestosterone to intiate sexual behaviour in the castrated male rat, Nature (Lond.) 227:964-965.

Phoenix, C.H., 1973, The role of testosterone in the sexual behavior of laboratory male rhesus, in *Primate Reproductive Behavior* (C.H. Phoenix, ed.) (Sym. Proc., IVth Int. Cong. Primatol., Portland, Ore., 1972; vol. 2), pp. 99-122, S. Karger, Basel.

Phoenix, C.H., Goy, R.W., Resko, J.A., and Koering, M., 1968, Probability of mating during various stages of the ovarian cycle in *Macaca mulatta*, Anat. Rec. 160:490 (abstr.).

Phoenix, C.H., Slob, A.K., and Goy, R.W., 1973, Effects of castration and replacement therapy on the sexual behavior of adult male rhesuses, J. Comp. Physiol. Psychol., 84:472-481.

Resko, J.A., and Phoenix, C.H., 1972, Sexual behavior and testosterone concentrations in the plasma of the rhesus monkey before and after castration, Endocrinology 91:499-503.

Schwartz, M., and Beach, F.A., 1954, Effects of adrenalectomy upon mating behavior in castrated male dogs, Amer. Psychol. 9:467 (abstr.).

Warren, R.P., and Aronson, L.R., 1956, Sexual behavior in castra-
    ted-adrenalectomized hamsters maintained on DCA, Endocrin-
    ology 58:293-304.

Whalen, R.E., and Luttge, W.G., 1971, Testosterone, androstenedione
    and dihydrotestosterone effects on mating behavior of male
    rats, Horm. Behav. 2:117-125.

THE BEHAVIORAL CONTROL OF REPRODUCTIVE PHYSIOLOGY [1]

N. T. Adler

Department of Psychology

University of Pennsylvania, Philadelphia

This article is dedicated to the late Professor D. S. Lehrman, whose work on the reproductive cycle of ring doves set the paradigm for the integrative study of reproductive behavior and physiology in animals.

I. BEHAVIORAL CONTROL OF REPRODUCTIVE PHYSIOLOGY: A COMPARATIVE OVERVIEW

The Nature of Behavioral Control

This paper analyzes the behavioral control of reproductive physiology in animals. Like the other articles on behavioral and social determinants of reproductive behavior in this volume, it treats social or behavioral events as the cause and reproductive physiology as the effect. This is the psychosomatic paradigm; it adds a dimension to the articles included elsewhere in the volume in which the physiological mechanisms controlling behavior are treated as the cause and behavior as the effect.

This article has two sections; this first part describes various behaviorally controlled reproductive processes, and the examples are drawn from a number of species. The purpose of this exercise is not just to catalogue some reproductive exotica, but rather to indicate the rich variety of behavioral interactions with physiological organization. The rest of the paper will deal specifically

[1] The preparation of this article and the research in the author's laboratory reported therein were supported by NIH grant HD-04522.

with behavioral and social control of reproduction in the rat.
This division was intended to support the somewhat polemical statement that in every mammalian species, and possibly in every animal
species, behavioral or psychological variables critically affect
some stage of reproductive physiology. To facilitate the discussion, Figure 1 presents a schematic diagram, a biological "time-
line," of the reproductive life-cycle in a "typical" female mammal.
The stages of the cycle (the center or time-line of the diagram)
are: ontogenetic development, adult sexual cyclicity, pregnancy,
and maternity. The endogenous factors, genetic influences acting
through neural and endocrine structures, are shown below the stages
of the cycle; the exogenous factors influencing reproduction are
indicated above the cycle.

The Development of Sexuality

The first epoch in the reproductive life of the organism is
the ontogenetic period in which the differentiation of the sexual
type (male or female) occurs. Several investigators have shown how
neonatal hormones affect this type of maturation; for example, in
mammals, the presence of androgen early in development predisposes
or "organizes" the adult organism to respond in a male-like fashion
(Clemens, this volume). In many animals, the nature of the early
hormone secretions is determined by the genetic material; presumably,
male mammals (with an XY chromosome complement) perinatally secrete

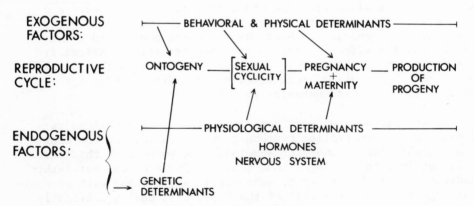

Fig. 1. The major stages in the reproductive life of a female mammal: endogenous events are shown under the time-line representing
the reproductive cycle; exogenous events are shown above.

enough androgen to masculinize the behavior and morphology of the adult. In some animals, however, the sex typing process also depends on the kind of social stimulation to which the animal is exposed (Robertson, 1972). For example, in the labrid fish, *Labroides dimidiatus*, the basic social unit is defined by the male territory which includes one mature male and a harem of some three to six females. In this group, the largest, oldest individual is the male, which dominates all of the females in the group; the larger, older females, in turn, dominate the smaller ones. If the male dies or is removed, sex reversal of the dominant female frequently occurs. Within a few hours after the dominant male's departure, this female begins to show male-like behavior (aggression). After two to four days, "she" begins to perform male courtship and spawning behavior. Within 14 to 18 days, the behavioral sex reversal sometimes extends to physiological functioning: "she" apparently can now release sperm. In this species, therefore, sex is not an invariable correlate of chromosomal complement; the presence or absence of the male determines the sex of the other animals.

In the coral reef fish, the maleness or femaleness of each fish can be determined by social stimulation. In other animals, the same holds true for the quantitative aspect of this development, that is, how quickly sexual development occurs. For example, the onset of puberty in mice is advanced by 20 days if the immature female is caged with an adult male (Vandenbergh, 1967). The underlying mechanism seems to be pheromonal, for puberty can also be slightly accelerated by housing females on bedding soiled by males (Vandenbergh, 1969).

The Ovarian Cycle

Once puberty has been reached, the adult sexual cyclicity of females begins. The ovarian cycles of mammals have several sequential phases: 1) an initial *follicular phase*, during which the ovarian follicles grow; 2) *ovulation*, when the mature follicle releases the egg; 3) the *luteal phase*, when the ruptured follicle is transformed into the corpus luteum, which in turn secretes progesterone. In different species, stimulation derived from behavior plays an important role in many, if not all, of these stages.

One of the most striking examples of behavioral control in the estrous cycle is seen in some species (e.g., cat, rabbit, ferret, mink, some species of mice, and perhaps tree shrews) in which ovulation is not spontaneous but depends upon stimuli delivered by the male during courtship and copulation (Rowlands, *passim*, 1966). Such behaviorally derived stimuli can also potentially influence ovulation even in species which normally ovulate spontaneously (Aron et al., 1966). For example, in female rats displaying normal estrous cycles, copulatory stimulation can advance the time when ovulation

will occur and can sometimes increase the number of eggs released
(Rogers, 1971; Rogers and Schwartz, 1972). Behavioral stimuli
exert an even stronger effect under some experimental conditions.
In female rats, for example, ovulation can be experimentally inhibited by placing these animals under conditions of constant light
(Brown-Grant et al., 1973; Hoffman, 1973) or by injecting a number
of pharmacological agents during a critical period of proestrus
(Everett and Sawyer, 1950). Female rats thus treated become "induced ovulators": they do not ovulate spontaneously but will do so
if they receive enough copulatory stimulation (Adler, 1973). Although the female rat is normally a "spontaneous ovulator," she
possesses the neuroanatomical and neuroendocrinal machinery for
induced ovulation.

In the mammalian ovarian cycle, once successful ovulation has
occurred, either of two sequences can follow. A female can go
through the luteal phase of the cycle and return to the start of
another estrous cycle, or she can become fertilized and enter the
prolonged luteal phase characteristic of pregnancy. (This option
of repeating the estrous cycle is the reason for putting brackets
around the stage labelled Sexual Cyclicity in Figure 1.) In the
females of some genera (e.g., primates), the spontaneous luteal
phase of each ovarian cycle is long; and there is enough progesterone to permit uterine implantation of a fertilized egg. If no
egg implants, the uterus, which has been developing under the influence of progesterone, sheds its inner lining, the endometrium, and
another cycle begins. (An apocryphal story attributes to William
Osler the statement that menstruation is the weeping and wailing of
a disappointed endometrium.) In the rat, the hamster, and several
species of mice, however, there is no spontaneous luteal phase of
any functional consequence (Alder, 1973); in these organisms, some
aspect of the male's copulation triggers the progestational state
underlying pregnancy. This process will be described in detail in
the second section of this article, which presents experiments on
reproduction in rats.

Pregnancy: the Gametes

The secretion of steroid hormones such as progesterone is, of
course, only one aspect of pregnancy. Another constellation of
events is also necessary, including sperm transport, fertilization,
and ovum development. These events are also affected by stimuli
derived from behavior. Some examples of such effects in rats will
be discussed in detail in the second section of the article. I will
confine myself here to an unusual case, a species of South American molly, *Mollinesia formosa*, which consists almost entirely of
females, each daughter being the genetic equivalent of her mother.
The females reproduce by a peculiar form of parthenogenesis, inas-
much as they do copulate with males, but males of another species

(Haskins et al., 1960; Hubbs, 1964; Kallman, 1962). If the female is prevented from copulating, she does not produce offspring. Copulation stimulates the eggs to develop, but there is no genetic union between sperm and egg. The male's sexual contribution is merely to trigger the gynogenetic development of *M. formosa*'s unfertilized eggs.

Maternity

The final stage of the reproductive sequence is, of course, maternity; the production and release of milk to the developing young is an essential component of mammalian reproduction. Lactation in the rat and rabbit have been extensively studied. In these species, both the production of milk (under the control of prolactin) and the reflexive delivery of milk to the suckling pup depend on neuroendocrine reflexes stimulated by the young (Rosenblatt and Lehrman, 1963; Grosvenor et al., 1970). Even before active maternity, the development of mammary gland activity in the rat is under the control of stimuli derived from behavior; in this case, the behavior is the female rat's self-licking. During pregnancy, a female prevented from licking her nipples shows reduced mammary development (Roth and Rosenblatt, 1968). After parturition, the control passes to the pups, which maintain the mother's capacity for lactation by the stimulation she derives from their suckling.

An almost universal law of biological organization is that any excitatory influence is accompanied by an inhibitory one. The maternal condition illustrates this nicely. While the young mammal is nursing, the hormonal condition induced by the suckling stimulus, by blocking the surge of luteinizing hormone, prevents the occurrence of ovulation and therefore of estrous cyclicity (Young, 1961). Marsupials carry such control of the reproductive cycle to extremes not seen in Eutherian mammals. The young red kangaroo, *Megaleia rufa*, is born in an immature state; it makes its way into the pouch and attaches to a nipple. The mother has already mated, ovulated, and become fertilized. The suckling of the pouch joey not only produces milk for itself but also sends the younger sibling (still a blastocyst) into embryonic diapause. For the duration of the older sibling's pouch life, which lasts hundreds of days, the embryo's growth remains suspended and does not resume development until the older sibling leaves the pouch (Sharman et al., 1966).

Triggers and Pumps: the Nature of Stimulus Control

Each of the reproductive changes described in the previous paragraphs illustrates the psychosomatic paradigm: behavior produces stimulation which alters the physiological status of the animal. Such behaviorally derived stimuli affect physiology in two

ways. The first is by *modulating* some ongoing physiological process, e.g., the advancement of puberty in female mice by the presence of a male-derived odor. This type of stimulus is analogous
to a pump which adds to or subtracts from an already operating
process. The second type of stimulus control, "triggering," is
somewhat stronger, e.g., the triggering of ovulation in some forms
by coital stimulation. Without the presence of a male, female mice
would eventually reach puberty; without copulation, however, the
"average" female rabbit or cat would not ovulate.

In a taxonomy of stimulus control, "trigger" stimuli would
differ from "pump" stimuli in several ways. Trigger stimuli act
rapidly, pump stimuli act over a longer period of time, in a tonic
fashion. Trigger stimuli cause a change of state in the organism
(e.g., ovulation, milk let-down); pump stimuli do not necessarily
change an organism's state. Rather, they can alter the timing or
intensity of a state (e.g., pheromonal stimuli synchronize estrous
cycles in female mice; Parkes and Bruce, 1961) or they can act
tonically to maintain or inhibit reproductive states (e.g., the
pups' suckling maintains milk production in the lactating mother
but inhibits ovulation).

The distinction between triggers and pumps is really the difference between stimuli that release a new response and stimuli that
exert tonic effects on organisms. This common distinction is made
in several areas of research. In the study of pheromones, Wilson
and Bossert (1963) developed the concept of "releaser" and "primer"
pheromones. In animal communication, Schleidt (1974) has developed
a theory of "tonic communication" to distinguish it from the phasic
forms of communication more commonly studied. In several areas of
psychology, the "releasing" and "motivating" effects of stimuli are
distinguished. In all of these cases, the distinction is meant to
point up the two basic ways of coordinating an organism with its
social and physical environment, priming by a pump and releasing
by a trigger.

II. *BEHAVIORAL CONTROL OF REPRODUCTION IN RATS*

The rest of this article will analyze the behavioral and social
control of reproduction in rats (*Rattus norvegicus*). By concentrating on one species, we can examine the multiple relationship between
behavior and physiology in a single system. Instead of evolutionary variety, the emphasis now is on mechanisms which integrate reproductive systems into one functional unit.

Male Behavior as a Stimulus for the Neuroendocrine Reflex

During copulation, male and female rats begin by investigating

each other, particularly in the anogenital region. After this preliminary relationship, the male quickly mounts and dismounts about
10 times. On some of the mounts, intromission occurs; and on the
final intromission, the male ejaculates, depositing sperm and an
enzymatic coagulate called the vaginal plug. After a five-minute
rest, the male resumes copulation (Beach, 1956). The broad functional question to be answered here is: how does the male's copulatory behavior affect the occurrence of a successful pregnancy in
the female? Since one of the most consistent features of this
copulatory pattern is the sequence of multiple intromissions, the
question arises: why does the male rat mount and dismount a number of times before ejaculating? To answer this question, we compared two sets of female rats (Adler, 1969; Wilson et al., 1965).
In a High Intromission Group, females were allowed to copulate with
the male undisturbed and received multiple intromissions followed
by ejaculation and the deposition of the sperm and vaginal plug in
the vagina. To produce a Low Intromission Group, we allowed male
rats to copulate with "stimulus" females. When the male was judged
ready to ejaculate, the stimulus female was removed and the experimental female substituted. If this experimental female received
an ejaculation preceded by no more than two intromissions, she was
classified as a Low Intromission Female. About 20 days after copulation, females in both groups were sacrificed and their uteri
examined for the presence of viable fetuses. Approximately 90% of
the females in the High Intromission Group were pregnant (n = 19);
however, only about 20% of the females in the Low Intromission
Group had viable fetuses (n = 10). Thus multiple intromissions
appear to be necessary for the induction of pregnancy.

How do multiple copulatory intromissions stimulate pregnancy?
Part of the answer is to be found in the organization of the female rat's four- to five-day estrous cycle. Unlike most primates,
the rat does not experience a functional luteal phase; that is,
the mature follicle ovulates but does not secrete enough progesterone to permit uterine implantation of a fertilized egg. If the
female is to become pregnant, some event must trigger the progestational state. Initially we hypothesized that the stimulation derived from multiple intromissions triggers a neuroendocrine reflex
which results in the secretion of progesterone. To test this hypothesis, we compared the number of females in both groups which
failed to show regular four-day estrous cycles (Adler, 1969; Wilson
et al., 1965). Almost 100% of the females in the High Intromission
Group (n = 9) failed to show four-day estrous cycles; whereas only
22% of the females in the Low Intromission Group failed to show cycles (n = 9). Since cessation of behavioral cyclicity is one of
the signs of progestational hormone secretion, we concluded that multiple intromissions stimulate the release of progesterone. Since the
proportion of females in each group that stopped showing behavioral
estrous cycles was the same as the proportion that had developing
pups in the previous experiment, multiple intromissions appeared to

stimulate the induction of pregnancy by stimulating the secretion of gestational hormones.

Several other findings support the hypothesis that the male's copulatory behavior facilitates progesterone secretion. Progesterone can maintain pregnancy in ovariectomized females (Talwalker et al., 1966), and exogenous injections of progesterone partially compensated for the detrimental effects of reduced intromission stimulation (Adler, 1969). Finally, females in a High Intromission Group had significantly more progesterone in their peripheral blood than either females in a Low Intromission Group or females receiving no copulatory stimulation (Adler et al., 1970). Within 24 hours after mating, the progesterone level in females of a High Intromission Group was twice that in females receiving no copulatory stimulation. The initial rise in progesterone may have been effected by a correlated decrease in the secretion of the 20-alpha-hydroxy-progesterone, which has a much weaker potential for maintaining pregnancy.

We also wanted to study the dynamics of this behaviorally initiated neuroendocrine reflex. The first step was to determine what aspects of male intromissions are responsible for initiating pregnancy. Since females in both groups received ejaculations with sperm and vaginal plugs, ejaculation could not account for the induction of the progestational state; on the other hand, ejaculation may be a necessary part of the stimulus. The results of several experiments, however, suggest otherwise. In one study, males were treated with guanethidine sulfate before being placed with females. These males could copulate but could not deposit any semen (sperm or plug material) during the ejaculatory response. Nonetheless, females mated to these males became progestational (Adler, 1969).

In another experiment, female rats were permitted a number of intromissions but no ejaculation (Adler, 1969). The effect of the intromissions on the incidence of the progestational state, as indicated by progressive cessation of behavioral receptivity, is illustrated in Figure 2. This is a "dose-response curve" in which the response is a physiological variable (occurrence of progestational state) and the stimulus is produced by behavior. The figure shows that as the number of intromissions increases, so does the proportion of females that become progestational. The female neuroendocrine system can "store" the effects of successive intromissions. Ejaculation does not seem a necessary component of the stimulus.

These results indicate that the multiple intromissions delivered during copulation provide stimulation which is both necessary and sufficient for the induction of the progestational state in female rats. The intromissions, however, have this effect only if they stimulate the vaginal-cervical area and to a sufficient depth. When the vaginocervical area of female rats is anesthetized before mating, the proportion of females becoming progestational is reduced (Adler, 1969). Related evidence comes from a study in which the rats' vaginas were sewn shut before copulation. The males mounted the females repeatedly (up to 50 or 60 times), but could not intromit; none of these females became progestational (Adler, 1969). Further evidence for the necessity of vaginocervical involvement comes from experiments in which the pelvic nerve, which contains afferent fibers from the vaginocervical area, was cut. Pelvic-neurectomized females did not become progestational after mechanical stimulation of the cervix (Kollar, 1953; Carlsson and DeFeo, 1965; Spies and Niswender, 1971).

The Mechanics of the Neuroendocrine Reflex

These experiments suggested that the necessary stimulus for induction of the progestational state is the intromission of the male penis into the vaginocervical area. Ejaculation is not necessary, but deep penetration is. Starting from the description of the necessary stimulus (activity in the pelvic nerve during copulatory intromissions) and some knowledge of the response (heightened progesterone secretion and decreased 20-alpha-hydroxy-progesterone release), we are trying to fill in the connecting links of this neuroendocrine reflex.

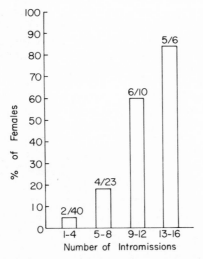

Fig. 2. Number of female rats showing cessation of behavioral receptivity and therefore, by inference, secretion of progesterone after different numbers of intromissions (from Adler, 1969).

Since the anatomical work mentioned above showed that cutting the pelvic nerve prevented the progestational response, J. Hutchison, B. Komisaruk, and I decided that an electrophysiological examination of the sensory fields of genital nerves would provide additional information on the neural organization underlying this neuroendocrine reflex. We recorded multiunit activity in the pelvic, pudendal, and genitofemoral nerves (Komisaruk et al., 1972). By stimulating various regions of the genital area, we located the sensory fields of each of these nerves. For example, the sensory field of the pudendal nerve extends from the base of the clitoral sheath to the base of the tail in the midline and laterally along the inner surface of the thigh. Responses were also elicited from the posterior outer surface of the thigh.

Except for the clitoral region, which is covered by short, fine abdominal fur, the abdominal sensory field of the pudendal nerve coincides closely with a triangular patch of the long, coarse yellowish fur of the dorsal body surface. There is a distinct boundary running across the lower abdomen where the coarse fur ends and the fine fur of the abdomen begins; the sensory field of the pudendal nerve rarely extends into this short fur region. Within the most sensitive region of the field, mechanical stimulation of the skin or deflection of as little as a single hair elicited a response.

One of the most striking features of the pudendal nerve response was that the total field size was significantly greater (33%) in castrated females given exogenous estrogen injections than in untreated controls (Komisaruk et al., 1972). A similar increase in field size was found by Kow and Pfaff (1973/74). The functional significance of this hormonal effect on a peripheral nerve is not yet understood, but it may facilitate the orientation of the female pudendum to the penis and thus the occurrence of intromission.

The sensory field of the genitofemoral nerve was complementary to the pudendal field. Except for overlap in the area of the clitoris, genitofemoral activity was stimulated from regions rostral to the pudendal field. Unlike those in the pudendal nerve, the units in the genitofemoral did not adapt rapidly. The sensory field of the pelvic nerve was the vagina, cervix, and rectum. Stimulation of these structures evoked a response in the nerve, but stimulation of the external perineum did not. The external sensory fields of the pudendal and genitofemoral nerves may be involved in the elicitation and orientation of lordosis (Komisaruk, this volume; Kow et al., this volume); whereas the pelvic nerve, responsive to internal stimulation, is needed to initiate the progestational response.

We are now in a position to trace the effects of the copulatory stimulus through the peripheral nervous system and into the central mechanism of the neuroendocrine reflex which results in progesterone secretion. As an initial strategy, we manipulated the parameters of the copulatory stimulus and correlated these manipulations with the percentage of females showing the progestational response. The purpose of the manipulations was to determine the storage capacities of this neuroendocrine system. As Figure 2 indicates, the greater proportion of females becoming progestational after an increased amount of intromittive stimulation indicates some sort of summation. Since each intromission lasts only 250 milliseconds (Bermant et al., 1969), the brief stimulation from each intromission must be stored and added to that from the succeeding ones to result in this cumulative curve.

To determine the limits of the storage capacity of this system, a second experiment was performed in which the rate of stimulation was varied (Edmonds et al., 1972). For a given group of females, both the number of intromissions permitted (two, five, or ten) and the interval between intromissions varied; the intromission-interval values ranged from the control rate of ad libitum copulation (approximately 40 seconds up to one hour. The results are summarized in Figure 3.

With 10 intromissions, 100% of the females became progestational. This result is especially striking because the intromissions could be spaced one every half hour without a diminution

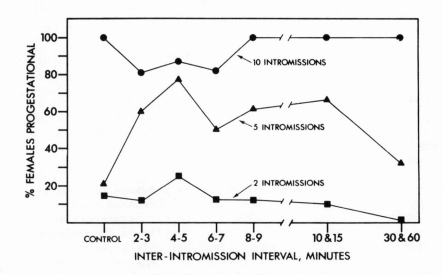

Fig. 3. Effects of different inter-intromission intervals on the induction of progestational hormone secretion in female rats (N = 213; from data in Edmonds et al., 1972).

in their effectiveness. The results for females receiving five
intromissions also indicate that spaced intromissions are as effective as intromissions delivered at the control rate and, in addition, that intromissions at the rate of one every four or five
minutes may be *more* effective in stimulating the progestational
response than the control rate. Ten intromissions at the rate of
one every half hour induced the progestational state in almost
100% of the females. One intromission delivered every half hour
provided a density of stimulation of only one part per 7,200. The
potency points to an exquisitely adapted form of neuroendocrine
integration by which species-specific stimuli (the multiple intromissions) are stored by an adaptively specialized neural mechanism.
The way the female rat's neuroendocrine system "translates" copulatory stimulation into an endocrine result is relevant to an understanding not only of neural storage capacities but also of behavioral function. This process suggests that wild rats do not
"need to" copulate at the accelerated rate of domestic animals in
small laboratory cages. I will return to this point later.

The operation of this entire pregnancy system can be concep-
tualized as a biological amplifier in which each stage lengthens
the temporal characteristics of the previous one. The brief penile
insertions lead to a central nervous system event, the duration of
which is undetermined but which may run for several hours or days
(Barraclough and Sawyer, 1959). Pituitary involvement may last 12
days (Pencharz and Long, 1933); the ovary secretes large amounts
of progesterone for 19 days (Morishige et al., 1973). The final
stage, uterine pregnancy, continues for more than 20 days. In this
sequence, the intromissions trigger the tonic secretion of progesterone.

The occurrence of such biological amplifiers points to another
difference between trigger and pump stimuli. For many triggering
stimuli, such as the copulatory stimulation which initiates progesterone secretion, some dimension of the response is amplified
with respect to the stimulus. In the case of the multiple intromissions, the duration of heightened progesterone secretion lasts
much longer than the copulatory stimuli that initiated the secretion in the first place.

There are other reproductive events which demonstrate an extended storage capability. The phenomenon of "delayed pseudopregnancy" is such a case (Everett, 1968); if proper parameters are
used, electrical stimulation of the brain results in the induction
of a progestational state in female rats, but only after a delay of
hours or days. Another example occurs in the hamster, in which
copulatory stimulation facilitates the subsequent parturition more
than 20 days after the stimulus (Diamond, 1972). Although exogenous progesterone injections can trigger the progestational response
and permit a normal pregnancy from artificial insemination,

parturition is not normal unless copulatory stimulation from a
male hamster is provided at the time of insemination. The precise
mechanism by which the stimulus is stored in the female's neuroendocrine system is not yet known precisely in any of these cases,
but phenomenologically they all represent the storage and amplifi-
cation of a triggering stumulus.

Unlike the effects of trigger stimuli, the effects of pump
stimuli are often reduced from the intensity of some dimension of
the pump stimuli; a great amount of stimulation is often needed
to maintain a physiological state. The odor of a male mouse must
be available for several days if it is to advance the date of the
first estrus in developing female mice (Vandenbergh, 1967). Briefly stated, trigger stimuli are amplified; pump stimuli are integrated.

Some Comparative Issues in the Induction of Pregnancy

The male rat's copulatory behavior, an important stimulus in
the social environment of the female rat, initiates a reproductive
event underlying pregnancy, progesterone secretion; and the neuroendocrine reflex just discussed mediates the transduction from one
event to the other (Adler, 1968, 1969, 1973; Dilley and Adler, 1968;
Chester and Zucker, 1970; Wilson et al., 1965). In other systems,
hormones also act as mediators relating the organism's physiological responses to ecologically important stimuli. Like the female
rat, the female mouse (Mus) and hamster experience no functional
luteal phases. In all of these forms, male copulation produces
the stimulation that triggers the secretion of progesterone (McGill,
1970; Diamond, 1968). In the mouse, however, the ejaculatory response seems to be the necessary stimulus; whereas in the hamster,
as in the rat, the preejaculatory behavior apparently is the trigger for the progestational state. Diamond (1970) has proposed the
concept of a "species-specific" vaginal code to emphasize that in
each form a relatively constrained range of stimuli triggers the
progestational state; the optimal stimulus parameters appear to be
possessed by the normal copulatory pattern of the male.

The Control of Sperm Transport in Rats

When first studying the copulatory influences on pregnancy in
rats, we concentrated on the function of multiple intromissions
in stimulating the neuroendocrine reflex that resulted in progesterone secretion and subsequent gestation. Since all of the female
rats in our experiment, in both the High and Low Intromission Groups,
had sperm and plug deposited in their vaginas, it was an easy assumption that sperm transport and fertilization were normal (Adler, 1969).
In one experiment, however, we checked the condition of ova in the
fallopian tubes of the female rats three days after copulation.

Females which had received many preejaculatory intromissions all had developing ova in the blastocyst stage (Fig. 4); whereas females which had received an ejaculation preceded by only one intromission or by none had unfertilized and degenerating ova (Adler, 1968, 1969). Moreover, females receiving two or more intromissions before ejaculation all had sperm in their uteri one hour after copulation (Fig. 4); the others did not (Adler, 1969). Multiple intromissions before ejaculation, therefore, have two functions: they trigger the neuroendocrine reflex which results in progesterone secretion, and they are necessary for the transport of sperm from the vagina through the tightly closed cervix into the uterine lumen. Without a few preejaculatory intromissions, there can be no sperm transport and therefore no fertilization.

A further requirement for sperm transport is the deposition of the vaginal plug (Blandau, 1945) in close juxtaposition to the cervix. Male rats surgically deprived of their seminal vesicles (and thus unable to deposit a vaginal plug) leave no sperm in the uterus after an ejaculation. In a preliminary anatomical investigation of the vaginal plug, Matthews found what might be called a "sperm cup" at the cervical end of the plug (Matthews and Adler, 1974). This cup is a semihollow cylinder extending a few millimeters into the solid plug material; it is rich with sperm and may function as a reservoir for the sperm for transfer into the cervix.

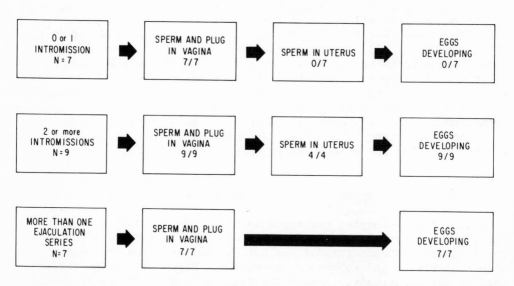

Fig. 4. Effect of different numbers of preejaculatory intromissions on sperm transport and fertilization in female rats (from Adler, 1969).

Given that multiple copulatory intromissions and deposition of a vaginal plug are both necessary for sperm transport, what is the mechanism by which they act? To begin this inquiry, we wanted to know the rate of transport of sperm into the uterus. We already knew that sperm do not pass into the uterus immediately after ejaculation; in a previous study, we had found that approximately eight minutes were required for maximal sperm passage in rats (Adler and Zoloth, 1970). In a series of experiments currently in progress, we sacrificed female rats at various times after they had received an ejaculation in order to determine the rate of sperm transport (Matthews and Adler, 1974). The values we obtained are shown in the following chart:

Minutes after Ejaculation	Average Number of Sperm x 10^5
0–2	6
2–4	155
4–6	215
6–8	458
45–60	459

The relatively long time period (6 to 8 minutes) required for complete sperm transport is compatible with the theory that the pre-ejaculatory intromissions stimulate uterine contractions, which in turn facilitate sperm transport through the cervix. Although this is still a speculation, we are currently examining the way in which the effect of the preejaculatory intromissions on sperm transport is mediated.

Another reason that the 6 to 8 minute transport-time is significant is that it demonstrates the necessity of a "behavioral refractory period." If the female receives cervical stimulation during the eight minutes or so after ejaculation, the transport of the sperm into the uterus is severely reduced from the expected or control values (Table 1). Both the experimenter's manual stimulation of the cervix and the male rat's postejaculatory intromissions reduced the total number of sperm in the uterus one hour after ejaculation (Adler and Zoloth, 1970).

Matthews is currently investigating the mechanism by which postejaculatory intromissions reduce sperm transport. He permitted each female in the experiment to receive a complete ejaculatory series and then in each case allowed a second male to give one intromission within the first eight minutes after that ejaculation. Approximately one hour after the postejaculatory intromission, the number of sperm in the uterine horn was estimated. He found that the

TABLE 1.   *The Effect on Uterine Sperm Content of Genital Stimu-*
           lation of Female Rats after Receipt of an Ejaculation[a]

Time between Ejaculation and Stimulation (min)	Females (no.)	Sperm Count x 10^5 (mean)	Females with Sperm Passage (%)		
			Normal	Reduced	None
Control					
(no stimulation)	14	446	86	14	0
Manual stimulation					
0–15	25	99*	16	40	44
15–30	7	288†	57	28	14
30–45	6	202†	50	33	17
45–60	4	464	100	0	0
Stimulation from five intromissions					
0–15	7	39*	28.5	43	28.5
15–30	4	464	100	0	0
30–45	6	363	100	0	0
45–60	6	341	66	0	33

[a]A given female's sperm count was reduced if the probability of obtaining such a count was < 0.05, given the control group as the
standard; *$p < 0.001$; †$p < 0.01$; †$p < 0.05$ (data from Adler and Zoloth,
1970).

effect of the single postejaculatory intromission was not as great
as those of five such intromissions. However, there was a reduction
in the number of sperm following the single postejaculatory intromission; and it seemed to halt sperm transport at the level attained
at the time the intromission occurred.

Matthews then compared the number of sperm in the uteri of two
groups of female rats. All females in both groups received a single
ejaculatory series. For females in one group, sperm content of the
uterus was estimated at two-minute intervals during the first ten
minutes after ejaculation, as I have already described. Each female
in the other group was given a single intromission within the first
ten minutes after ejaculation, and the number of sperm in her uterus
was estimated an hour later. Females in the first group sacrificed
at 0–2 min after ejaculation had on the average 6 x 10^5 sperm in
their uterus; females in the second group given a single intromission 0–2 min after ejaculation and then dissected an hour after

ejaculation had $\underline{0.6 \times 10^5}$ sperm. For 4-6 min after ejaculation, the corresponding values for the two groups were $\underline{215 \times 10^5}$ and $\underline{316 \times 10^5}$ sperm; and the values for 8-10 min after ejaculation were $\underline{368 \times 10^5}$ and $\underline{395 \times 10^5}$.

After a single postejaculatory intromission, the vaginal plug was still in place, but loosened on its dorsal aspect from its attachment to the cervix. Putting these findings together, we arrive at this tentative scheme describing sperm transport: The multiple preejaculatory intromissions seem to prime the female so that upon ejaculation, sperm can be transferred from the vagina to the uterus. The preejaculatory intromissions alone are not sufficient to ensure sperm transport; the vaginal plug must also be deposited. The close approximation of the plug to the cervix assures an adequate supply of sperm to be transported and may provide cervical stimulation necessary for that transport. The complete transport of the sperm requires approximately six to eight minutes; if during that time further copulatory intromissions cause the penis to strike the vaginal plug and disrupt its close approximation to the cervix, then sperm transport is disrupted.

This potential disruption of sperm transport may have functional significance for the organization of reproduction in this species. Theoretically, one male could "cancel" the effects of another male's copulation. If the second male begins copulating too soon after the previous male has completed an ejaculatory series, his multiple intromissions may prevent the first male's sperm from reaching the uterus, presumably by dislodging the plug. The second male's sperm would then be deposited upon his ejaculation and could fertilize the ova. To test this possibility, we arranged a series of mating tests involving albino female rats (Adler and Zoloth, 1970). An albino female was allowed to mate first with an albino male and was then with a pigmented male for a second ejaculatory series. Paternity could be established soon after the pups are born since pups sired by albino fathers are light colored and have clear eyes, whereas pups of pigmented fathers have dark skins and pigmented eyes. If the substitution of a pigmented for an albino male is made within the first 15 minutes after the initial ejaculation, approximately 60% of the offspring are pigmented (Table 2). If, however, 45 to 60 minutes elapse between copulations, only 23% of the offspring are pigmented. Therefore, a male rat can in large part prevent the insemination of a given female by another male that had copulated previously (Adler and Zoloth, 1970).

Thus, for effective sperm transport, the female rat must have no cervical stimulation for several minutes after receiving an ejaculation. What then prevents a male which has just ejaculated from resuming his intromission during this critical time? We suggest that it is the male's postejaculatory refractory period that prevents him from resuming copulation too soon (Adler and Zoloth,

TABLE 2. *Effect of Different Mating Sequences of Two Male Rats on the Paternity of the Offspring* [a]

Time between Ejaculation and Intromission (min)	Females (no.)	Pups (no.)	Offspring (%) Produced by Ejaculation of Each Male	
			First ♂	Second ♂
Albino ♂ first, pigmented second				
0-15	5	62*	34	66
15-30	5	59	54	46
30-45	4	57†	88	12
45-60	5	60†	77	23

[a] Statistical evaluation by the binomial test; *p< 0.025; †p< 0.001 (from Adler and Zoloth, 1970).

1970). The male rats in our laboratory require an average of 4.5 minutes after ejaculating before they deliver the first intromission of their next ejaculatory series. Since three intromissions on the average are required to dislodge the vaginal plug totally (Lisk, 1969) and since sperm transport is relatively complete within six to eight minutes, the postejaculatory refractory period is of the correct magnitude to permit effective sperm transport into the uterine lumen.

The hypothesized function of the postejaculatory refractory period may have ecological significance for rodent population dynamics. Under conditions of crowding, laboratory rats show persistent social pathology (Calhoun, 1962a). One type of behavioral abnormality is a kind of pansexual behavior in which males mount at a much higher rate than usual. In these crowded colonies, reproduction decreases, partly because of the pathological increase in mounting. Even where crowding is not a problem, colonies of rats with low social rank have more males than those of higher social rank and a reduced number of pups (Calhoun, 1962b). One of the reasons for this abnormal reproductive performance may be copulatory interference with sperm transport.

Reproductive processes are also blocked in other species: the odor of strange males inhibits implantation in female mice (Parkes and Bruce, 1968); sperm are rejected by the bursa copulatrix of Drosophila females during interspecific mating (Dobzhansky et al., 1968); and prolonged auditory stimulation can reduce fertility in rats (Zondek and Tamari, 1967).

The Dual Control of Reproductive Behavior

If the postejaculatory refractory period of the male rat is an integral part of the copulatory process leading to sperm transport and successful pregnancy, its occurance may be the result of an active physiological process of a type that normally ensures a relevant behavioral inhibition. Several experiments now point to the operation of just such an active inhibitory process during the postejaculatory interval, when the male rat displays a pattern of "tonic immobility" (Dewsbury, 1967). Along with this behavior pattern, the hippocampal and cortical EEG display the kind of spindling and slow wave electrical activity (Kurtz and Adler, 1973) that often signals physiological inhibition (Gellhorn, 1967) (Fig. 5). Another feature of the postejaculatory interval which indicates its active nature is the 22 KHz vocalization emitted by the male during his "absolute refractory period." One function of this call may be to prevent the female from physically contacting him (Barfield and Geyer, 1974). On the basis of behavioral and electrophysiological indices like these, Kurtz has developed an opponent-process model to describe the male rat copulatory pattern (Kurtz and Adler, 1973). The performance of the successive intromissions culminating in

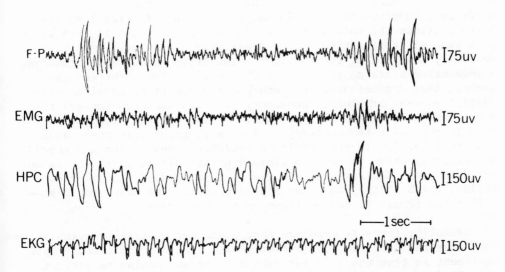

Fig. 5. Rest after ejaculation--Rat 53. Cortical tracing shows high amplitude spindles of 16 Hz, while hippocampal activity is irregular, slow, and high amplitude. Heart rate is 6.5 beats per second. F-P = frontal and parietal cortices; EMG = electromyogram from neck muscle; HPC = dorsal hippocampus; EKG = electrocardiogram.

ejaculation is induced by a positive or appetitive system; the electrophysiological correlate of this appetitive state is theta activity recorded from the hippocampus. The behavioral refractoriness of the postejaculatory interval is under the control of an inhibitory system reflected in the spindles recorded from the cortical EEG. Barfield and Geyer have demonstrated a 0.95 correlation between this inhibitory EEG spindling and the emission of the 22 KHz vocalization.

The control of the male rat's copulatory pattern by linked excitatory and inhibitory mechanisms illustrates the general principle of dual systems controlling autonomic-somatic integration (Gellhorn, 1967). In his investigations of hypothalamic functioning, Hess coined the terms "ergotrophic" and "trophotrophic" control to refer to these systems; but his neologisms applied to sympathetic and parasympathetic activity in the autonomic nervous system (see Gellhorn, 1967). Since somatic activity is also involved, however, "action-oriented" and "vegetative" seem to be preferable terms. In the case of the male rat's sexual behavior, it is adaptive that a vegetative state be induced in the male after he ejaculates. If he were sexually active too soon after his ejaculation, he could disrupt the transport of his own sperm in the female (see preceding section).

A pattern of multiple control is also applicable to the female rat's sexual behavior. The male rat's initial sexual stimulation elicits the lordosis reflex (see the articles of Kow et al. and Komisaruk, this volume, and Diakow, 1974). The elicited lordosis posture allows the male to achieve intromission, which in turn triggers several reproductive events underlying pregnancy. As a kind of secondary adaptation, the cervical stimulation concurrently induces a vegetative state on the part of the female. Komisaruk and co-workers have demonstrated the remarkable property of cervical stimulation whereby it blocks locomotory reflexes and even sensory responses to normally painful stimuli. Since cervical stimulation from intromission is necessary to induce pregnancy, it seems adaptive that the female rat should be inhibited from moving and should be less responsive to certain classes of stimuli while receiving an intromission. Komisaruk even suggests that lordosis is an extensor reflex, which is the classical approach reflex and which would stabilize the female in her position on the substrate.

Finally, there is evidence that prolonged copulatory stimulation can, over the long run, inhibit or reduce female receptivity. This statement is true not only for rats but for an amazing variety of animals including hamsters, guinea pigs, turkeys, lizards, fruit flies, and some cockroaches (Hardy and DeBold, 1972; Carter and Schein, 1971; Goldfoot and Goy, 1970; Schein and Hale, 1965; Crews, 1973; Manning, 1967; Barth, 1968). Sexual behavior in the female is thus a product of opposing systems. Again, in the female rat, the male's copulatory behavior stimulates the lordosis reflex, a classic part of sexual receptivity, at the same time that it inhibits her movement. Thus in the sexual encounter, the female is in a negative state, at least during intromission. Over a longer period, continued copulatory stimulation terminates the receptive "mood" (presumably after the function of copulation has been achieved) and brings the female out of the vegetative state conducive to copulation. The relations between male and female behavior and reproductive physiology are illustrated schematically in Figure 6.

The Copulatory Pattern of Wild Rats

If the copulatory pattern of the male rat is truly adaptive, then some of the features we have discovered in the laboratory should prevail among feral animals. In order to examine the ecological significance of this behavior-physiology interaction and to study mammalian sexual behavior in a more naturalistic framework, Martha McClintock has begun investigating the sexual behavior of wild rats under seminatural conditions. By means of slow-motion videotape recording, she has recorded for periods of 24 hours the sexual behavior of wild and domestic rats in 7 ft x 7 ft cages which contained a maze-like burrow system. The major results of this study are now clear: the wild rat, like the domestic rat, shows a copulatory pattern characterized by multiple (spaced) intromissions, an ejaculation, and a postejaculatory refractory period. How this pattern influences pregnancy remains to be determined.

One of McClintock's major findings is that the rats' sexual behavior is more extended than that of domestic animals in the laboratory. There were more intromissions in the ejaculatory series, and the interval between intromissions and the postejaculatory

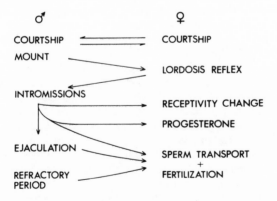

Fig. 6. Schematic representation of the intersection between male and female rats during copulation.

interval were both longer. Sometimes, one series would last an hour
The comparative numbers of intromissions preceding ejaculation are
illustrated in the chart below:

Female

		Domestic	Wild
Male	Domestic	5.3	17.8
	Wild	3.3	13.8

The strain of the female is much more significant in controlling
the number of intromissions than that of the male. As determined by
an analysis of variance, the female strain in this experiment accounted for 58% of the variance; the male strain was not significant, nor was the interaction effect. In the evaluation of
the control of sexual behavior, the particular level of biological
integration being analyzed must be specified. For an individual
pair the question of whether the female or the male is more important in controlling the intromission frequency has little meaning,
even if the specific stimuli involved in copulation can be analyzed
and the qualitative contribution of each animal to the interaction
be determined. At the level of the population, however, the effects
of sex (or strain) can be quantitatively determined and analyzed
statistically.

Summary

My discussion of the reproductive physiology of the female rat
has been based almost entirely on the functions of the male's copulatory behavior. In terms of the time-line (Fig. 1) presented at
the beginning of the article, a female rat progresses from the state
of estrus into pregnancy only if the male's behavior triggers certain critical physiological events. First, his copulation, in particular the multiple intromissions, triggers a neuroendocrine reflex
which results in the secretion of progesterone in sufficient amounts
to permit uterine implantation of the fertilized ova and subsequent
development. Second, these same multiple intromissions are necessary for the transport of sperm from the vagina into the uterus after
ejaculation. Third, the male refractory period after ejaculation is
also necessary for sperm transport since cervical stimulation during
the first few minutes after sperm deposition inhibits transport.
Both the behavioral events themselves and their temporal organization are important. The multiple intromissions before ejaculation
can be separated over time and still induce the progestational state;
and subsequent intromissions cannot follow an ejaculation too quickly
without disturbing sperm transport. It is by means of this deli-
cately coordinated behavioral pattern that reproduction in this

species is adaptively integrated.

ACKNOWLEDGMENTS

The author wishes to thank Drs. H. Gleitman, J. Levy, B. Rosner, P. Rozin, M. Seligman, and R. Solomon for their invaluable discussions and criticism of the material presented here.

REFERENCES

Adler, N.T., 1968, Effects of the male's copulatory behavior in the initiation of pregnancy in the female rat, Anat. Rec. 160:305.

Adler, N.T., 1969, Effects of the male's copulatory behavior in successful pregnancy of the female rat, J. Comp. Physiol. Psychol. 69:613-622.

Adler, N.T., 1973, The biopsychology of hormones and behavior, in Comparative Psychology: A Modern Survey (D. Dewsbury, ed.), Chapter 9, pp. 301-343, McGraw-Hill, New York.

Adler, N.T., and Zoloth, S.R., 1970, Copulatory behavior can inhibit pregnancy in female rats, Science 168:1480-1482.

Adler, N.T., Resko, J.A., and Goy, R.W., 1970, The effect of copulatory behavior on hormonal change in the female rat prior to implantation, Physiol. Behav. 5:1003-1007.

Aron, Cl., Asch, G., and Roos, J., 1966, Triggering of ovulation by coitus in the rat, Int. J. Cytol. 20:139-172.

Barfield, R.J., and Geyer, L.A., 1974, The ultrasonic postejaculatory vocalization and the postejaculatory refractory period of the male rat, J. Comp. Physiol. Psychol. (in press).

Barraclough, C.A., and Sawyer, C.H., 1959, Induction of pseudopregnancy in the rat by reserpine and chloropromazine, Endocrinology 65:563-571.

Barth, Robert H., Jr., 1968, The comparative physiology of reproductive processes in cockroaches. Part I. Mating behaviour and its endocrine control, in Advances in Reproductive Physiology, (A. McLaren, ed.), Vol. 3, pp. 167-207, Logos Press, London.

Beach, F.A., 1956, Characteristics of masculine sex drive, Nebraska Symposium on Motivation, (M. R. Jones, ed.), vol. 4, pp. 1-32, University of Nebraska Press, Lincoln.

Bermant, G., Anderson, T., and Parkinson, S., 1969, Copulation in
    rats. Relations among intromission duration, frequency, and
    pacing, Psychonom. Sci. 17:293-294.

Blandau, R.J., 1945, On factors involved in sperm transport through
    the cervix uteri of the albino rat, Amer. J. Anat. 77:253-272.

Brown-Grant, K., Davidson, J.M., and Greig, F., 1973, Induced ovu-
    lation in albino rats exposed to constant light, J. Endocrin.
    57:7-22.

Calhoun, J.B., 1962a, Population density and social pathology,
    Sci. Amer. 206(2):139.

Calhoun, J.B., 1962b, The Ecology and Sociology of the Norway Rat,
    Public Health Service Pub. No. 1008, U.S. Public Health Serv-
    ice, Bethesda, Md.

Carlson, R.R., and DeFeo, V.J., 1965, Role of the pelvic nerve vs.
    the abdominal sympathetic nerves in the reproductive function
    of the female rat, Endocrinology 77:1014-1022.

Carter, C.S., and Schein, M.W., 1971, Sexual receptivity and ex-
    haustion in the female golden hamster, Horm. Behav. 2:191-200.

Chester, R.V., and Zucker, I., 1970, Influence of male copulatory
    behavior on sperm transport, pregnancy, and pseudopregnancy
    in female rats, Physiol. Behav. 5:35-43.

Crews, David, 1973, Coition-induced inhibition of sexual receptiv-
    ity in female lizards (Anolis carolinensis), Psychol. Behav.
    11:463-468.

Davidson, J.M., Smith, E.R., and Bowers, C.Y., 1974, Effects of
    mating on gonadotropin release in the female rat, J. Endocrinol.
    (in press).

Dewsbury, D.A., 1967, A quantitative description of the behavior of
    rats during copulation, Behaviour XXIX:154-178.

Diakow, C., 1974, Sensory bases for mating behavior in mammals, in
    Advances in the Study of Behavior, vol. V (D.S. Lehrman, R.A.
    Hinde, and J.S. Rosenblatt, eds.), Academic Press, London/New
    York (in press).

Diamond, M., 1968, Induction of pseudopregnancy in the golden ham-
    ster, J. Reprod. Fert. 17:165-168.

Diamond, M., 1970, Intromission pattern and species vaginal code
    in relation to induction of pseudopregnancy, Science, 169:
    995-997.

Diamond, M., 1972, Vaginal stimulation and progesterone in relation to pregnancy, Biol. Reprod. 6:281-287.

Dilley, W.G., and Adler, N.T., 1968, Postcopulatory mammary gland secretion in rats, Proc. Soc. Exp. Biol. Med. 129:964.

Dobzhansky, T., Ehrman, L., and Kastritsis, P.A., 1968, Ethological isolation between sympatric and allopatric species of the Obscura group of Drosophila, Anim. Behav. 16:39-87.

Edmonds, S., Zoloth, S.R., and Adler, N.T., 1972, Storage of copulatory stimulation in the female rat, Physiol. Behav. 8:161-164.

Everett, J.W., 1968, Delayed pseudopregnancy in the rat, a tool for the study of central neural mechanisms in reproduction, in *Perspectives in Reproduction and Sexual Behavior* (M. Diamond, ed.), pp. 25-32, Indiana University Press, Bloomington.

Gellhorn, E., 1967, *Principles of Autonomic-Somatic Integration. Physiological Basis and Psychological and Clinical Implications,* University of Minnesota Press, Minneapolis.

Goldfoot, D.A., and Goy, R.W., 1970, Abbreviation of behavioral estrus in guinea pigs by coital and vagino-cervical stimulation, J. Comp. Physiol. Psychol. 72:426-434.

Grosvenor, C.E., Maiweg, H., and Mena, F., 1970, A study of factors involved in the development of the exteroceptive release of prolactin in the lactating rat, Horm. Behav. 1:111-120.

Hardy, D.F., and DeBold, J.F., 1972, Effects of coital stimulation upon behavior of the female rat, J. Comp. Physiol. Psychol. 78:400-408.

Haskins, C.P., Haskins, E.F., and Hewitt, R.E., 1960, Pseudogamy as an evolutionary factor in the poeciliid fish *Mollienesia formosa*, Evolution 14:473-483.

Hoffman, J. C., 1973, The influence of photoperiods on reproductive functions in female mammals, in *Handbook of Physiology - Endocrinology* (Sect. 7; R.O. Greep and E.B. Atwood, eds.) Vol. II, Pt. 1, pp. 57-77, American Physiological Society, Washington, D.C.

Hubbs, C., 1964, Interactions between a bisexual fish species and its gynogenetic sexual parasite, Bull. Texas Mem. Mus. 8:1-72.

Kallman, K.D., 1962, Population genetics of the gynogenetic teleost *Mollienesia formosa* (Girar), Evolution 16:497-504.

Kollar, E.J., 1953, Reproduction in the female rat after pelvic nerve neurectomy, Anat. Rec. 115:641-658.

Komisaruk, B.R., Adler, N.T., and Hutchison, J.B., 1972, Genital sensory field: Enlargement by estrogen treatment in female rats, Science 178:1295-1298.

Kow, L.-M., and Pfaff, D.W., 1973/74, Effects of estrogen treatment on the size of receptive field and response threshold of pudendal nerve in the female rat, Neuroendocrin. 13:299-313.

Kurtz, R.G., and Adler, N.T., 1973, Electrophysiological correlates of copulatory behavior in the male rat, J. Comp. Physiol. Psychol. 84:225-239.

Lehrman, D.S., 1965, Interaction between internal and external environments in the regulation of the reproductive cycle of the ring dove, in *Sex and Behavior* (F.A. Beach, ed.), pp. 355-381, John Wiley & Sons, New York.

Lisk, R.D., 1969, Cyclic fluctuations in sexual responsiveness in the male rat, J. Exp. Zool. 171:313-326.

Matthews, M., and Adler, N.T., 1974, Copulatory effects on sperm transport (in preparation).

Manning, A., 1967, The control of sexual receptivity in female *Drosophila*, Anim. Behav. 15:239-250.

McGill, T.E., 1970, Induction of luteal activity in female house mice, Horm. Behav. 1:211-222.

Morishige, W.K., Pepe, G., and Rothchild, I., 1973, Serum luteinizing hormone, prolactin and progesterone levels during pregnancy in the rat, Endocrinology 92, 5:1527-1530.

Parkes, A.S., and Bruce H.M., 1961, Olfactory stimuli in mammalian reproduction, Science 134:1049-1054.

Pencharz, R.I., and Long, J.A., 1933, Hypophysectomy in the pregnant rat, Amer. J. Anat. 53:117-139.

Robertson, D.R., 1972, Social control of sex reversal in a coral-reef fish, Science 17:1007-1009.

Rodgers, Charles H., 1971, Influence of copulation on ovulation in the cycling rat, Endocrinology 88:433-436.

Rogers, C.H., and Schwartz, N.B., 1972, Diencephalic regulation of plasma LH, ovulation, and sexual behavior in the rat, Endocrinology 90:461-465.

Rosenblatt, J.S., and Lehrman, D.S., 1963, Maternal behavior of the laboratory rat, in *Maternal Behavior in Mammals* (H.L. Rheingold, ed.), pp. 8-57, John Wiley & Sons, New York.

Roth, L.L., and Rosenblatt, J.S., 1968, Self-licking and mammary development during pregnancy in the rat, J. Endocrinol. 42: 363-378.

Rowlands, I.W. (ed.), 1966, *Comparative Biology of Reproduction in Mammals* (Zool. Soc. London Sym. No. 15), Academic Press, London/New York.

Schein, M.W., and Hale, E.B., 1965, Stimuli eliciting sexual behavior, in *Sex and Behavior* (F.A. Beach, ed.), pp. 440-483, John Wiley & Sons, New York.

Schleidt, W.M., 1974, Tonic communication: continual effects of discrete signs in animal communication systems, J. Theoret. Biol. (in press).

Sharman, G.B., Calaby, J.H., and Poole, W.E., 1966, Patterns of reproduction in female diprotodont marsupials, in *Comparative Biology of Reproduction in Mammals* (Zool. Soc. London Sym. No. 15; I.W. Rowlands, ed.), pp. 205-229, Academic Press, London/New York.

Spies, H.G., and Niswender, G.D., 1971, Levels of prolactin, LH, and FSH in the serum of intact and pelvic-neurectomized rats, Endocrinology 88:937-943.

Talwalker, P.K., Krähenbühl, C., and Desaulles, P.A., 1966, Maintenance of pregnancy in spayed rats with 20-α-hydroxypregn-4-ene-3-one and 20-β-hydroxypregn-4-ene-3-one, Nature (Lond.) 209:86-87.

Vandenbergh, J.G., 1967, Effect of the presence of a male on the sexual maturation of female mice, Endocrinology 81:345-349.

Vandenbergh, J.G., 1969, Male odor accelerates female sexual maturation in mice, Endocrinology 84:658-660.

Wilson, E.O., and Bossert, W.H., 1963, Chemical communication among animals, Rec. Prog. Horm. Res. XIX:673-716.

Wilson, J.R., Adler, N.T., and LeBoeuf, B., 1965, The effects of intromission frequency on successful pregnancy in the female rat, Proc. Nat. Acad. Sci. 53:1392-1395.

Young, W.C., 1961, The hormones and mating behavior, in *Sex and Internal Secretions*, 3rd ed. (W.C. Young, ed.), Vol. 2, pp. 1173-1239, Williams & Wilkins, Baltimore.

Zondek, B., and Tamari, I., 1967, Effects of auditory stimuli on
    reproduction, in *Effects of External Stimuli on Reproduction*
    (Ciba Found. Study Group No. 26; G.E.W. Wolstenholme and
    M. O'Conner, eds.), pp. 4-16, Little, Brown, Boston.

MALE DOMINANCE AND AGGRESSION IN JAPANESE MACAQUE REPRODUCTION[1]

G. Gray Eaton

Oregon Regional Primate Research Center, Beaverton

Dominant males have been reported to mate more frequently than lower-ranking males in multimale troops of baboons (DeVore, 1965), rhesus monkeys (Carpenter, 1942; Kaufman, 1965), and Japanese macaques (Tokuda, 1961-2; Hanby et al., 1971). Such behavior would appear to have selective advantages for fostering genetic traits that contribute to the phenotype of high-ranking males, e.g., perhaps a low threshold for instituting attack, a high tolerance of pain, and a high social intelligence. However, Rowell (1966) has argued that dominance in caged baboons is a social construct that is maintained by the behavior of lower-ranking animals and that female primate sexual receptivity is not necessarily related to ovulation. Therefore, she concluded it would be difficult for traits that contribute to dominance to be selected through reproductive success.

Nevertheless, with some primate species it is possible to analyze how particular types of aggressive interactions contribute to dominance and to correlate that dominance with patterns of reproductive behavior. DeVore, for example, has performed such an analysis with free-ranging troops of savanna baboons in South Africa (*Papio ursinus*) and Kenya (*Papio anubis*), (Devore, 1965; Hall and DeVore, 1965). Unfortunately, these animals did not arrange themselves in neat linear order. On the contrary, two or three males formed a dominant central hierarchy by cooperating during aggressive interactions with other males. The males in the central hierarchy then tended to copulate more frequently than those outside even

[1]Publication No. 667 of the Oregon Regional Primate Research Center

287

though the central males did not necessarily outrank the other males in individual conflict situations.

On the other hand, Saayman (1970) did not report a correlation between the dominance rank and mating success of adult male chacma baboons (P. ursinus) studied in South Africa. Although these copulated more frequently than subadult males, Saayman observed that fighting between males for females was rare. He reported that one of the three adult males in the troop, a toothless old male, twice lost his consort to the same more physically dominant adult male without overt fighting.

In two artificially formed troops of rhesus monkeys that had been transplanted from India to Cayo Santiago, Puerto Rico, Carpenter (1942) reported a high correlation between male dominance rank and the frequency of copulation. All the males were ranked on an aggressiveness score, but neither the criteria for dominance nor the data were presented in this pioneering field study of primate sexual behavior. In a subsequent study of the same macaques, Kaufman (1965) also reported a positive correlation between male dominance and the amount of breeding activity. Dominance rank was determined by displacement at food and water sources and the exchange of aggressive and submissive postures and calls, but again neither the criteria for dominance nor the actual data were presented.

For a number of years, Tokuda (1961-2) periodically studied the Koshima Island troop of Japanese macaques and reported male rank to be positively correlated with the frequency of copulatory behavior. He, too, failed to report how rank was determined; but he did observe that with few exceptions males occupying the geographic center of the troop were high ranking and did most of the mating during the breeding season.

In a subsequent study of a confined troop of Japanese macaques, Hanby et al. (1971) reported that high-ranking males were shown to ejaculate more frequently than middle- or low-ranking males when the data were grouped into these three categories. On the other hand, the differences between these groupings were not significant when the mean number of mount sequences, mounts, days active, or number of female partners was compared. How the males were divided into the three dominance categories was not reported, but these workers took the individual ranking of the males from another study that had arranged every troop member in a linear order matrix through the frequency of successful unidirectional attacks (Alexander and Bowers, 1969).

As part of a longitudinal study of reproductive behavior, frequency counts of aggressive and sexual interactions have been made in a confined troop of Japanese macaques (Eaton, 1971, 1972, 1973). This paper provides additional information on how male

aggression, dominance, and sexual behavior are related by analyzing
these data collected during an annual breeding season of the Oregon
troop of Japanese macaques.

The troop was captured near Hiroshima in 1964 and now inhabits
an 8000 m^2 (two-acre) grassy corral at the Oregon Regional Primate
Research Center. Confinement has not markedly affected patterns of
social (Alexander and Bowers, 1967), aggressive (Alexander and Roth,
1971), or sexual behavior (Hanby et al., 1971). The macaques are
not handled except for an annual two- to three-day roundup for
general husbandry purposes sometime after the March to July birth
season. The troop is observed for the first three hours after sun-
rise, five days a week, by two or three trained observers who record
on tape or checklist the frequency of the behaviors defined below.
The data reported here were collected during an annual breeding
season (8-16-71 to 3-2-72), which was defined as the period between
the first and last observed ejaculations. During this period there
were 21 adult males (4 years and older), 36 adult female (3 years
and older), 26 subadult males, and 24 subadult females.

Four patterns of aggressive behavior were recorded (cf. Alexander and Roth, 1971) in terms of increasing severity: 1) *threat*,
lunging toward another animal, gaping or earflattening, or emitting
a "woof" vocalization; 2) *chase*: pursuit accompanied by threats;
3) *punish*: biting of brief duration that did not injure, and/or
pinning to the ground, striking, leaping on, or pulling fur or body
parts; 4) *assault*: biting that injured or was prolonged and accompanied by vigorous head movement while the teeth were gripping the
victim. If more than one type of behavior occurred during a single
sequence, only the most severe was scored, e.g., if one animal
chased, caught, and punished another, only a *punish* was scored. With
each incident of aggression, one of three forms of response was also
recorded: 1) *submission*: moving away, grimacing, cringing, or
crouching; 2) *stand-off*: submission not apparent or submissive gestures interspersed with aggressive actions; 3) *ignore*: no aggressive
or submissive behavior in response to the attack. The criteria used
to form the males' dominance rank involved only aggressive interactions that were unambiguously followed by *submission* (cf. Sade,
1967). Cases in which more than one attacker was involved were also
ignored in this analysis.

The dominance hierarchy was formed by constructing a matrix of
all the adult males of the troop (N = 21) and by entering the frequency of each animal's attacks met by submission on each of the
others. The ranks were then renumbered so as to obtain a minimum
number of reversals of attack and submission behaviors between
individuals in the linear order. Ties in position in the new hierarchy were broken by giving precedence first to the male in the tied
pair that had more frequently attacked the other male, then if a pair
was still tied, to the male that had been attacked by a fewer number

of other males, and finally, if the tie persisted, to the male that had attacked the greater number of other males.

The basic units of sexual behavior recorded were adult hetero-sexual mount series, which were defined as two or more mounts in less than five minutes by the same pair of animals. Mounts were single or double foot clasps with thrusts. The occurrence of intro-mission could not be ascertained. Each series was assigned at the time of the first mount to one of four categories of initiation: 1) the male approached from more than 10 to 15 m or had followed the female; 2) the female approached from more than 10 to 15 m or had followed the male; 3) both approached or had walked together or were seated less than 5 m apart without other animals nearby; 4) initia-tion unknown. After a series was initiated, one of six terminations was recorded: 1) ejaculation, recognized by a pause in thrusting and the rigid posture of the male or by semen on the female's perineum; 2) male movement at least 10 to 15 m away and failure to return to mount within five minutes; 3) female movement at least 10 to 15 m away and failure to return to be mounted within five minutes; 4) separation of both partners for more than 10 to 15 m and failure to return to mount within five minutes; 5) failure of a mount to occur within five minutes even though the pair remained less than 10 to 15 m apart; 6) termination unknown (either observations ceased while a pair was still mounting, or a pair that had been mounting was observed to have separated).

The dominance hierarchy of the 21 adult males, compiled from 646 attacks in which submissive behavior was observed, is illustrated in Table 1. The sums across the bottom row are the number of males that attacked the male that heads each column minus those males that had been more frequently attacked by the male in question; e.g., although 17 males attacked 1721, one of them (2191) was attacked an equal number of times by 1721 and so was not included in the sum-mation for 1721. The sums in the right hand column are the number of males that were attacked by the male at the left of each row minus those that had more frequently attacked the male in question; e.g., 1721 attacked six males, but was given credit for only two (1764 and 1735) because the other four had attacked him more fre-quently than he had attacked them.

Of the 646 attacks in Table 1, 560 were *chases*, 49 were *punishes*, 18 were *threats*, and 14 were *assaults*. Most of the reversals in Table 1 were from *chases*. There were no reversals involved with *assaults*, only one with *punishes* (1721 punished 1768), and four with *threats*.

The sexual behavior of the males is given in Tables 2 and 3. Within each table, the males are ranked according to the hierarchy shown in Table 1. Table 2 summarizes the mount series terminations for each of the six categories discussed earlier. Each male was

TABLE 1. *Dominance Rank of Adult Male Japanese Macaques Compiled from 646 Attacks in which Submissive Behavior was Observed*[a]

Dominance Rank	Male Number	Field Mark	△	∧	▽	B	3	E	α	ς	ʟ	I	h	d	π	F	ϕ	ᴺ	ʌ	7	9	V	X	No. of Males Attacked[b]
1	1723	△		8	2	0	10	3	3	0	1	0	0	0	2	0	8	0	1	1	1	1	2	13
2	1751	∧			7	5	3	2	1	0	2	3	15	5	6	0	4	2	4	1	3	3	1	17
3	1728	▽				9	9	1	1	0	4	0	16	10	5	1	20	3	5	0	1	12	3	15
4	1766	B					5	6	1	6	1	6	6	4	5	3	5	3	3	2	2	13	2	16
5	1734	3						9	1	2	1	12	12	5	6	2	11	0	0	1	4	13	0	13
6	2378	E							1	4	1	0	1	1	1	0	0	1	0	1	1	6	1	9
7	1725	α								0	1	1	1	3	3	0	0	1	1	1	1	6	0	8
8	3223	ς									1	21	2	1	0	0	0	0	0	0	0	1	0	4
9	1719	ʟ										1	1	1	1	3	1	1	9	0	1	13	2	9
10	3302	I											1	1	3	0	2	0	0	0	1	3	0	5
11	2191	h						2						3	6	1	10	1	8	3	2	9	1	10
12	1768	d									1				6	1	2	0	2	3	0	3	1	7
13	1756	π														1	4	2	1	5	3	3	0	5
14	1736	F								1							0	0	0	5	3	6	1	4
15	1741	ϕ									5							8	5	1	0	6	5	5
16	1735	ᴺ								1									1	0	0	0	0	1
17	1716	ʌ									1						1			6	3	33	6	4
18	1733	7													1				1		1	1	1	4
19	1753	9																				4	3	2
20	1721	V							2				3	1	2					2	1		4	2
21	1764	X									1											1	1	2
Attacked by No. of males[c]			0	1	2	2	4	5	5	6	7	4	9	8	11	7	10	10	11	10	13	16	14	

[a] Entries below diagonal line indicate reversals of dominance rank.
[b] Minus those males that had more frequently attacked the male in question.
[c] Minus those males that had been more frequently attacked by the male in question.

TABLE 2.   *Male Dominance Rank and Breeding Season Totals of the*
Frequency of Mount Series Terminations in a Confined
Troop of Japanese Macaques [a]

		Frequency of Mount Series Terminations					
Male Number	Dominance Rank	Ejaculations	Male Leaves	Female Leaves	Both Leave	Pauses	Unknowns
1723	1	67	9	3	0	9	4
1751	2	40	18	4	1	66	14
1728	3	10	16	3	2	2	0
1766	4	10	2	3	0	1	3
1734	5	47	29	12	2	24	12
2378	6	1	0	0	0	0	1
1725	7	38	7	4	1	17	7
3223	8	4	0	1	1	4	0
1719	9	10	21	2	3	7	2
3302	10	4	5	1	0	0	0
2191	11	25	11	3	0	24	4
1768	12	17	6	1	2	17	2
1756	13	49	11	4	1	47	12
1736	14	3	9	3	2	7	1
1741	15	34	17	8	2	52	8
1735	16	48	36	14	5	83	11
1716	17	19	7	4	0	33	3
1733	18	16	4	1	0	3	2
1753	19	1	0	0	0	0	0
1721	20	110	52	13	3	76	17
1764	21	4	6	5	1	4	0

[a]See text for definitions of dominance rank, breeding seasons and
mount series terminations.

ranked in each of the categories, and a Spearman rank order correlation coefficient was then computed between dominance rank and rank
within each category (Siegel, 1956). Male 1723, for example, was
ranked Number 1 in the dominance hierarchy and Number 2 in frequency
of ejaculation; whereas male 1721 was ranked Number 20 in the dominance hierarchy and Number 1 in frequency of ejaculation. The
resulting correlation between dominance rank and frequency of ejaculation was low and not significant ($r = 0.097$, $N = 21$, $p > .05$), like
all the correlations between dominance rank and other terminations
including the total series frequency given in Table 3.

To analyze further the relation between dominance rank and
sexual behavior in Table 2, each individual male's score on each of

TABLE 3. *Male Dominance Rank and Breeding Season Totals of the*
Frequency of Mount Series Initiations in a Confined Troop
of Japanese Macaques[a]

Male Number	Dominance Rank	Frequency of Mount Series Initiations				
		Male	Female	Both	Unknown	Total
1723	1	42	10	20	20	92
1751	2	23	8	87	25	143
1728	3	19	1	3	10	33
1766	4	6	2	6	5	19
1734	5	34	11	46	35	126
2378	6	1	0	0	1	2
1725	7	11	8	33	22	74
3223	8	1	3	3	3	10
1719	9	26	0	8	11	45
3302	10	3	1	2	4	10
2191	11	11	5	32	19	67
1768	12	6	4	22	13	45
1756	13	8	11	72	33	124
1736	14	11	0	5	9	25
1741	15	15	7	75	24	121
1735	16	33	8	99	57	197
1716	17	3	3	41	19	66
1733	18	5	0	10	11	26
1753	19	0	1	0	0	1
1721	20	55	25	128	63	271
1764	21	11	2	2	5	20

[a] See text for definitions of dominance rank, breeding season and
mount series initiations.

the terminations was expressed as a percent of his own total mount
series; e.g., for male 1721 who had 110 ejaculations out of 271
mount series, 40.6 percent of his mount series ended in ejaculation.
Therefore, although he was ranked first in frequency of ejaculation,
he was ranked seventh in percent of series ending with ejaculation.
However, even when ranked in this manner, none of the sexual behavior
categories were significantly correlated with dominance rank.

    The mount series initiations given in Table 3 were analyzed
in the same manner as the terminations in Table 2. Dominance rank
did not significantly correlate with sexual behavior that was categorized by how a mount series was initiated. Again it made no
difference whether each male's frequency of initiations counts or

the percentage of his own total of mount series initiations were
ranked and correlated with dominance.

The lack of correlation between male dominance rank and sexual
behavior in this troop of Japanese macaques contrasts with some of
the previously discussed reports of this phenomenon in other baboon
and macaque troops. Several explanations are possible, ranging
from species differences, through cultural differences within
species, to different interpretations of dominance rank by research-
ers. Moreover, because the most dominant males tend to occupy the
geographical center of baboon and macaque troops, their behavior
is more likely to be recorded than that of lower ranking males on
the periphery of the troop.

The fact that in the wild lower-ranking males do live on the
periphery of troops may be relevant. In the Oregon macaque troop
the fence restricts the outward movement of the low-ranking males,
and they therefore have access to all the females most of the time
since the concentric distribution of the troop occurs only during
feeding when the animals are quite hungry. However, the fact that
three years earlier Hanby et al. (1971) had observed in this same
troop a correlation between groups of dominant males and their frequency of ejaculation argues that there may be cultural differences
between troops and that such patterns of behavior can change over
time within a troop.

It is, of course, possible that the dominant males actually do
most of the inseminating and the low-ranking males waste their
gametes before and after conception has taken place. However,
without a reliable test for paternity, it is difficult to test this
hypothesis. There is some variation in reported gestation periods
for Japanese macaques (*Macaca fuscata*), e.g., Hanby et al. (1971),
165 days; Tanaka et al. (1970), 180 days; Tokuda (1961-2), 150 to
170 days; so it would appear futile to attempt to determine paternity by counting back from birthdate. However, a preliminary statistical analysis involving the males that copulated with each of
the females before she gave birth suggests that high-ranking males
did not inseminate proportionately more females than did lower-
ranked males. A more complete analysis is in progress and will be
the subject of a later report.

Although the dominance hierarchy shown in Table 1 is not
absolutely linear, the percentage of reversals in which a lower-
ranked male more frequently attacked a higher-ranked male is very
small (22 out of 646, or 3%). Informal observations of position
displacement (without overt aggressive behavior) during feeding have
repeatedly validated the five top-ranked males' positions in the
hierarchy and those of many of the lower-ranked males. Males ranked
2 through 5 could be classified as "subleaders" (Kawai, 1965; Koyama,
1967) since they tended to dominate most of the adult females of the

troop, which in turn dominated the remainder of the adult males. The top five males also "punished" and "assaulted" more frequently than other males. However, other subdivisions such as "middle" or "low" ranking were not obvious from the data or from general observations of the troop.

Some of the high-ranking males did harass other males by chasing away the females they were mounting. However, as Hanby et al. (1971) reported, this behavior did not appear to keep the pair apart; and as shown in Table 2, neither "male," "female," or "both" terminations correlated inversely with dominance rank, a situation that would have been the case if dominant males significantly interrupted the mating activities of lower-ranking males.

Popular writings on the evolution of human behavior have made it fashionable to characterize human beings as "naked" pongidae and have emphasized dominance and aggression when reviewing studies of nonhuman primates. However, many species of nonhuman primates, including those most recently separated from human ancestral stock, show relatively little aggression that is related to dominance or sexual behavior (chimpanzee: Goodall, 1968; gorilla: Schaller, 1963). Moreover, the data presented in this paper indicate that in this troop of Japanese macaques there is little or no relation between male dominance rank and mating behavior. Furthermore, the two contrary facts—that such a relation does exist in some troops of baboons and macaques under some conditions and that it did not exist in this troop at this time—serve to emphasize the plasticity of individual and group behavior that characterizes all primates.

ACKNOWLEDGMENTS

This research was supported in part by Grants HD–05969, RR–05694, and RR–00163 from the National Institutes of Health, United States Public Health Service. I thank Jens Jensen for writing the computer programs used in the analyses and Kurt Modahl and Eileen Eaton for their technical assistance.

REFERENCES

Alexander, B.K., and Bowers, J.M., 1969, Social organization of a troop of Japanese monkeys in a two-acre enclosure, Folia Primatol. 10:230-242.

Alexander, B.K., and Bowers, J.M., 1967, The social structure of the Oregon troop of Japanese macaques, Primates 8:333-340.

Alexander, B.K., and Roth, E.M., 1971, The effects of acute crowding on aggressive behavior of Japanese monkeys, Behaviour 39:73-90.

Carpenter, C.R., 1942, Sexual behavior of free ranging rhesus monkeys (*Macaca mulatta*); II. Periodicity of estrus, homosexual, auto-erotic and non-conformist behavior, J. Comp. Psychol. 33:143-162.

DeVore, I., 1965, Male dominance and mating behavior in baboons, in *Sex and Behavior* (F.A. Beach, ed.), pp. 266-289, John Wiley & Sons, New York.

Eaton, G.G., 1973, Social and endocrine determinants of simian and prosimian female sexual behavior, in *Primate Reproductive Behavior* (vol. 2., Symp. Proc. IVth Int. Cong. Primat., Portland, Ore., 1972, C.H. Phoenix, ed.), pp. 20-35, S. Karger, Basel.

Eaton, G.G., 1972, Seasonal sexual behavior and intrauterine contraceptive devices in a confined troop of Japanese macaques, Horm. Behav. 3:133-142.

Eaton, G.G., 1971, The Oregon troop of Japanese macaques, Primate News 9:4-9.

Goodall, J.V.L., 1968, The behaviour of free-living chimpanzees in the Gombe Stream Reserve, Anim. Behav. Monogr. 1:161-311.

Hall, K.R.L., and DeVore, I., 1965, Baboon social behavior, in *Primate Behavior: Field Studies of Monkeys and Apes* (I. DeVore, ed.), pp. 53-111, Holt, Rinehart and Winston, New York.

Hanby, J.P., Robertson, L.T., and Phoenix, C.H., 1971, The sexual behavior of a confined troop of Japanese macaques, Folia Primatol. 16:123-143.

Kaufman, J.H., 1965, A three-year study of mating behavior in a free-ranging band of rhesus monkeys, Ecology 46:500-512.

Kawai, M., 1965, On the system of social ranks in a natural troop of Japanese monkeys: II. Ranking order as observed among the monkeys on and near the test box, in *Japanese Monkeys: A Collection of Translations* (K. Imanishi and S.A. Altmann, eds.), pp. 87-104, S.A. Altmann, Edmonton.

Koyama, N., 1967, On dominance rank and kinship of a wild Japanese monkey troop in Arashiyama, Primates 8:189-216.

Rowell, T.E., 1966, Hierarchy in the organization of a captive baboon group, Anim. Behav. 14:430-443.

Saayman, G.S., 1970, The menstrual cycle and sexual behavior in a troop of free ranging Chacma baboons (*Papio ursinus*), Folia

Primatol. 12:81.

Sade, D.S., 1967, Determinants of dominance in a group of free-
    ranging rhesus monkeys, in *Social Communication Among Primates*.
    (S.A. Altmann, ed.), pp. 99-114, University Chicago Press,
    Chicago.

Schaller, G.B., 1963, *The Mountain Gorilla: Ecology and Behavior*,
    University Chicago Press, Chicago.

Siegel, S., 1956, *Nonparametric Statistics*, McGraw-Hill, New York.

Tanaka, T., Tokuda, K., and Kotera, S., 1970, Effects of infant
    loss on the interbirth interval of Japanese monkeys, Primates
    11:113-117.

Tokuda, K., 1961-2, A study on the sexual behavior in the Japanese
    monkey troop, Primates 3:1-40.

THE REPRODUCTIVE BEHAVIOR OF MINORITY GROUPS IN THE U.S.A.

Harley L. Browning

Population Research Center

The University of Texas at Austin

In any consideration of the reproductive behavior of minority groups, the first question that naturally comes to mind is what do we mean by a minority group. Since women increasingly are identifying themselves as a minority, one could state that all reproduction is by members of a minority group. Rather than being dismissed as facetious, this statement merits serious consideration. It seems to me quite evident that reproduction in this and other countries would be appreciably different if women enjoyed the same status as men, within as well as outside the context of the family. Indeed, as Bumpass (1973) in attempting to account for the recent decline in fertility has stated "...motherhood has been the last major vestige of ascribed status in a modern industrial society. A primary consequence of complete fertility control is to place motherhood more squarely in competition with other social roles." Norman Ryder (1973) also has repeatedly made much the same point. Although I believe that women can be considered as a minority group within the context of the study of reproductive behavior, such an inclusive definition is not appropriate for this paper.

In defining a minority group, I do not propose to go into all the problems and the many sociological ramifications. For my purposes, a minority group is a relatively small population within a country which is identified, both by its members and by others, as a distinct social group and which has experienced discrimination of one kind or another. The case of Catholics is a difficult one. They have experienced discrimination, but because of size (Catholics constitute 26% or more of the total population), varied origins (Irish, Italian, etc.), and group identity and cohesiveness, I prefer to exclude them from further consideration.

On the basis of the above definition, we can identify as minorities blacks, Mexican Americans, Puerto Ricans, and native Americans, along with Jews, Japanese Americans, and Mormons. (I exclude
Appalachian whites because of doubts about their self-definition
as a minority and because of the nature of the discrimination directed against them.) What is striking about this listing is that
it includes groups with quite different socioeconomic statuses (SES)
as well as different fertility behavior. Cross-classifying SES
by fertility, one finds the following distribution:

Socioeconomic Status

	+	**−**
+ Fertility	Mormons	Blacks, Mexican Americans, Puerto Ricans, Native Americans
−	Jews, Japanese Americans	

Only one cell, low SES with low fertility, is vacant because I
know of no minority group of any appreciable size in the U.S.A.
with these characteristics. (Hippies might qualify; but the status
of members of the group is often transitory, and there are doubts
about their low SES.) The combination of high socioeconomic status
and low fertility of the Jews has been widely known, but a similar
pattern holds for Japanese Americans (Uhlenberg, 1972). Perhaps
most interesting is the situation of the Mormons. Despite the
lack of abundant data to document the assertion, they appear to be
a rare instance of a group that has managed to maintain high levels
of fertility and at the same time reach high levels of education
and income. Blacks, Mexican Americans, Puerto Ricans, and native
Americans are all characterized by high fertility and particularly
low socioeconomic levels.

Thus we find minority-group status associated with both low and
high levels of fertility. I do not propose to provide a theoretical framework that will adequately account for all types of
minority fertility, a problem that has preoccupied Goldscheider
(1971). Indeed, I shall confine my discussion to just two minority
groups of low socioeconomic status and high fertility, blacks and
Mexican Americans, with some comparison with the Anglo majority.
After a brief consideration of the fertility characteristics of
the two groups, I will discuss whether there is an "ethnic" component in their fertility behavior that is distinguishable from
socioeconomic factors. I will also consider within-group fertility

differences and how to account for them. Because of my greater
familiarity with Mexican Americans, most of my attention is devoted
to this group. And since much of my concern will be with the
problematics of "ethnicity" as a factor in differential fertility
and with accounting for within-group fertility differentials, my
presentation will be frankly speculative, concentrating more on
problem-stating rather than on problem-solving.

BLACK FERTILITY

Any discussion of black fertility and its change over time must
rely upon the authoritative work of Reynolds Farley. In his book,
Growth of the Black Population (1970), and in his long article
prepared for the Commission on Population Growth and the American
Future, "Fertility and Mortality Trends Among Blacks in the United
States" (1972), he has provided comprehensive historical and contemporary coverage.

Two main conclusions are warranted from his historical analysis.
First, black fertility, as measured for the period 1850-1970, has
been substantially higher than that of whites, except for a period
during the late 1920's and '30's when black and white rates closely
converged and for a time black fertility dipped even lower than
that of whites. Second, the direction of movement for blacks and
whites was quite similar over more than 120 years. Both groups
had a long decline, accentuated by the depression, followed by a
baby boom. Currently, both groups are experiencing marked fertility declines which began in the late '50's. In other words,
despite the quite different socioeconomic circumstances of blacks
and whites, they still seem to have responded in much the same way
to the broader socioeconomic forces affecting fertility.

Perhaps the parallelism is most strikingly demonstrated by the
data on changes in birth expectations for whites and Negroes during
the short period 1965-1972 (Table 1). The patterns of change for
each of the four age groups are quite similar. Within the span of
only seven years, there is a remarkable drop of 27% for white wives
aged 25-29 and 29% for Negro wives. Negro wives in each age group
still expect more children than white wives, but for the 18-24 age
group the difference has diminished to only one-tenth of a child.

Norman Ryder (1973) casts further light on these differentials
and their changes by his analysis of wanted and unwanted births,
drawing on the 1965-1970 National Fertility Surveys. Black fertility was 40% greater than white fertility between 1961 and 1965,
but from 1966 to 1970 the difference declined to only 10%. The
main reason for this narrowing was the ability of black wives to
reduce substantially their unwanted births. Unwanted births for
white wives declined from 0.16 in the 1961-1965 period to 0.09

TABLE 1.   Total Births Expected by Wives Age 18 to 39 Years*

Respondents	Age in Years			
	18-24	25-29	30-34	35-39
White				
1965	3.1	3.4	3.6	3.4
1972	2.3	2.5	2.9	3.2
Difference, 1965 to 1972	-.8	-.9	-.7	-.2
% Change	-26%	-27%	-19%	-6%
Negro				
1965	3.4	4.0	4.4	4.1
1972	2.4	2.8	3.7	4.0
Difference, 1965 to 1972	-1.0	-1.2	-.7	-.1
% Change	-29%	-29%	-16%	-2%

*Source: Birth Expectations and Fertility: June, 1972 (Current Population Reports, Series P-20, No. 248, April, 1973)

in the 1966-1970 period; the comparable figures for black wives were 0.49 to 0.20. Thus, nearly three-quarters of the black fertility decline was due to greater control over unwanted births.

This remarkable short-term decline was made possible by much easier access to and effective use of contraceptives. Ryder shows that whereas only 15% of the black married women were using either the pill or the IUD in 1965 (compared with 29% of whites), by 1970 this figure had jumped to 40%, above the 38% of white women.

In a sense this decline in fertility should not have been unexpected, for in earlier national surveys, the 1960 survey reported by Pascal K. Whelpton et al. (1966) and the 1965 study of Ryder and Westoff (1971), black, currently married women consistently reported that they *desired* fewer children than did their white counterparts (except for those with elementary educations) but that they *expected* to have more (except for black women with college educations) than white married women. Evidently the widespread availability and usage of the "new" contraceptives have made it possible to approximate more closely the desired family size.

Parenthetically, the above figures demonstrating that married black women want both to restrict their births and are increasingly effective in doing so are in striking contrast with the statements of black nationalist leaders who endorse a rapidly growing black population and who see any diminution in the present rate of growth as "genocide." Evidently they are having about as much effect on the black population as Catholic leaders who urge Catholics to abstain from all forms of contraception except the rhythm method. It is also unlikely that pronatalist statements by Chicano leaders are having any noticeable effect on Mexican-American fertility.

Table 2 shows the latest available information on differentials for whites and Negroes by two age groups, 15 to 24 and 25 to 34. In some instances, the differentials by age are narrowing (particularly in the South and nonmetropolitan areas); and in others they are widening somewhat (as in those with high school education, partial or complete).

MEXICAN AMERICANS AS A MINORITY

When we turn from a consideration of black fertility to that of Mexican Americans, we are confronted by a dearth of information and considerable difficulties in even defining the group. Parenthetically, the U.S. Bureau of the Census has wrestled with this problem for many decades. (See Hernandez, Estrada, and Alvirez, 1973, for an excellent review of the definitional problem.) Mexican Americans are not to be confused with the Spanish-speaking or the Spanish-heritage population of the U.S.A. The latter was reported at 9.6 million in the 1970 census. The *1973 Manpower Report of the President* refers to these people as "the country's second largest minority group" (page 85); but it is questionable whether those with a Spanish-speaking heritage form a true sociological group since they include persons of Mexican, Puerto Rican, and Cuban birth or parentage, located in different areas of the country: Mexicans in five Southwestern states, Puerto Ricans in three Mid-Atlantic states, and Cubans in Florida. Table 3 compares these three populations with whites and Negroes. The Spanish-speaking populations differ as much among themselves as they do in comparisons with white and black populations.

Bradshaw and Bean (1972) assembled fragmentary data on Mexican-American reproductive behavior for the last 120 years. In the area of San Antonio, Texas, in 1850, the fertility of married women with Spanish surnames and spouse present was a third higher than for other women. They conclude that Mexican-American fertility, like black fertility, has recently followed the national trends. In a more recent article (1973), these authors conclude that " . . . these data provide little evidence to support the thesis that the fertility levels of the two populations have substantially converged

*TABLE 2.   Children Ever Born per 1,000 Women, Ages 15 to 25 and 25
to 34 Years, White and Negro (June 1972)**

	Ages 15 to 24		Ages 25 to 34	
	White	Negro	White	Negro
Region				
Total U.S.	339	391	1,971	2,287
South	401	412	1,930	2,557
Metropolitan	306	386	1,871	2,179
Non-Metropolitan	411	406	2,205	2,754
Years of School				
Less than 8	578		2,660	
		341		3,487
8	317		2,642	
High School: 1-3	288	355	2,733	2,840
4	474	519	2,005	2,002
College: 1-3	200		1,620	
		220		1,537
4 +	105		1,092	
Work Status				
In Labor Force	190	361	1,465	1,979
Not in Labor Force	518	418	2,372	2,763

*Source: Birth Expectations and Fertility: June, 1972 (Current
Population Reports, Series P-20, No. 248, April 1973)

TABLE 3. *Selected Characteristics by Ethnic Status for Five Minority Groups in the United States 1970**

Characteristics	White	Negro	Mexican American	Puerto Rican	Cuban
Percent of families with female head	9.0	27.4	12.8	26.7	9.6
Percent with own children under 6 years	25.5	30.9	40.3	44.2	27.0
Number of children born per 1,000 women age 35 to 44	2,888	3,489	3,895	3,269	2,206
Percent of wives working some time during year	50.0	63.3	38.9	33.5	49.9
Percent of women working with children under six	28.4	47.6	29.8	16.6	38.6
Median family income (1971)	10,672	6,440	7,486	6,185	9,371

*Source: *Manpower Report of the President*, Chapter 4, 1973; data, except for family income, are taken from the 1970 census.

from 1950 to 1970..." In fact, as they point out, the 1969 age-adjusted fertility measures for women of Mexican origin and the 1970 measures for Spanish-speaking or Spanish-surname women are nearly the same as those for the 1950 Spanish-surname population. They also note that whereas Mexican-American fertility generally paralleled Anglo fertility between 1950 and 1970, "period fluctuations in the former have not been sharp as in the latter." They then speculate whether the relative insensitivity of Mexican-

American fertility to socioeconomic conditions is due mainly to
discriminatory practices or "to cultural orientations in the population supportive of having large families."

As far as Bradshaw and Bean can determine, socioeconomic indicators such as education, occupation, and income are related to
fertility in these women in much the same way as they are in white
and black populations

Other factors more closely tied to reproduction may help to
explain the higher fertility among Mexican-American than among
Anglo women. For one thing, Mexican-American women marry at an
earlier age than do Anglo women; however, the difference between
their medians is only one year. This is offset to some extent by
the fact that fewer Mexican-American women get married or have their
spouses present than Anglo women. In the five south-western states
in 1969, only 65.7% of the women of Mexican origin age 15 to 44 were
ever married compared with 74.3% for Anglo women. Only 55.3% of
Mexican women age 15 to 44 were married and had their spouse
present; whereas the corresponding figure for Anglo women was 65.4%.
Bradshaw and Bean estimate that the total number of children born
to Mexican-American women is roughly 7% less than if they had
"accessibility" to husbands in the same proportion as Anglo women.

Unfortunately, there are no surveys of the total Mexican-
American population comparable to those of white and black populations reported in the National Fertility Surveys, so I must draw
upon the 1969 Austin Family and Fertility Survey of 348 couples.
Even though it is a restricted sample, many of the findings of this
survey probably have wider applicability to the total Mexican-
American population. (1) About 24% of first births were premari-
tally conceived, a figure not much different from that for the
whole United States. (2) The mean length of first-birth intervals
for postmaritally conceived first births was 16.9 months for the
Austin Mexican-American women compared with 19.2 months for white
women and 18.0 for black women. It was anticipated that once
married Mexican-American women would have their first births
a good deal sooner than white women, but the 2.3 month difference
is not large. (3) Although more than 90% of the Austin sample
reported themselves as Catholic, about 86% had used some type of
contraception, the pill much more than the rhythm method. Indeed,
by 1969 65% of the Austin women had tried the pill. (4) If one
divides the Austin Mexican-American women into three marital cohorts, it is evident that the more recent cohorts were more likely
to use contraceptives, and to use them more effectively. As a
whole, however, the Austin women were not very successful contra-
ceptors.

After their review of the specific practices affecting the
fertility of the Mexican-American sample, Bradshaw and Bean (1972)

concluded: "The practices and circumstances examined, whether considered separately or together, do not offer a satisfactory explanation of the differences between Mexican-American fertility and
fertility of other white women."

THE INDEPENDENT EFFECT OF ETHNICITY ON DIFFERENTIAL FERTILITY

To what extent can the fertility behavior of a minority group
and its variance with that of the white majority of the U.S.A. be
interpreted as a consequence of ethnic characteristics, i.e.,
their values, belief systems, and behavioral patterns, rather than
of their differential socioeconomic position? Before getting into
a general discussion of this question, let us briefly consider
some of the results of empirical studies.

After reviewing the evidence on the relationship between education and fertility for five groups (whites, nonwhites, Japanese
Americans, Jews, and Roman Catholics), Goldscheider and Uhlenberg
(1969) stated: "... under given social and economic changes and
concomitant acculturation, the insecurities and marginality associated with minority group status exert an independent effect on
fertility..." Continuing along these lines, Sly (1970) addressed
the same problem in examining white-nonwhite fertility differentials with 1960 census data. On the basis of an analysis-of-
variance design, he concluded that regional and socioeconomic
characteristics were important, but not ethnicity. He argued for
the distinction between "structural" (education, occupation, and
income) and "cultural" (norms, values, beliefs, and life-styles)
assimilation.

Roberts and Lee (1973) raised objections to the procedures
used in both of the above surveys (e.g., failure to control for
factors such as age at first marriage, employment status of women,
and the validity of "nonwhite" to identify a minority group). They
based their own analysis on the 1960 1-in-100 Public Use Sample for
three groups in the five south-western states: 1) persons with
Spanish surnames, 2) other whites (Anglos), and 3) Negroes. Using
an analysis-of-variance design, they found that cumulative fertility measures (ever-married women age 40 to 49) "provide very strong
support for the independent effects argument. In every comparison
ethnicity exerts a powerful effect independent of income, education, occupation, and residence." In a current fertility analysis
(ever-married women age 20 to 29), ethnicity is the only variable
that has statistical significance.

Using the same data set for the same year and for the same
three ethnic groups in the Southwest, Bean (1973) concentrated on
the relationship between income and fertility. His regression
analysis of the data revealed "considerable variation by ethnicity

in the processes of family formation." He finds that "Income as
here measured bears a positive, though not exceptionally strong,
relation to fertility for Anglos and a negative relation for Mexican Americans and Blacks." But what is perhaps more interesting,
Bean finds differences between Mexican Americans and blacks. Using
the concept of "relative" income, which is the ratio of actual to
"estimated" income (the latter derived from the husband's occupation, education, and age), he finds relative income effects for
Anglos and blacks, but not for Mexican Americans. He suggests
that the lack of this effect for the latter group is related to
reference group theory. "The economic attainment of Mexican Americans can be relatively low and still be higher than would have
been likely if they had lived in Mexico." For blacks, an excess
of actual-over-estimated income "may well indicate extraordinary
efforts to overcome social discrimination. Such compensatory
behavior is likely to be facilitated by lower fertility, a possibility consistent with the negative relationship between relative
income and fertility among blacks."

Whatever the numerous technical problems inherent in distinguishing between "structural" and "cultural" factors, I believe
the various studies reviewed above provide support for the assertion that ethnicity has an independent effect. As Goldscheider
(1971) notes, however, "The isolation of minority group status as
an independent factor in fertility is only a starting point. The
significant yet unresolved issues are how and under what conditions
minority group status depresses fertility and what dimensions of
minority group identification affect fertility." His concern with
what "depresses" fertility is peculiar to his concern with Jewish
fertility, but the statement can be applied equally to the problem
of what elevates minority fertility.

Ryder (1973), who recently posed the problem in what is probably the broadest possible context, the demographic transition,
should be quoted at some length because his statement is forceful,
pithy, and unequivocable.

A single framework encompasses most theories about
the sources of differences in fertility. The transition from high to low fertility experienced by all
modern countries is customarily explained by the
transformation of the social structure implicit in
industrialization, in brief, the movement from a
traditional to a rational way of life. The same is
true of fertility differences. The more the process
of modernization has affected a group, the lower its
fertility. Modernization enhances the importance of
urban industry relative to rural agriculture, and the
importance of the individual relative to the family.
The child is transformed from a source of labor in the

family enterprise to a consumer of education. Modern
urban life releases the woman from bondage to the home.
This process occurs sequentially through the social
ranks, with those of higher status being the first
to adopt the new ways and profit from them. The application of rationality in the reproductive sphere is
highly correlated with its application elsewhere; each
in turn depends on the extent to which there is confidence in future rewards for present effort.

Fertility can be expected in subpopulations where
tradition and religion are important, because of their
emphasis on the family vis-a-vis the individual, the
father vis-a-vis the child, the husband vis-a-vis the
wife, and the spiritual vis-a-vis the material world.
These conditions are most likely in rural society.
Yet it would be unwise to stress too much the economic exploitation of children and women in the pre-
modern world. Children and family are alternatives
to material success both as sources of satisfaction
and as primitive social security systems.

Except for the committed Catholic, the high-
fertility subpopulations have been the Southern blacks,
the Southwest Mexican Americans, the Appalachian whites,
the Puerto Ricans, and the American Indians. All of
these groups are poor, uneducated, and rural in origin.
All belong to separate, isolated, highly visible
minorities and all are excluded from the dominant
culture and its resources, and thus from full participation in the process of modernization that produces
low fertility. They lack the skills necessary to
succeed in many areas of life, including the prevention
of unwanted children. Perhaps a life of uncertainty
and futility, reinforced by that other-worldly orientation that is the traditional consolation of the destitute
and deprived, makes planning for the future seem a
pointless exercise. Perhaps also the exclusion from
the dominant subculture is accompanied by the development
of subcultural norms that help reduce the cost of minority membership in the short run, but at the cost of
reinforcing the boundaries and limitations imposed by
the majority culture....

The view of fertility differences presented here
is essentially time-bound. The essence of the position
is that groups that still have higher-than-average
fertility have not yet become modern. The inference
is that, as they come to participate more fully and
equally in modern life, and as the secular values of

primacy of the individual--particularly the child and
woman--permeate the entire social system, their fertility will also decline....

Central to Ryder's interpretation is the "modernization"
process, whereby in the process of economic development people move
from "a traditional to a rational way of life." As I understand
him, economic development creates the conditions that provide for
structural assimilation, with the shift from manual to nonmanual
positions and the general upgrading of the labor force making for
more "modern" positions. But modernization for Ryder also has a
prominent value and belief system component. Basically, this
statement means a shift from a family-and-kin-centered orientation
to a situation where "the secular values of primacy of the indi-
vidual--particularly the child and the woman--permeate the entire
social system." Ryder suggests that minority groups who are
"excluded from the dominant culture and its resources" usually
develop "subcultural norms that help reduce the cost of minority
membership," and I interpret this to mean positive sanctions for
large families. Linking this idea to the Goldscheider formulation,
we can say that those groups most desirous of being assimilated
and becoming modern will translate the "insecurities of minority
group membership" into lower-than-average fertility, whereas those
who are "excluded" will translate their "insecurities" into higher-
than-average fertility.

Ryder understandably is reluctant to include the high-fertility
"committed Catholic" with the low-status high-fertility subpopulations. Without departing from his basic framework, he shifts his
emphasis somewhat in accounting for the "desire for relatively
more children, and of acquiescence to restrictions on effective
fertility regulation" on the part of committed Catholics to their
"emphasis on family solidarity as the bedrock of the social order."

Ryder's formulation is probably the most general one that
can be made, linking as it does minority fertility to the demographic transition; and in the main it does catch the essence of
fertility change within the last hundred years or so. However,
some aspects make it less suitable or helpful in accounting for
the fertility of specific minority groups and how it changes over
time. Basically, the concept of the modernization process is a
linear model of change, that is, groups move from a more traditional to a more modern mode of existence over time. As the members of a minority group overcome various restrictions on participation in the larger society, the fertility differential between
them and the dominant group will be reduced.

Consider this formulation from the historical perspective of
American blacks. Looking at the trend of black and white fertility

in the U.S.A. since 1850 (setting aside measurement problems), we
find that the movements of the two groups generally have been
parallel. However, the gap between the two has widened and
narrowed at various points. The widest gap about 1880 was followed
by a narrowing until the 1920's and early '30's, when black fertility dropped briefly below that of whites. Then the gap widened
in the '50's, only to narrow once again the the late '60's and
early '70's.

The pronounced fertility fluctuations themselves, as well as
the substantial changes in the magnitude of black-white fertility
differences, do not seem to fit the modernization model very well.
Did blacks become more modern from 1880 to 1930 and then become
more traditional during the 1950's? Or did they ever have a
"traditional" culture in the U.S.A., in the sense of the long-term,
largely autochthonous development that is implicit in most discussions of "traditional" societies? The conditions of slavery
and the attitudes and practices of the slave owners regarding
reproduction can scarcely be called a "traditional" orientation.
Partly because of its sweeping generalizations, Ryder's formulation gives the impression of a rather monolithic adherence either
to the traditional or to the modern pattern. He does, however,
note that modernization "occurs sequentially through the social
ranks, with those of higher status being the first to adopt the
new ways and profit from them." Goldscheider sees the problem
somewhat differently (1971): "...the quality of minority group
cohesion and integration becomes a key axis of fertility hetero-
geneity *within* the minority group. In this sense differentials
within subgroups must be interpreted. The degree of minority
group integration and the accentuated marginal position between
acculturation and structural separation will determine the behavior
patterns of minority members, all other things considered, vis-a-
vis the majority community."

FERTILITY DIFFERENCES WITHIN A MEXICAN-AMERICAN POPULATION

In minority group fertility analysis, most of the attention
has been devoted to the extent of between-group differences, in
this instance between Mexican Americans and Anglos, rather than of
within-group differences. As a result, certain minorities are
characterized as high-fertility groups; and sufficient attention
has not been given to the variation in fertility levels within the
group. Guided perhaps by orientations that assume relatively
uniform socioeconomic conditions and/or a high degree of adherence
to cultural norms, writers have often stopped with the group
average and failed to appreciate the diversity within it.

I want to pursue this point by examining the situation of
Mexican Americans. I am not sufficiently acquainted with the
situation of blacks, particularly their cultural circumstances, to

deal with them. Because Mexican Americans originate from a society
with high fertility and are over whelmingly Catholic in their religious preference, it is easy to generalize in stereotypical
statements about their value system. But to examine the differences within this group in terms of family size preferences would
seem to be much more worthwhile. To do this, let us again draw
upon the Austin Family and Fertility Survey, the study of 348
couples carried out in 1969.

The Austin survey is far from ideal for our purposes, since
it is of only one community and even then is not a representative
sample of Austin. We took only couples married at least three
years, with spouse present and the wife no older than 35. The
sample was drawn to overrepresent those with higher education. In
one-third of the couples, the husband had completed 12 or more
years of schooling, a high level of attainment for Texas. Because
of this bias, the sample underestimates the average family size
of the population as a whole.

Let us first consider the matter of religion. Ryder, you
will recall, says that "high fertility can be expected in subpopulations where tradition and religion are important..." David
Alvirez (1972) drew upon the Austin survey to determine the relation between religion (specifically Catholicism, for more than 90%
of the sample reported themselves Catholics) and fertility. He
constructed two models of religion, an Institutional Model and a
Religiosity Model. The first was based on attachment to the
Catholic Church as such: marriage in the Church, attendance at
Mass, and reception of Communion. The second, which is often used
by sociologists to determine the more personal commitment of the
individual to religion, included a self-appraisal of the individual's religious status: praying at home, making a petition to a
saint or the Virgin Mary, etc.

After several tests of these models, Alvirez found at best
only a very weak relationship between both models and fertility.
The association was a bit stronger for women than for men as is, I
think, understandable. The interesting thing is that these
findings contrast with those reported by Westoff and associates
and by Ryder and Westoff (1971) and based on national data. In
their book based on the 1965 survey, Ryder and Westoff say,
"Various indices of religiousness among Catholics, including both
measures of formal associational involvement and community involvement, reveal the expected direct relationship with fertility."

What is there to distinguish the Mexican-American Catholics
from the other American Catholics? A number of factors may provide an explanation:

(1) Whereas in the North, the Catholic religious structure has

been dominated by European, specifically Irish, elements, in the Southwest there has been more Latin and "folk" Catholicism.

(2) There is relatively less opportunity for Mexican-American Catholics to attend Catholic schools, and this has lessened the amount of indoctrination they have received on the teachings of the Church, specifically in regard to birth control. As far as we can tell, most Mexican Americans, especially those of lower socioeconomic position, do not interpret their adoption of the pill or other chemical or mechanical contraceptives as really affecting their status as "good" Catholics.

What this leads up to is the thesis that Mexican Americans as a group have high fertility not so much because they are Catholic but because they are Mexican-American in culture and orientation. This does not mean that no elements of Catholicism have penetrated the general culture of the group. On the contrary, the Church has been particularly important in reinforcing values which favor the role of motherhood and the positive regard for large families. But it is very difficult to extricate the specifically Catholic doctrines and practices from the general Mexican-American culture that favors large families. Thus, whether one is a Mexican-American Catholic or a Protestant has little predictable consequence for fertility differentials. David Alvirez (1973) divided the Austin sample into three groups: both spouses Catholic; one Catholic and one Protestant; and both Protestant. There was no significant variation in wanted completed family size among the three groups.

Finally, the orientation that sees the general culture as fostering large families and the intertwining of "religion" and "tradition" can be examined in terms of Mexican data. Since the Mexican-American culture has much of its roots in Mexico, the hypothesis that Catholic practices there would have a weak relationship to fertility differences should be examined. In a study of the fertility of women in Mexico City, Benitez Zenteno (1970) found that attendance at Mass, which would be one indicator of attachment to religion, had almost no relationship to the mean number of children. He concludes:

> One would expect that the highest attendance at religious services would be associated with a greater number of live-born children, given the position of the Church toward the use of contraceptive methods and/or the limitation of the family. However, according to the sample, this is not so and generally among the women who attend such services the number of children they have had is independent of attendance at religious services.

Thus the evidence from Mexico is also consistent with the other bits and pieces of data that we have pulled together to try to help explain the high fertility pattern of Mexican Americans.

Returning to the consideration of the differences in wanted family size within the Austin sample, the investigators obtained the number of children wanted by each of these married couples by adding the number of children born up to the time of the interview, but ignoring mortality, to the number of additional children wanted by the respondent husbands and wives. (In a few cases this procedure meant subtraction, for the respondents reported they already had more children than they wanted.)

The 796 respondents wanted a mean of 4.2 children, i.e. about one child more than was wanted by the white population of the U.S.A. for periods before the data of survey in 1969. But what is the distribution behind this mean? Classifying the response into three groups (the fact that none of the respondents wanted no children probably reflected the nature of the sample) produced the following distribution:

No. of Children Wanted	Percent of Sample	Mean No. of Children Wanted
Small (1 or 2)	15	1.9
Medium (3 or 4)	52	3.7
Large (5 or more)	33	6.5

Thus a small percentage (15%) of the total number of Mexican Americans sampled wanted small families. At the other extreme, a third wanted large families (five or more children can be rated large by any criterion). Thus by U.S. standards, there is considerable variation in wanted family size, even within a high-fertility group.

At this time, the reasons for the choices of small or large families by these respondents cannot be explained. Neither of the two extreme hypotheses proposed warrants much support. The first, the ethnic homogeneity hypothesis, assumes that the value orientations and ethnic identification of Mexican Americans are sufficiently similar to make any differences in wanted family size attributable to chance factors. (Even though the Austin Survey included a large number of items designed to tap the ethnicity variable, no clear factor has emerged.) The second and opposite hypothesis assumes that family size reflects a fundamental cleavage within the Mexican-American population; i.e., those with large-family preferences differ systematically from respondents with small-family preferences on a wide scale of indices: income,

parental background, language preference, and spouse interaction.
In the Austin sample, some indices were predictably related, others
were related inversely to predictions.

In any event, whatever the difficulty of satisfactorily explaining the differences between the large- and small-family choice
groups, it is important that those who direct family planning programs be aware of this variation and take it into account in trying
to "reach" members of this population. Clearly, the most suscep-
tible couples to such efforts are those who desire a small family;
they should, therefore, be the most willing and motivated to
accept family planning assistance. The large intermediate group
wanting three or four children are also candidates for assistance
since some of them are likely to overrun their preferred family
size. The last group, probably at least a third of Mexican-American
couples, cannot be expected to respond favorably to overtures from
family planning clinics. These couples are having large families
because they *want* large families, and in a democratic society
their desires should be respected.

Over a period of time, however, the relative importance of
these three groups may change. Indeed, the dramatic decline in
wanted family size among younger women has manifested itself among
Mexican-American women. But since in a society moving toward a
stationary state a family of three or four children must be labeled
a large family, for the foreseeable future the Mexican-American
population will continue to have a substantial proportion of couples
with large families. Meanwhile, social scientists must try to
arrive at a better understanding of the nature of minority fertility
than they now possess.

REFERENCES

Alvirez, D., 1973, The effects of formal church affiliation and
    religiosity on the fertility patterns of Mexican-American
    Catholics, Demography 10:19-36.

Bean, F.D., 1973, Ethnic variations in the income-fertility rela-
    tionship (unpub. paper presented at the annual meeting of the
    Population Association of America, New Orleans, April 26-28,
    1973).

Benitez Zenteno, R., 1970, Fecundidad, in *Dinamica de la poblacion
    de Mexico,* publication of the Centro de Estuidios Economicos
    y Demograficos, El Colegio de Mexico, Mexico, D.F.

Bumpass, L.L., 1973, Is low fertility here to stay?, Family Plan-
    ing Perspectives 5:67-69.

Bradshaw, B.S., and Bean, F.D., 1972, Some aspects of the fertility of Mexican-Americans, in *Demographic and Social Aspects of Population Growth,* (C.F. Westoff and R. Parke, Jr., eds.), Vol. I., U.S. Government Printing Office, Washington, D.C.

Bradshaw, B.S., and Bean, F.D., 1973, Trends in the fertility of Mexican Americans, 1950–1970, Soc. Sci. Quart. 53:688–696.

Farley, R., 1970, *Growth of the Black Population,* Markham, Chicago.

Farley, R., 1972, Fertility and mortality trends among Blacks in the United States, in *Demographic and Social Aspects of Population Growth* (C.F. Westoff and R. Parke, Jr. eds.), Vol. I, pp. 111–138, U.S. Government Printing Office, Washington D.C.

Goldscheider, C., 1971, *Population, Modernization, and Social Structure,* Little, Brown and Co., Boston.

Goldscheider, C., and Uhlenberg, P.R., 1969, Minority group status and fertility, American Journal of Sociology 74:361–372.

Hernandez, J., Estrada, L., and Alvirez, D., 1973, Census data and the problem of conceptually defining the Mexican American population, Social Science Quarterly 53:671–687.

Roberts, R.E., and Lee, E.S., 1973, Minority group status and fertility revisited (unpub. paper presented at the annual meeting of the Population Association of America, New Orleans, April 26–28, 1973).

Ryder, N. B., 1973, Recent trends and group differences in fertility, in *Toward the End of Growth: Population in America* (C.F. Westoff, ed.), Prentice-Hall, Englewood Cliffs, N.J.

Ryder, N.B., and Westoff, C.F., 1971, *Reproduction in the United States: 1965,* Princeton University Press, Princeton, N.J.

Sly, D., 1970, Minority group status and fertility: An extension of Goldscheider and Uhlenberg, Amer. J. Sociol. 76:443–459.

Uhlenberg, P.R., 1972, Demographic correlates of group achievement: contrasting patterns of Mexican-Americans and Japanese-Americans, Demography 9:119–128.

United States Bureau of the Census, 1973, Birth expectations and fertility: June, 1972, in *Current Population Reports,* Ser. P-20, No. 248 (April), U.S. Government Printing Office, Washington, D.C.

United States Department of Labor, 1973, *Manpower Report of the President*, U.S. Government Printing Office, Washington, D.C.

Whelpton, P.K., Campbell, A.A., and Patterson, J.E., 1966, *Fertility and Family Planning in the United States*, Princeton University Press, Princeton, N.J.

FERTILITY BEHAVIOR OF AMERICAN WOMEN

Larry L. Bumpass

Center for Demography and Ecology

The University of Wisconsin, Madison

The recent low fertility of U.S. women has been much heralded. The crude birth rate for 1972 was 15 (below that of the Depression) and the period total fertility rate is below replacement. The crucial question is whether this low fertility is merely a transitory product of delays in marriages and births or whether basic long-run changes have occurred. Although predictions are highly vulnerable, I think recent declines reflect a revolution in the fertility regime. I will outline the argument upon which that conclusion is based[1] and then report briefly on findings which are relevant to points in that argument and concern the relationship between employment and fertility expectations among recently married women.

The explanation suggested here is to some extent a particular historical account of the achievement of low fertility in the U.S. The historical process through which replacement fertility has evolved has obviously differed from that described here for a number of developed countries. However, I suspect the context of the maintenance of low fertility (described here as a modern fertility regime) may be more similar across modern societies than the historical paths to replacement fertility.

The structure of the argument is complex but at its core is the concept of a revolution in fertility control values. The rapid diffusion of the pill brought not simply a quantitative increase in efficacy but rather a qualitative leap that changed the rules under which fertility decisions are made. As as innovation that affected directly the lives of individuals, the oral contraceptive diffused

---

[1]This section was originally published as an essay (Bumpass, 1973).

throughout the population at an unprecedented rate (Ryder, 1972a), vastly improving contraceptive protection and separating contraception from sexual activity. As a result, significant proportions of the population came to expect fertility control to be both *complete* and *unobtrusive*. Indeed, I think that the rapid diffusion may be attributable as much to the sexual unobtrusiveness of the pill as to its greater efficacy; but the ensuing revolution in fertility derives from the latter.

The essential characteristic of this revolution is that childbearing can now be voluntary in a radically different sense than before. Under the previous fertility regime, women could not confidently plan a lifetime of childlessness nor even the prevention of unwanted fertility. Obviously some women could and did avoid unwanted fertility, but no individual could be sure of that outcome in advance. Traditional effective contraceptives markedly delayed and reduced in number accidental conceptions but did not prevent these altogether. Among women using traditional "effective" methods to prevent unwanted pregnancy, nearly half fail within a five-year period (Ryder and Westoff, 1971), partly as a consequence of the coital relatedness of these methods. The gestalt that accompanied this regime was such that the adult role expectations of women were structured around motherhood, and the separation of work and home in industrial society buttressed this role structure. Cultural values with respect to fertility were, in part, rationalizations of the inevitable.

The position taken here is that the diffusion of the pill set in motion a field of forces leading to both the reduction of fertility and the establishment of a new fertility regime in which the pill occupies an important but not exclusive position. Given the levels of ill-timed and unwanted fertility under the old regime, the most obvious effect of the pill on fertility is through improved efficacy[1]. This effect is really the actualization of fertility goals that existed even among baby boom cohorts. However, far more significant in the long run are the expectations precipitated by pill use that fertility control can and ought to be both total and sexually unobtrusive. These expectations are viewed as reducing fertility through: 1) facilitating the adoption of other effective means of preventing unwanted births (the intrauterine devide or IUD, sterilization, and abortion) and 2) leading to reductions in the number of children intended by American women.

The IUD shares fully in the modern contraceptive values and has gained increasingly rapid acceptance since its introduction

---

[1] In the early 1960s, about one-half of all pregnancies were accidental; one-fifth were reported by the mothers as unwanted ever (see Bumpass and Westoff, 1970).

in the mid-1960s. By 1970, about six percent of women of childbearing age were using this method (Ryder, 1972a); in time, the
IUD may rival the pill and thus may moderate the role of sterilization and abortion in the immediate future.

Contraceptive sterilization has been both generally disapproved
and relatively infrequent in the population. The last half of the
sixties witnessed a dramatic reversal in this pattern with emerging
majority approval and an accelerating incidence of sterilization
(Bumpass and Presser, 1972). By 1970, more than one in every six
U.S. couples who had had all the children they wanted had been
sterilized for contraceptive reasons; and among women over 30,
sterilization was the most prevalent contraceptive method (Westoff,
1972). These changes are generally interpreted as responses to
the new contraceptive values and to concern over long-term administration of the oral contraceptive. Women completing desired
childbearing around age 25 face about 20 years of risk of unwanted
pregnancy and thus must decide whether to continue with the pill
for that length of time, to return to more cumbersome, less effective methods, or to seek another highly effective unobtrusive
method. If my interpretation is correct, changes in sterilization
attitudes and practices are a nice example of the realignment of
values in one area to adjust to a new milieu introduced by technological innovation.

A similar interpretation is placed on changes in the area of
abortion, though here the roots of change seem much more complex
(Bumpass and Presser, 1973). With growing expectations of complete
fertility control, I would predict the erosion of social rationalizations of fertility. Myths that halo even accidental pregnancy
may weaken when such an event is no longer seen as inevitable. In
addition, of course, changes in attitudes and practices with respect to abortion have been responsive to concern with the health
consequences of illegal abortions, the changing legal situation,
and concern with the role of accidental pregnancy in limiting the
achievement of women. However, this latter factor may be seen as
a consequence as well as a cause in these trends; i.e., new fertility control values have set the stage for popular receptivity
to equal opportunity themes by making commitment to nonfamilial
roles a realistic option.

Although the increasing prevalence of each of these methods
further reduces unwanted fertility, it is this last point that is
central to the reduction of fertility goals. The potential for
complete fertility control makes childbearing, when and if it
occurs, a matter of choice in an ultimate sense that never before
existed. For the first time, motherhood itself is fully a matter
for rational evaluation. Since it need not be rationalized because
inevitable, costs as well as virtues must be weighed. The literature is replete with the psychological and emotional costs of

motherhood. What is contended here is not that these costs have
just been discovered; but that they have become more relevant to
fertility decisions and, in addition, increasingly salient in the
context of growing concern for equal opportunity for women. As
nonfamilial opportunities become more equal, the opportunity costs
of childbearing escalate. Obviously, childbearing roles offer
much that is rewarding; but these rewards are probably not experienced equally by all women. Motherhood has been the last major
vestige of ascribed status in a modern industrial society. A primary consequence of complete fertility control is to place motherhood more squarely in competition with other social roles. As
fertility becomes more a matter for *decision*, the focus of decision
progressively rests on the planning of pregnancy rather than on
its prevention. The decision to have a child must be weighed not
only against the direct social, psychological, and economic costs
of children but also against the loss of the wife's earnings[1]; and
intrinsic satisfaction with her job.

Of course, deeply seated values ascribing motherhood to all
women will not disappear over night. What is more remarkable is
that we have experienced our recent declines in fertility in the
context of strong pronatalist values (Blake, 1972). Indeed, that
such rapid change is occurring under the influence of what has been
described as "coercive pronatalism" suggests that the staffing of
parental roles will become problematic for a society in which
young girls are less socialized to ascribed motherhood.

I see two major facets of the decline in fertility during this
transition period. The replacement of cohorts is a major mechanism
of social change (Ryder, 1965), and I expect succeeding cohorts of
young women to hold progressively less to ascriptive values of
motherhood. As new cohorts are socialized in a modern fertility
control society with an awareness of alternatives to motherhood,
aggregate cohort fertility goals will be lower,[2] perhaps with
many women remaining childless. For cohorts socialized under the
old fertility regime, I expect final cohort fertility to be lower
than intended, i.e., I expect the experience of fertility control
to lead to smaller family sizes than the women themselves say they
want. The extension of birth intervals through the reduction of
accidental pregnancy should reduct total cohort fertility by

---

[1]Wive's earnings play an important role in family life styles. In
a majority of the families in which both husband and wife have in-
comes, the wife's income represents over a fifth of total family
income (see Sweet, 1973).
[2]Fertility expectations among young white women under 24 ranged
from 3.2 to 3.1 over the years 1955-1965, but had dropped to 2.2
by 1972 (U.S. Bureau of the Census, 1972).

increasing the "risk" time available for the reduction of intentions (Friedman et al., 1965). Many women, finding increasing gratification in their jobs and freedom from the intensive demands of infant childcare, will become reluctant to return to childbearing. Others will simply discover that they have passed an age, or age of youngest child, beyond which they no longer wish to become reinvolved in infant care, even if ideally they would "like another." (I expect the same process to result in many members of the younger cohorts remaining childless, though few such women now tell us that they desire to do so.)

My interpretation does not rest on universal diffusion of modern methods but on the diffusion of the gestalt associated with these methods. The transition is still in process. Indeed, we have not yet recorded the effect of the recent Supreme Court decision on abortion; and I think much of the rapid fertility decline since 1970 is attributable to liberal legalized abortion practices in a few states (Bumpass and Presser, 1973).

It is important to note that our fertility-related population problems are not over. Indeed, they may only have begun. For while fertility may be expected to average below replacement for successive cohorts, its time pattern may fluctuate markedly. With near complete control, we may experience very deep troughs after the "bad years" and rather high peaks after the "good" ones as delayed marriages and births are made up. As Norman Ryder (1972b) has repeatedly emphasized, the consequences of varying cohort sizes for the life chances of their members and for the structure of social institutions (most obviously, education) are far more profound than moderate differences in the overall growth rate.

Among the consequences of these changes has been a marked convergence in the contraceptive experience and fertility expectations of subgroups of the population. Ryder and Westoff report that differentials in recent fertility between Catholics and non-Catholics and between blacks and whites closed by more than half a birth.

The relationship between employment and fertility gains increased interest in the context outlined here. The topic is extremely complex because of the difficulty of assigning causal priority in the often reported negative relationship. In cross-section, women who have worked the most may have limited their fertility in order to work; however, it is also plausible that they were simply able to work more because of fertility that was lower for other reasons. The findings reported here are only a subset of the analyses now under way of employment and fertility on the basis of the 1970 National Fertility Study data, a national probability sample of 6700 women. In the context of recent fertility changes, I will focus on those women who were first married

in the five years preceding the 1970 survey (N = 1409). Since the
great majority of these women intend future childbearing, the use
of intended family size enables us to examine the relationship of
future fertility goals to early employment experience and attitude
for this cohort of recently married women. We will consider
briefly the employment histories of these women and a number of
attitudes relevant to working and having children. More detailed
discussion of the results will be included in my chapter of a
monograph on the 1970 study now being prepared by Ryder and
Westoff.

Dummy variable regression analysis is used to adjust addi-
tively for the effects of differing education and age at marriage
as well as for other variables in the analysis. The first column
of both tables included here shows the unadjusted category deviations from the grand mean; the following columns present the net
effect for each variable after the effects of other variables have
been controlled in an additive regression model.

EMPLOYMENT

Several authors have theorized that labor force experience
before marriage is an important socializing experience, giving
young women alternative skills and role definitions to the mother
role which might later compete with that role.

The data in the first panel of Table 1 support that hypothesis
only among blacks. Expected fertility is slightly lower among
white women who worked before marriage than those who did not, but
this difference is largely attributable to the higher education
and later age at marriage of women employed full-time before
marriage. However, among blacks, the lower fertility of women
employed before marriage persists, net of education and age at
marriage, at over half a child lower than women who did not work.

The relationship of current employment to fertility is difficult to interpret. Among recently married women, those who are
currently employed are likely to be a lower-fertility subset since
women who have not recently given birth are more likely to be
working. Whatever the causal ordering, white women who are currently working expect about .2 fewer children than those who are
not. There is no differential fertility by current employment
among recently married black women. The more interesting point in
the second panel of Table 1 is that women who indicate reasons
that are not primarily financial expect .2 fewer children than
other working women among non-Catholics and .1 fewer among Catholics. Nonfinancial reasons for anticipating future employment are
related to .1 fewer fertility among non-Catholics but to .3 higher
fertility among Catholics.

The effect of employment in the first birth interval on total fertility intentions can be examined with less ambiguity than current employment by focusing on women who recently gave birth to their first child.

We might expect employment after marriage to be more relevant than premarital employment to the definition of adult and marital female roles in ways competitive with childbearing. Such employment not only provides opportunity for the development of skills and interests outside the home but also may lead to adjustments in marital role patterns that incorporate the wife's employment, including a life style partially dependent upon the additional income. We find that among women bearing a first child in the three years preceding interview, work in first interval is associated with lower fertility by .24 among white non-Catholics, .46 among white Catholics, and .31 among blacks, net of other factors. With education, age at marriage, and the length of the first interval controlled, the adjusted deviations from the grand mean are:

	Net deviation	N
White non-Catholic		
Worked in first interval	.08	325
Did not work	.16	155
Mean: 2.411		
White Catholic		
Worked in first interval	−.12	142
Did not work	.34	52
Mean: 2.744		
Black		
Worked in first interval	−.14	45
Did not work	.17	43
Mean: 2.438		

Given the problems of causal inference in cross-sectional analyses of employment and fertility, the attitude data in Table 2 lend further confidence that the relationship includes causal effects of orientations on fertility. The first item is the least relevant to this concern but is included for its own interest. Wives who report their husbands as disapproving of their working expect more children than other wives. Controls for education, age at marriage, and employment reduce the effect somewhat; and it persists only among blacks after the other attitude variables are controlled.

TABLE 1. *Differentials in Total Intended Family Size by Respon-*
dent's Work History and Reasons for Work, for Currently
Married Women under 45 Years of Age, First Married 0-4.9
Years Before Interview, by Race and Religion: Gross and
Net Deviations from Grand Mean

Characteristics	Deviations from Grand Mean			
	Gross	Net[1]	Net[2]	N
Whether wife had full-time **employment before marriage**				
WHITE NON-CATHOLIC				
Yes	-.05	-.00	.01	615
No	.11	.01	-.01	268
WHITE CATHOLIC				
Yes	-.04	-.02	-.02	262
No	.14	.07	.07	72
BLACK				
Yes	-.24	-.21	-.21	121
No	.41	.36	.35	71
Reasons for current **or future employment**				
WHITE NON-CATHOLIC				
Currently working:				
non-financial	-.38	-.34	-.23	101
financial	-.12	-.11	-.03	273
Expects future work:				
non-financial	-.01	-.02	.03	99
financial	.20	.15	.12	170
Neither	.16	.17	.04	240
WHITE CATHOLIC				
Currently working:				
non-financial	-.14	-.17	-.21	47
financial	-.09	-.09	-.12	84
Expects future work:				
non-financial	.12	.13	.21	34
financial	-.10	-.15	-.08	70
Neither	.18	.21	.19	99

TABLE 1. (continued)

	Deviations from Grand Mean			
Characteristics	Gross	Net[1]	Net[2]	N
BLACK				
Currently working:				
non-financial	**	**	**	15
financial	-.00	.05	.17	71
Expects future work:				
non-financial	**	**	**	15
financial	.10	.04	-.07	69
Neither	-.02	-.05	-.14	22

Grand Means

WHITE NON-CATHOLIC: 2.38

WHITE CATHOLIC: 2.76

BLACK: 2.61

Net[1] = net of wife's age at marriage and education.

Net[2] = net of wife's age at marriage, education, reasons for working (second section of table) or work before marriage
(first section), number of weeks worked in 1970, husband's
attitude to wife's work, whether wife thinks the man should
be the achiever, whether wife thinks mother's working harms
preschoolers, and age youngest child should be before
mother works full time.

** N less than 20 cases

TABLE 2.   *Differentials in Total Intended Family Size by Attitudes toward Employment and Childbearing for Currently Married Women under 45 Years of Age, First Married 0-4.9 Years before Interview, by Race and Religion: Gross and Net Deviations from Grand Mean*

| Characteristics | Deviations from Grand Mean | | | | |
	Gross	Net[1]	Net[2]	Net[3]	N
Husband's attitude toward wife's work					
WHITE NON-CATHOLIC					
Disapproves	.13	.12	.07	.05	536
Other	-.20	-.18	-.11	-.08	347
WHITE CATHOLIC					
Disapproves	.06	.07	.04	.01	214
Other	-.10	-.13	-.17	-.01	120
BLACK					
Disapproves	.13	.10	.13	.11	116
Other	-.20	-.15	-.19	-.17	76
Whether respondent thinks mother's working is harmful to preschoolers					
WHITE NON-CATHOLIC					
Agrees	.09	.09	.08	.03	645
Other	-.25	-.24	-.21	-.09	238
WHITE CATHOLIC					
Agrees	.06	.07	.05	.03	232
Other	-.14	-.16	-.12	-.02	102
BLACK					
Agrees	.01	.02	.03	.00	104
Other	-.01	-.02	-.04	-.01	88
Age youngest child should be before mother works full-time					
WHITE NON-CATHOLIC					
0-4 & any age	-.14	-.20	-.19	-.11	175
5-9 & school age	-.04	-.03	-.04	-.04	535
10-15+ & out of high school	.23	.26	.27	.21	161
WHITE CATHOLIC					
0-4 & any age	-.17	-.25	-.24	-.17	61
5-9 & school age	.00	.00	.00	.01	210
10-15+ & out of high school	.14	.22	.21	.18	62

TABLE 2. (continued)

| Characteristics | Deviations from Grand Mean | | | | |
	Gross	Net[1]	Net[2]	Net[3]	N
BLACK					
0-4 & any age	.04	.00	-.02	-.00	104
5-9 & school age	-.06	.00	.04	.02	79
10-15+ & out of high school	**	**	**	**	8
Whether respondent agrees "it is better if the man is the achiever"					
WHITE NON-CATHOLIC					
Agrees	.12	.11	.10	.07	645
Other	-.31	-.31	-.26	-.19	238
WHITE CATHOLIC					
Agrees	.10	.10	.10	.08	247
Other	-.29	-.29	-.27	-.23	87
BLACK					
Agrees	.06	.05	.02	.01	137
Other	-.16	-.12	-.06	-.03	55

Grand Means

WHITE NON-CATHOLIC: 2.38

WHITE CATHOLIC: 2.76

BLACK: 2.61

Net[1] = net of wife's education and age at marriage

Net[2] = net of wife's education, age at marriage, work before marriage, work history and intentions, and number of weeks worked in 1970

Net[3] = net of wife's education, age at marriage, work before marriage, work history and intentions, number of weeks worked, and all other variables in Table 2

** N less than 20 cases

The next two items relate directly to how the respondent feels about working and children and the final item to sex-role separation more generally. Among these recently married white women, the beliefs that working is harmful to preschoolers and that women ought to wait until their children are older before working are clearly related to lower expected fertility. The differences are quite large in the context of current fertility levels and persist net of age at marriage, education, and employment history. The belief that a mother's employment is harmful to preschoolers is associated with .29 lower expected fertility among non-Catholics and .17 lower expected fertility among Catholics. Respondents were asked the following question:

> Suppose a married women has had all the children she wants, is offered a good job, and can arrange for child-care. What age should her youngest child be before she takes the job on a *full-time* basis?

When we compare women giving responses of under five years with those indicating a child should be over 10, the difference in intended fertility is .46 among non-Catholics and .45 among Catholics net of the background variables and remains over .3 net of background and other attitude variables.

"Nontraditional" responses to these questions (particularly to the latter items) indicate an orientation toward female employment as other than a residual role. That is, women who say a women should return to work before her children reach school age are asserting, in part, that children should not rule out other adult female roles; and such women anticipate substantially lower fertility.

We approach attitudes toward female roles more generally in the final item. It is of interest that three-quarters of this sample of recently married women agree with the statement, "It is better for everyone involved if the man is the achiever outside the home and the woman takes care of the home and family." However, disagreement with this traditional position is associated with markedly lower expected fertility among whites. Again controlling for background factors, the difference is .36 for non-Catholics and .37 for Catholics; and these differentials generally persist when the other attitude variables are included in the model. It is of interest that the preceding differentials in fertility by role attitude are not found among the black women in our sample.

We began this discussion with a speculative interpretation in which technological changes in the fertility control regime were seen as moving the population towards lower fertility along a number of routes in addition to higher contraceptive efficacy. It

was argued that qualitative leaps in fertility control would facilitate rapid changes in the views of American women with regard
to alternatives to motherhood as a primary female role. Although
the evidence remains sketchy and merely suggestive and the argument
still highly speculative, we have seen that employment experience
in the first birth interval and, in particular, attitudes toward
female roles are strongly related to lower fertility goals among
recent brides and mothers in 1970. Over the next few years, we
hope to examine more systematically the notion that a genuine revolution is occurring in the U.S. fertility regime.

REFERENCES

Blake, J., 1972, Coercive pronatalism and American population
policy, Preliminary Paper No. 2, International Population and
Urban Research, University of California, Berkeley.

Bumpass, L.L., 1973, Is low fertility here to stay?, Family Planning Perspectives 5:67-69.

Bumpass, L.L., and Presser, H.B., 1972, Contraceptive sterilization
in the U.S.: 1965 and 1970, Demography 9:531-548.

Bumpass, L.L., and Presser, H.B., 1973, The increasing acceptance
of sterilization and abortion, in *Toward the End of Growth:
Population in America* (C.F. Westoff, ed.), pp. 33-46,
Prentice-Hall, Englewood Cliffs, N.J.

Bumpass, L.L., and Westoff, C.F., 1970, The 'perfect contraceptive'
population, Science 169:1177-1182.

Freedman, R., Coombs, L., and Bumpass, L., 1965, Stability and
change in expectations about family size: a longitudinal
study, Demography 2:250-275.

Ryder, N.B., 1965, The cohort in the study of social change, Amer.
Sociol. Rev. 30:843-861.

Ryder, N.B., 1972a, Time series of pill and IUD use: United States
1961-70, Studies in Family Planning 3:233-240.

Ryder, N.B., 1972b, A demographic optimum projection for the United
States, in *Demographic and Social Aspects of Population
Growth: The Commission on Population Growth and the American
Future Research Reports* (C.F. Westoff and R. Parke, Jr., eds.),
Vol. I, pp. 605-622, U.S. Government Printing Office, Wash-
ington, D.C.

Ryder, N.B., and Westoff, C.F., 1971, *Reproduction in the United*

States: *1965*, Princeton University Press, Princeton, N.J.

Sweet, J.A., 1973, *The Employment and Earnings of Married Women,*
    Seminar Press, New York (see especially Chapter 6).

U.S. Bureau of the Census, 1972, Birth Expectations and Fertility:
    June 1972, in *Current Population Reports,* Ser. P-20, No. 240
    (September), U.S. Government Printing Office, Washington,
    D.C.

Westoff, C.F., 1972, The modernization of U.S. contraceptive
    practices, Family Planning Perspectives 4:9-12.

HUMAN SEXUALITY AND EVOLUTION

Frank A. Beach

Department of Psychology

University of California, Berkeley

INTRODUCTION

My first experiment on reproductive behavior was published in 1937, and since that time most of my research has dealt with the neural, hormonal,and experiential control of sexual activity in various species of birds and mammals. In the course of nearly four decades I have authored, coauthored, or edited three books dealing with sexual behavior. Two years ago I decided to capitalize upon my experience as an experimentalist and author by teaching a new undergraduate course entitled "Human Sexuality." The result was very nearly a total disaster.

At first the class listened patiently to my lectures about mating in animals and sexual differentiation in the human embryo; but as the weeks passed it became clear that most students regarded my approach as irrelevant to the central issues, which they defined as "psychological explanation" of human sexuality. I wanted to develop a theoretical framework within which various aspects of human sexuality could be objectively interpreted. The students wanted instant analysis of their personal sexual experiences.

To me, the problem seemed to develop a line of approach to the entire subject of human sexuality that offered some hope of reconciling my objectives with the needs and expectations of the students. I have not solved that problem, but the essay that follows represents a first step in this direction. It is a child of my own frustration, and as an exercise in self-therapy it has proven helpful. Whether it has any additional merit I leave to the judgement of my readers.

SEX VS. SEXUALITY

The arguments summarized here are neither original nor profound, but even so they eluded me until three orienting conclusions gradually became clear. (1) All laymen and nearly all psychologists and psychiatrists implicitly and unconsciously distinguish between male-female and masculine-feminine as separate, dyadic categories. (2) The fundamental issues underlying most treatments of sexuality center upon masculinity and femininity and have relatively little to do with maleness and femaleness as such. (3) Nearly all of the problems that are traditionally dealt with in theories of sexuality are much more relevant to the behavior of men and women than to that of males and females of other species.

When I had gotten this far in my thinking, I began to suspect that I might achieve further insights by starting from a working hypothesis that sexuality is a product of the evolution of *Homo sapiens*. If this were true, however, if sexuality were a product of evolution, it almost surely would be an evolutionary "emergent." That is to say, it would represent a new level or form of organization not apparent in previous stages of evolution, one involving saltatory change from preexisting levels that could not be predicted from a knowledge of the characteristics or qualities of those levels. It is my present feeling that human sexuality is about as closely related to the mating behavior of other species as human language is related to animal communication, a relationship that is distant indeed.

My synoptic survey of facts and theories that follows is organized in terms of the ontogenetic history of the individual and the phylogenetic history of the species. Individual development is subdivided into the epigenesis of maleness and femaleness on the one hand and of masculinity and feminity on the other.

The term *epigenesis* is chosen deliberately to emphasize the fact that development of structure and of behavior involves a process of gradual diversification and differentiation in which new traits, characters, or attributes actually develop out of initially undifferentiated ones. Ontogeny is not merely a process of accretion in which new features are superimposed on those present in the preceding stages. It is a progression from one level of organization to the next, in the course of which new and quite different characteristics, relationships, and functions appear at successive levels. As a matter of fact, epigenesis might be thought of as "emergent ontogenesis."

EPIGENESIS OF MALE AND FEMALE

In every sexually reproducing species a fundamental difference

between males and females is permanently and irreversibly established at the moment of fertilization. In all mammals the haploid
nuclei of the paternal and maternal gametes fuse, and the diploid
nucleus of the zygote contains either two X chromosomes or one X
and one Y. This means that the nucleus of every somatic cell in
the individual produced from that zygote will contain either two
X's or an X and a Y chromosome which are direct descendants of
the original sex chromosomes.

As epigenesis progresses through embryonic and fetal stages,
the process of sexual differentiation results in the appearance of
more and more differences between males and females. For example,
at an early point in embryonic life the gonads of both sexes contain cortical and medullary components. In XX embryos the cortex
gives rise to an ovary, and in XY embryos the medulla differentiates into a testis. A new sex difference has emerged as a consequence of the original difference in chromosomal balance.

Hormones secreted by the embryonic testis have a directive
effect upon the subsequent growth and development of accessory
male and female sex structures. Embryos of both sexes possess primordial tissues which can develop into male or female sex access-
ories; but in the genetic male, testicular hormone stimulates the
Wolffian ducts to differentiate into the epididymis, vas deferens,
and seminal vesicle, while the Mullerian ducts fail to develop.
In female embryos there is no testicular hormone, and the embryonic
ovary is inactive. Since the Wolffian derivatives are not stimulated and the Mullerian system is not suppressed, the latter differentiates into the fallopian tubes, uterus, and median vagina.

After these changes have taken place, male and female organisms are distinguishable in terms of differences in the sex
chromosomes, the structure of the gonads, the secretion or lack of
secretion of gonadal hormone, and the anatomy of the sex accessor-
ies. Figuratively speaking, the gap between male and female
steadily widens as development proceeds.

In anticipation of conclusions to be developed later, it is
important to emphasize here that these successively appearing
differentiating characters will persist throughout the entire existence of the individual and will contribute to the development
of additional sexual dimorphisms at much later stages of ontogeny.
For example, the uterus, which differentiates in the female embryo, eventually will be essential to menstruation; and beginning
at menarche, the occurrence of menstrual cycles will indirectly
affect the development of some of the psychological characteristics
which differentiate adolescent girls from boys of the same age.

Sex differences mentioned thus far are established in the
embryo. Additional ones appear during the fetal stage of development.

Compare, for example, an XY fetus bearing testes, epididymis, and vas deferens with an XX fetus possessing ovaries, tubes, and uterus. The gonads of the XY fetus secrete testosterone (which is different from the hormone produced earlier by the embryonic testis); whereas those of the XX fetus are endocrinologically inactive. Testosterone contributes to the development of still further divergences between males and females. In the male, primordial tissues of the urogenital sinus differentiate to form a penis and a scrotum. In the female the same anlagen give rise to a vagina and a clitoris. There are now at least five separate characters distinguishing males from females.

Sex differences in reproductive anatomy are striking and easily observed; but others, more difficult to discern, are at least equally important in terms of their effects upon the eventual development of sex-related behavior patterns. I refer to differences in the male and female brain which are established during prenatal life. The existence of such differences in animals has been established by experiments involving exposure of unborn females to stimulation by testicular hormone and by observing their sexually dimorphic behavior as adults. Numerous studies have shown that prenatal androgen treatment decreases female behavior and increases the incidence of mating responses typical of the male (Gorski, 1971). If male animals are deprived of the testosterone normally secreted by their own testes, their sexually dimorphic behavior in adulthood is more like that of females than is the behavior of normal males; and they show clear deficiences in characteristic male behavior (Beach, 1971).

Behavioral changes induced by altering the endocrine environment during development provide at best presumptive evidence for sex differences in the brain, but recent discoveries prove the existence of neuroanatomical differences in brains of male and female rats (Field and Raisman, 1973). The differences, which are present at birth, can only be detected by electronmicroscopy and are restricted to one small area in the hypothalamus. Nevertheless their significance is difficult to exaggerate, for the region in which they occur is involved in mediating sexually dimorphic mating behavior. Furthermore, brains of genetic females can be made to develop like those of males by exposing the females to androgen in utero; and brains of males castrated at birth develop postnatally to resemble those of normal females.

All of the sex differences mentioned to this point have been qualitative rather than quantitative. The variables involved are discontinuous rather than continuous, but it is important to emphasize that quantitative differences between males and females represent the operation of continuous variables.

With very rare exceptions, every newborn child can be assigned unequivocally to one of two populations designated as male or female.

The variables in terms of which these populations differ are numerous, and for most measures there is marked interindividual variability within both populations. In fact, for the majority of sex-
related traits, differences between male and female can only be
expressed in terms of means and standard deviations; and there is
appreciable overlap between the distributions of the two populations.

This, of course, does not make such sex differences unimpor-
tant, but it is important to recognize their relative nature. For
example, at birth female infants are, *on the average*, 1/2 to 1 cm
shorter and 300 gm lighter than males. Newborn females are, *on the
average*, four to six weeks more advanced than males in skeletal
maturation. Infant females are, *on the average*, less muscular,
less active, and slightly more sensitive to pain and more irritable
than males. When the inventory of congenital sex differences is
examined *in toto*, it appears that the only characters for which
males and females are qualitatively dichotomous are those in which
dichotomy is essential for reproductive function, e.g., testis vs.
ovary, uterus vs. epididymis, or penis and scrotum vs. clitoris
and vagina.

Many differences between human males and females have been established by the time of birth, and additional differences continue
to emerge through the childhood year; but sex differentiation is
markedly accelerated during the period of adolescence. Some of the
more obvious changes at this time involve development of the secondary sex characters. In girls, the externally observable changes
include growth of the breasts, development of female habitus with
broadening of the hips, onset of menstruation, etc. Analogous
changes in boys involve change in pitch and timbre of the voice,
growth of the penis, occurrence of seminal ejaculation, gradual
appearance of facial hair, and eventual beard development.

Less obvious but equally significant changes occur in various
physiological functions. Figure 1 shows that in adolescence the
muscular strength of boys increases to a much greater degree than
that of girls, and Figure 2 demonstrates similar differences in
blood pressure and blood hemoglobin. Marked increase in the secretion of testicular hormone (Fig. 3) contributes to these and other
changes which characterize male adolescence. The appearance of
female secondary sex characters is linked to the beginning of the
cyclic secretion of estrogen and progesterone by the ovaries.
Taken in combination, the dimorphic secondary sex characters reflect
a new stage of divergence between males and females. The final
degree of separation is achieved even later when females experience
pregnancy, parturition, and lactation.

The essential features of the epigenesis of male and female,
so sketchily summarized here, seem to me to be the following.
Starting from an original difference in chromosomal balance, male
and female individuals pass through a succession of steps or stages

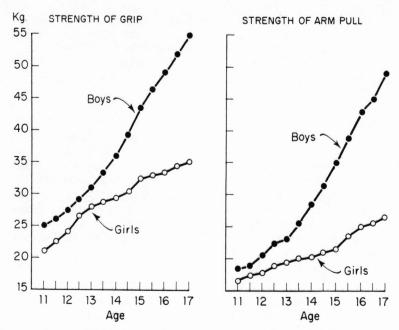

Fig. 1. Sex differences in muscular strength at different ages (based on data presented in Tanner, 1962).

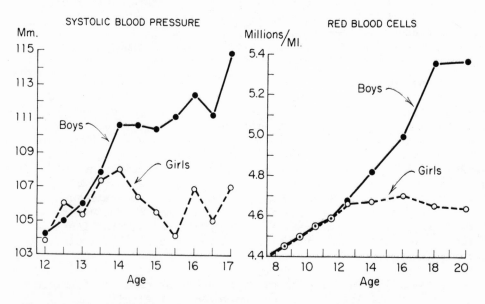

Fig. 2. Sex differences in blood pressure and erythrocyte count at different ages (based on data presented in Tanner, 1962).

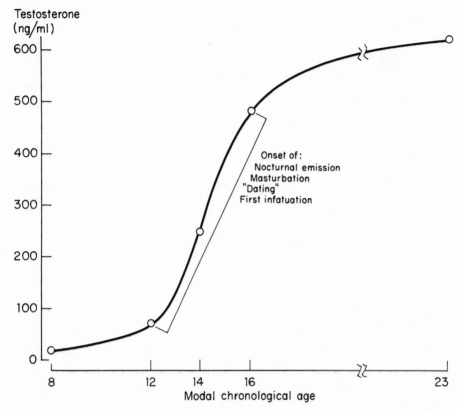

Fig. 3. Levels of plasma testosterone in human males at different ages (August et al., 1972; data referring to behavior taken from Ramsey, 1943, and Kephart, 1973.)

of development, at each one of which new differences between the sexes emerge. Some are mutually exclusive; whereas others are expressions of continuous variables which produce quantitative differences. Sexual differentiation is best viewed as a developmental progression from one level to the next. At each level new organization appears which was not discernible at earlier levels. The essential process is not one of addition or accretion but of organic growth and differentiation. I believe that the development of sexuality is also epigenetic and is in various ways analogous to the development of sex.

EPIGENESIS OF MASCULINE AND FEMININE

Most of the basic differences between male and female develop between fertilization and birth. In the space of nine months, the

foundation is laid for the eventual expression of the male-female
dichotomies found in the neonate as well as those arising in the
course of the individual's lifetime. In contrast, differences
between masculine and feminine do not even begin to develop until
the second or third year after birth; and their full maturation
occupies a span of fifteen to twenty years. When we transfer our
attention from male and female to masculine and feminine, we are
shifting from the analysis of differences in sex to the analysis
of differences in sexuality.

The concept of sexuality involves two components, gender role
and gender identity. Two gender roles are defined by every human
society. They constitute a composite of all those behavioral traits,
attitudes, and emotional characteristics that together define and
differentiate the masculine and feminine members of that society.
Some dimensions of sexuality are very much the same in all societies,
but others exhibit extreme intersocietal variability. The evolution of gender roles will be considered in the final section of
this essay. In this section, I wish to analyze the epigenesis of
masculine and feminine, or the development of gender identity.

Gender identity pertains to each person's feelings and convic-
tions regarding his or her own sexual makeup. More specifically,
it refers to the individual's concept of his or her own masculinity
or femininity. This is an exceedingly important dimension of the
more general self-concept, which in turn plays a vital role in overall psychological development. If a society is to function effectively, it must establish and maintain viable gender roles. If
individual members of a society are to function effectively, they
must develop integrated and stable gender identities.

I have already suggested that the development of the individual's gender identity progresses through successive stages, taking
new form or organization as new levels emerge. The principal
stages recognized and utilized by every society are infancy, childhood, adolescence, and maturity. As defined by most societies,
gender roles in early childhood are relatively simple, i.e., they
involve few variables; and they become increasingly complex and
more finely differentiated at each successive stage of postnatal
ontogeny.

As already noted, all societies lay the foundation for the development of gender identity by assigning every new infant to one
of two gender roles. Now, despite the fact that the majority of
male-female differences are quantitative rather than qualitative
and involve extensive overlap between the sexes, all societies di-
vide their newborn members into two discontinuous classes; and this
exceedingly influential decision is based upon one dichotomous character, the genital anatomy. The significance of this fact is that
the sex role assigned at birth will have a powerful and continuing

influence on the development of the individual's gender identity.
The implicit and intuitive definition of male and female as discontinuous and mutually exclusive categories leads inevitably to a
bipolar, unidimensional concept of masculine and feminine. This
concept violates the biological facts, so to speak, and in some
societies, including our own, can do mischief by encouraging com-
partmentalized and opposing stereotypes of masculinity and femininity which create serious problems not only for the society but for
many individuals as well.

Sex role ascription has no significance for the newborn infant,
but for the society it predetermines which one of two programs of
training will be used in the sexualization of that individual over
the next fifteen or twenty years. In contrast to sex role assignment, the actual start of gender identity development coincides
more or less with the onset of verbal learning. Disregarding the
simple conditioning of which neonates and even fetuses may be
capable, we can date the emergence of gender identity from the start
of language acquisition.

Development of gender identity depends upon two categories of
learning. The child must learn the distinguishing characteristics
of the gender roles of both sexes and must also learn to behave in
a manner consonant with his or her assigned gender role. The growing child begins to learn about sex differences as soon as he begins
to use and understand language. As a first step each male learns to
identify himself as a boy and not a girl although at first this
means no more than saying, "I am Johnny."

By three years of age or earlier most middleclass American
children can correctly state their own sex; and at four years they
can identify the sex of dolls of various ages, relying principally
upon differences in hair style, clothing, etc. One group of three-
year olds was required to assemble cutouts of male and female figures separated into three vertical sections consisting of head,
trunk, and below-the-trunk. When the figures were distinctively and
differentially attired, most children performed successfully; but
when the cutouts were nude,many children had difficulty identifying
the sex of the section bearing the genitals. Heads were correctly
matched with trunks; but errors with respect to the below-the-trunk
section were made by 88% of the three-year olds, 69% of the four-
year olds, and 31% of the six-year olds (Katcher, 1955). Other
studies have shown that American children do not achieve a clear
understanding of genital differences until they are five to seven
years old even when there has been express parental instruction.

An important concept which young children master with difficulty is that of gender constancy. In one study, children from four
to eight years old were shown pictures of a boy and a girl and were
asked whether the girl could become a boy if she so desired. Most

four-year olds were sure that she could do so if she would cut her hair, change to boys' clothes, and play boys' games. Only children six years and older were confident that even if she made these changes, the child in the picture would still be a girl (Kohlberg, 1966).

Certain aspects of adult gender roles are learned fairly early in life. These include not only such simple items as sex-related differences in dress but also various aspects of adult daily routine. For example, in most middleclass families five-year-old children know that fathers go to work in the daytime, whereas mothers stay at home or go shopping. They may also know that older brothers play football and older sisters engage in different forms of recreation.

The essential point at issue here is the fact that a growing child's concept of male and female gender roles is not easily and quickly acquired. Instead it is built up gradually over a number of years as a result of observation and instruction. Even so, this learned discrimination by itself is not sufficient to insure development of the individual's gender identity. It is not enough for a little boy to learn how men and women behave or how his society expects and desires him to behave. He must learn to govern his behavior so that it conforms to those expectations. Furthermore, he must, in most respects at least, *enjoy* fulfilling his assigned gender role. Successful development of gender identity involves three requirements: 1) to know how to behave in accordance with gender role; 2) to want to behave in the appropriate manner; and 3) to be able to execute the appropriate behavioral patterns. Achieving concordance of all three variables is what gender identity "is all about." Incongruities between any two can interfere with the epigenesis of masculinity or femininity.

Fortunately, but not at all fortuitously, in all societies many of the major dimensions of childhood gender roles are correlated with basic male-female differences characteristic of the human species. Such differences are reflected in the fact that in all societies which have been carefully studied boys are more active, venturesome, self-assertive, and physically aggressive than girls. Girls tend to be more sedentary, more dependent, and less aggressive. These sex-related differences are manifest in early childhood. For example, one review of observations of American children indicates that "...in virtually every relevant study of preschool and school-age children, aggressive behavior has been found to be more frequent among boys than girls. Boys also show more negative attention-seeking and antisocial behavior than girls. Even in their fantasies... boys indicate a greater preoccupation with aggressive themes" (Mussen, Conger, and Kagan, 1969).

Descriptions of behavior of children in numerous other societies reveal a marked intercultural commonality of such early sex differences. For instance, in his account of one Melanesian society, Davenport observes that boys of six years and younger characteristically engage in a great deal of rough-and-tumble play, gang fighting, and other forms of vigorous social interaction. Little girls in contrast play quietly in small groups or alone (Davenport, 1965).

Societies in general appear to have intuitively recognized certain male-female differences, formalized them to some extent, and incorporated them in the stereotyped gender roles for the two sexes. The way in which initial sex differences in aggression are modified and moulded in our society is indicated in the following quotation.

> Aggressive behavior is an accepted component of traditional masculine behavior (i.e., sex-typed behavior) but not of feminine behavior. Aggression in girls typically meets with more punishment than it does in boys, and the role-models young girls choose are less likely to be overtly aggressive. For these reasons young girls who are aggressive will gradually learn to inhibit aggressive manifestations, while boys have more freedom to express their hostile feelings and will continue to manifest aggressive behaviors (Mussen, Conger, and Kagan, 1969, p.341).

These authors note elsewhere that American boys are encouraged to behave aggressively in certain situations and condemned when they fail to do so.

In the Melanesian society studied by Davenport, strict gender-role training begins as soon as children can walk. Little boys are allowed to run freely all over the village, forming age-stratified gangs. They range about, "fighting amongst themselves and getting into mischief, always under the close but permissive surveillance of adults;" little girls, in contrast, "are virtually never separated from their mothers or older sisters while the latter are engaged in household and garden work before sunup to past sundown" (Davenport, 1965, p.195).

These examples could be multiplied almost indefinitely, but they will suffice to illustrate one hypothesis I wish to suggest concerning differences between males and females and between cultural definitions of masculine and feminine. Using aggression as an example, the hypothesis begins with the reasonably conservative assumption that before or in the absence of any social training, punishment, or reinforcement, human males and females differ in their tendencies to interact actively, vigorously, and aggressively with other individuals. It is further assumed that the strength of these tendencies varies considerably among individual members of

both sexes, although the average value for the male population significantly exceeds that for the female population. In other words,
the situation is that which obtains with respect to sex differences in all continuously varying traits. The picture is one of
two distribution curves with reliably different means but a considerable degree of overlap.

Starting from this base, society imposes a distinction in gender roles which embodies the original male-female difference but
eventually exaggerates it by reducing or eliminating the area of
bisexual overlap. Little boys who tend to be less aggressive than
the average for the male population are encouraged to display more
aggressivity. Little girls who show more aggression than the average for the female population are punished and encouraged to behave
less aggressively.

To the extent that social rewards and punishments can modify
aggressive behavior, the net result of the foregoing process will
be to produce a kind of regression toward the mean in both male
and female populations, which in turn increases the apparent bi-
modality of the two distributions. Society has, so to speak, used
the gender role differences which it defines in the first place as
a wedge to drive between the two populations and thus artificially
widen the distance between them. Society did not create the original difference, but society did seize upon that difference and exploit it for purposes that will be discussed in connection with the
evolution of human sexuality.

I suggest that this hypothesis is germaine to many gender-role
differences that have wide cross-cultural generality. Consider,
for example, sex differences in the games children play. Many of
them reflect or exploit underlying and unlearned male-female differences in strength, agility, physical endurance, and other biological variables. In numerous societies around the world, boys
throw at targets, run races, catch balls, climb trees, and in general prefer activities for which most males are physically better
suited than most females by virtue of their superior control over
large muscle groups, better eye-hand coordination in gross movements, etc.

Because of the concordance between childhood activities which
societies designate as masculine or feminine and congenital sex
differences which render these activities selectively suited to
the average boy's and the average girl's abilities and interests,
most members of both sexes tend to enjoy the actualization of their
sex roles, practice them frequently, and become proficient in their
performance. All of this constitutes and contributes to the development and growth of a masculine or feminine gender identity, but
the repertoire of human male-female differences is not confined to
physical capacities or sensory-motor coordinations.

Observations of children in different societies reveal that boys are more curious than girls about their environment and are in general more likely to ask questions starting with "why" or "how." Boys tend to be more analytical than girls and independent in solving problems (Smith, 1933). Boys are also more persistent. In one experiment, children were given two puzzles, only one of which they were able to solve. When later allowed to return to one puzzle, boys tended to choose the one they had failed; whereas more girls preferred to repeat their earlier success (Crandall and Rabson, 1960). Among American children of school age, more boys than girls prefer the study of science and mathematics; and more girls than boys prefer literature. In IQ tests, adolescent girls tend to excel on verbal measures; whereas boys of similar age are superior on quantitative and spatial problems (Kagan, 1970).

Earlier in this essay I mentioned the existence of a neuro-anatomical sex difference in the brains of newborn rats. The demonstration of emotional and intellectual differences between boys and girls raises the age-old question about possible differences between male and female brains in our own species. Although most scientists today would probably withhold judgment on this issue, it is amusing to recall that no such uncertainty plagued the minds of many authorities less than a century ago. The following excerpt is taken from a medical treatise entitled *Cerebral Hyperaemia* written by William A. Hammond, M.D., in 1895.

> Certainly, my experience goes to establish the fact that the study of mathematics is bad for the average young woman's mind. I have repeatedly had instances of cerebral hyperaemia under my charge occurring in young ladies of from fifteen to seventeen years of age, in whom it was directly induced by the study of calculus, spherical trigonometry, and civil engineering. I have now the care of a young lady, sixteen years of age, in whom the disease came on rapidly, in consequence of long-continued and close application to the solution of a mathematical problem. But so long as there are ambitious women who want their sex to study all the subjects men do, I suppose civil engineering will be responsible for many hyperaemia brains in young girls (Hammond, 1895, pp.71-72).

Leaving aside the validity of Dr. Hammond's theories concerning the etiology of cerebral hyperaemia in adolescent girls, it is a fact that gender role specifications and the components of gender identity undergo several important changes beginning at the time of puberty. Some of these are correlated with the new physiological and anatomical differences which characterize the onset of adolescence and serve further to distinguish males from females.

Representative sex differences have been illustrated in Figure 1, which reveals the increasing disparity between boys and girls with respect to muscular strength, and Figure 2, which shows differences in the oxygen-carrying capacity of the blood (red cells) and the ability of the circulatory system to deliver fresh blood to the sensory and motor mechanisms for behavior (systolic blood pressure). These differences are functionally related to some of the gender role assignments discussed in the final section of this essay. They are also differences which at the time of their development exert indirect effects upon the epigenesis of gender identity. For example, the adolescent boy's concept of his own masculinity is affected by changes in his physical capacities which are apparent to himself and to his associates.

Behavioral changes associated with puberty and with the increase in testosterone (Fig. 3) include occurrence of the first seminal ejaculation at a median age of 13.8 years and the beginning of masturbation at about 12 years. Nocturnal emissions appear at this time and are most frequent between the ages of 12 and 16 years (Ramsey, 1943). Events of this nature serve to focus the boy's attention upon genital aspects of sexuality and contribute to the development of a masculine gender identity.

In many societies the occurrence of menarche is associated with additional qualifications of the feminine gender role, and at the same time the beginning of menstruation contributes new dimensions to the adolescent girl's gender identity. Certainly, changes in society's attitudes and expectations play an important role in modifying adolescent gender identity; but the major physiological changes which occur at the same time must also be considered. For example, the relatively sudden increase in the secretion of ovarian hormones is not without effect upon behavioral tendencies. Psychologists have compared the social behavior and interest patterns of pre- and postmenarchial girls of the same chronological age. Postmenarchial girls were more interested in daydreaming, personal adornment, and the display of personal and social activities with the opposite sex (Stone and Barker, 1939).

The last item is especially significant. The emergence of strong heterosexual attraction is a very common element in the development of adolescent gender identity for both sexes. The broadening of interpersonal associations to include heterosexual activities is characteristic of this stage of psychosexual ontogenesis. Many societies indirectly acknowledge the fundamental change by restructuring gender roles between late childhood and early adolescence. Rules governing sex-related behavior of adolescents may range from active encouragement of dyadic heterosexual liaisons on the one hand to sweeping prohibitions against social contact between teen-age males and females on the other (Ford and Beach, 1952). Both extremes reflect implicit recognition of the

fact that important and powerful changes in sexuality are to be
expected during puberty.

In our own and many other societies, one striking manifesta-
tion of pubertal changes in gender identity is referred to as "falling in love." One study of more than 1,000 American men and women
(Kephart, 1973) indicated that for most people three important
components of "romantic love" are: 1) strong emotional attachment
toward a person of the opposite sex; 2) tendency to think of this
person in an idealized manner; and 3) pronounced physical attraction, "the fulfillment of which is reckoned in terms of touch."

Eighty-four percent of the respondents recognized romantic
love in terms of their own experience; and of these 90 percent
differentiated between love and infatuation, although they felt
the discrimination could only be made in retrospect. The average
age for beginning dating was 13 years in both sexes. For males
the first infatuation had occurred on the average at 13.5 years
and the first love affair at 17.5. Corresponding means for females were 13 and 17 years.

It seems to me we need not look far for the biological correlates of this new and emergent level of gender identity development. No one would assume that there is a one-to-one relationship
between the physiological alterations occurring at puberty and the
psychologically complex processes of falling in love. On the
other hand, it can scarcely be pure chance or entirely the result
of social conditioning that the timing of an initial, sustained
upsurge of heterosexual attraction in adolescents of both sexes
coincides so closely with the marked rise in secretion of gonadal
hormones.

The foregoing discussion of the epigenesis of masculinity and
femininity is obviously incomplete, and important features of psychosexual differentiation have necessarily been omitted. The
essential points are that gender identity is a fundamental variable
in the broader concept of human sexuality, that it develops in the
individual over a long period of time under the influence of socially structured programming based upon each society's gender roles,
and that some of its dimorphic components, though certainly not all
of them, are closely associated with and develop out of biological
differences between the human male and female.

PHYLOGENESIS OF SEXUALITY

The final section of this essay is concerned with interrelationships between sexuality and human evolution. In the two
previous sections, I outlined the main features of sexual differentiation and the differentiation of one of the components of

sexuality, gender identity. I tried to show that certain salient
differences between masculinity and feminity can be directly or
indirectly linked to correlated differences in maleness and femaleness. The other component of sexuality is gender role, and gender
identity is defined in terms of gender role. Masculine and feminine
gender roles differ in detail from one society to the next, but
there is nevertheless considerable cross-cultural similarity.

My discussion is addressed to the following questions. Where
did gender roles come from? How does it happen that basic concepts of masculinity and femininity are fundamentally so similar
from one culture to the next? What are the biological functions
and social values of separate gender roles for males and females?
I wish to emphasize at the outset that what I now present is
neither a theory to explain the evolution of *Homo sapiens* nor an
attempt to account for all aspects of human sexuality. It is
nothing more than a working hypothesis, a preliminary attempt to
interpret masculine and feminine gender roles as both products and
determinants of human evolution.

Evolutionary change depends upon genetic variation and natural
selection, and the genetic unit upon which selection operates is
not the individual but the gene pool represented by members of an
interbreeding population. Individuals do not evolve; populations
do. If it is to evolve, a population must survive; and the key to
survival is reproductive success of the population rather than of
the individual.

The traits or characteristics specific to any species include
not only anatomical and physiological features but also emotional
tendencies, behavioral capacities, and intellectual capabilities.
Characters favored and preserved by natural selection tend to be
those which make the greatest contribution to the survival and
reproduction of the interbreeding population. Evolution involves
continual and reciprocal interaction between the group genotype
and the results of its phenotypic expression, and this includes a
"feedback effect" of behavior upon the genotype. Behavioral
patterns or tendencies which maintain or increase the probability
for survival and reproduction of the group will be preserved and
perpetuated. Counteradaptive patterns of behavior will in the main
fail the test of natural selection and will not be incorporated
into the species genotype.

According to this point of view, the total behavioral repertoire of any species can be regarded as a mechanism or aggregation
of mechanisms reflecting that species's successful solution of problems of survival and reproduction. Species-specific behavior is
adaptive for the simple reason that it has been repeatedly and
continuously tested and moulded by the forces of natural selection.

Different species have different kinds of behavior for the same reason that they have different types of teeth, limbs, digestive systems, or skeletons: because these anatomical, physiological, and behavioral characters have contributed to and been preserved in the evolution of the species.

One behavioral characteristic man shares with all other primates is that of being a highly social animal, and organized patterns of group living can contribute in various and significant ways to both survival and reproduction. Human society is properly classified as an adaptive mechanism, and this implies that our species-specific patterns of interindividual behavior are what they are because they have withstood the test of natural selection and have contributed to the evolution of our species.

If it is assumed that the human form of social organization has been both a means and a product of man's evolution, then it should be possible to analyze the adaptive function of those forms of behavior which characterize human society; and one ubiquitous characteristic of all human cultures is the dichotomous categorization of their members on the basis of sex and the assignment of different gender roles to males and females. The questions therefore arise as to how sexuality and society are interrelated and in what ways sexuality has affected and been affected by the structuring, preservation, and perpetuation of human society.

Any serious attempt to answer such questions soon reveals that sexuality is an essential element in the very warp of human social life. It is true but trivial that without *sex* the human species could not survive for more than one generation. It is immeasurably more significant to recognize that without *sexuality* society in its human form could neither survive in the future nor have evolved in the past.

Before attempting to document these sweeping conclusions, I must enter one caveat. In speaking of human society, I do not refer to present-day American or Europe or even the Egyptian dynastic societies of three thousand years B.C.E. My analysis deals with human social organization as it existed some eight or ten thousand years ago before the domestication of plants and animals and the invention of agriculture. The reasons for this restriction are reflected in the following quotation.

> Even 6000 years ago large parts of the world's population were nonagricultural, and the entire evolution of man from the earliest populations of *Homo erectus* to existing races took place during the period in which man was a hunter. The common factors that dominated human evolution and produced *Homo sapiens* were preagricultural. Agricultural ways of life have dominated

less than 1 percent of human history and there is no
evidence of major biological changes during that period of time. . . . The origin of all common characteristics must be sought in preagricultural times
(Washburn and Lancaster, 1968, pp.213-214).

Preagricultural man was not only a hunter, he was also a
gatherer. In fact even when game was plentiful at least seventy
to eighty percent of his diet probably consisted of plant material.
In addition to being a hunter-and-gatherer early man was, as
already noted, a social animal; and as his prehuman, primate pro-
genitors evolved into group-living forms, their evolution had to
include development of certain basic controls over interindividual
behavior. For social groups to survive and reproduce, intragroup
conflict and competition must be modulated; and in addition the
social group must serve adaptive functions above and beyond those
provided by the behavior of its members acting independently.

As a hunting-and-gathering species, preagricultural man probably lived in small groups, rarely exceeding 20 to 50 individuals
(Goldschmidt, 1959). Paleontological evidence suggests that early
man killed, butchered, and brought to the group's living place a
variety of animals, including species sufficiently large and dan-
gerous that they could only have been overcome by concerted and
coordinated action by a number of individuals (Washburn and Lancaster, 1968). Coordination of hunting necessitated not only
momentary cooperation but planning in advance and some form of
leadership. It seems reasonable to assume, therefore, that the
social structuring of primitive man's life involved foresight,
cooperation, and at least temporary hierarchical distribution and
acceptance of authority.

Once a kill had been made and meat brought to the living site,
there had to be some form of food sharing, a type of behavior
which is rare in any living primate except between a mother and her
offspring. Sharing was a necessity to insure the survival of nonhunting members of the group who were in all likelihood chiefly
females, infants, and subadults. Although females probably did
not hunt regularly, they were far from being an economic liability
to the social group. It was primarily the females who gathered
seeds, berries, roots, and other vegetable foods. These also were
collected at the living place and shared with other members of the
group.

Several theorists have emphasized the importance of sex differences in the reproductive function as factors that influence the
evolution of human social systems (Zuckerman, 1932, Chance and
Mead, 1953); the foregoing reconstruction of prehistoric man's
society points to an additional dimension in the foundations for
differentiation of gender roles. Such differentiation was not based

solely upon differences in sexual and parental functions but also
upon sexually dimorphic economic specialization coupled with sex
differences in physical attributes. Women are hypothesized to
have been gatherers rather than hunters because successful hunting
depended upon masculine strength and endurance and perhaps even
upon certain emotional tendencies, such as less fearfulness and
greater willingness to venture far from the safety of the home base.
Of course, the more sedentary role of females was also directly
associated with the restrictive effects of pregnancy, the necessity of remaining with the young during lactation, and possibly a
stronger tendency toward nurturant behavior in general.

It is a general rule of organic evolution that a genetic mu-
tation, which must of course occur in one individual, will gradually
spread and invade the group genotype provided it does not reduce
the potential of the bearers for survival and reproduction and has
no deleterious effects upon perpetuation of the gene pool. Furthermore, among such mutations the ones most likely to be preserved
by natural selection are those which most directly contribute to
the survival and reproduction of the interbreeding group.

It is here suggested that when man evolved into a hunter-and-
gatherer, he exploited, so to speak, preexisting sex differences
which are easily observed in nonhuman primates today. Males are
stronger, more aggressive, dominant, and powerful than females.
They have greater physical endurance, are less fearful, and more
investigatory and hence more likely to venture into and explore
unfamiliar environments. Of course, nonpregnant and nonlactating
females possess all of the same traits and capabilities, but on
the average they are less strongly developed in the female population than in the male. Both females and males possess other
characteristics which are just as important to group survival and
are more highly developed in the feminine sex.

Now, as the hunting-and-gathering way of life developed in the
protohuman primate, the aforementioned sex differences assumed
new functional significance. Male genotypes that were above average in promoting those characteristics specifically related to
effective performance of the hunter role were especially adaptive
from the point of view of group survival. Within the female population, natural selection favored perpetuation and dissemination
of those gene patterns which contributed most to behavior consonant
with nonhunting, with gathering, with remaining near the home base,
etc. The basic notion suggested here is that, irrespective of sex
differences in reproductive function, human evolution could easily
have involved selection pressures contributing to gradual and progressive genetic divergence between males and females that increased
the ease and proficiency with which men and women adopted and performed those sex-related gender roles having adaptive value for the
reproducing population.

Reconstructions of the life of prehistoric man are not products of armchair theorizing. They are based in part upon such archeological evidence as large collections of elephant, rhinoceros, or baboon bones with flint arrows embedded in them and signs of prolonged human occupancy nearby. They are founded also upon intensive study of the few surviving contemporary societies which subsist on the basis of a hunting-and-gathering economy. Sahlins (1959), who has compared twelve such societies, believes that their social organization closely resembles that of preagricultural man.

All of these peoples live in relatively open groups which maintain friendly relations with each other. All such societies practice division of labor by sex, and the sharing of food and other items is universal. As a matter of fact, such sharing is regarded by Sahlins as a *sine qua non* of the human condition:

> Food sharing is an outstanding functional criterion of man. Among all hunters and gatherers there is a constant give and take of vital goods through hospitality and gift exchange. Everywhere, generosity is a great social virtue. Also general is the custom of pooling large game among the entire band, either as a matter of course, or in times of scarcity (Sahlins, 1959, p.66).

Perhaps the most central aspects of human sexuality and gender role are those pertaining to interindividual relationships within the family, and human family structure probably evolved *pari passu* with the evolution of the species. Heterosexual pair bonding for reproductive purposes is well known in fishes, birds, and some types of mammals; but it is rare in primates. Very few species of monkeys form stable family groups; and no examples have been described for man's nearest living relatives, the chimpanzee and gorilla. Seen against this background of anthropoid behavior, *Homo sapiens* stands out as a striking exception. There is not today and probably has never been a human society which lacked the essential elements of family structure.

In his generalized description of present-day hunting-and-gathering peoples and their social structure, Goldschmidt (1959) stresses the facts that they live mostly in small bands and that each band is divided into families or hearth groups consisting of a marital couple and their immature and unmarried children. Sahlins' (1959) painstaking analysis of twelve primitive societies reveals that their social organization always and without exception includes the formation of domestic units through marriage. He adds the following important note: "In the domestic economy of the family there is constant reciprocity and pooling of resources."

Evolution of the nuclear family may well have constituted an essential feature of man's phylogenetic history. In fact, differentiation of the family unit within the large social group probably represents a diagnostic character of *Homo sapiens*. If this is true--if the family evolved while man was evolving as a hunting-and-gathering, group-living, tool-making, language-using, culture-transmitting primate--what factors were responsible? Family structure is not essential for individual survival within the group nor for the production of young. It is not even necessary for rearing of young by their mothers or for the protection of young by adult male members of the society. All of these necessities are met in modern primate groups which include no family structure.

One unique aspect of human evolution is reflected in Sahlins' reference to the domestic economy of the family and his observation that this includes "constant reciprocity and pooling of resources." For a theory of human sexuality which necessitates selective advantage conferred by emergence of the family, Sahlins' statement contains a clue; and this clue has been followed up by Washburn and Lancaster (1968).

They point out that in nonhuman primates mother-young subgroups are stable and long lasting, and in addition the social relations within the entire group are ordered by positive affectional ties and by the strength of personal dominance. Both sorts of socializing influences persist in human society in which mother-young bonds are intensified by prolongation of the dependency period. In human society, however, a new element is added. This is the element of economic reciprocity, and economic reciprocity creates an entirely new set of interpersonal bonds. In preagricultural society, men hunted, women gathered, and subsequently they shared each other's spoils. The conclusion is: "According to this view the human family is the result of the reciprocity of hunting, the addition of a male to the mother-plus-young social group of the monkeys and apes" (Washburn and Lancaster, 1968).

Economic reciprocity is only one factor which might help explain evolution of the human family. Others include the ubiquitous tendency and even necessity to form strong and enduring emotional bonds with particular individuals within the social group and, in all likelihood, a specific need for heterosexual bonding which includes but also exceeds the need for sexual gratification.

We have now seen that in the course of human evolution the gradual differentiation of masculine and feminine gender roles served various adaptive functions including those of reproductive and economic significance, but still another function of such behavioral dimorphism was related to the neotony which characterizes man's ontogenetic development. The young of our species are much slower than those of any other primate in attaining a state of physical independence. What is at least equally important is the

fact that attainment of complete social or behavioral maturity is
an even more gradual process. This is true in large measure because developing the capacities and techniques for adult roles in
human society necessitates many years of learning and experience.

The young of preagricultural man had to acquire specific skills
involved in the "survival behavior" of his species, but he also had
to learn how to "be human." This included learning many patterns
of interindividual behavior essential to the preservation and perpetuation of his society. Central to such social maturation was
learning to fulfill the respective gender roles which over hundreds
of thousands of years had slowly evolved, differentiated, and been
preserved because they conferred adaptive values upon the human
social group. Learning to meet the demands of the masculine gender
role would have been facilitated by the possession of emotional
traits and intellectual capacities differing at least in degree
from those contributing selectively to rapid and efficient learning
of the feminine gender role. One result could be gradual selection
favoring genetic linkages in which female reproductive functions
were tied to a bias in interests and capacities fitting the individual for maximal efficiency in learning and living the feminine
gender role. Precisely the opposite pattern of selection may have
operated on the XY genotype in an evolving population. Regardless
of the validity of such speculations, it is clear that the evolution
of the family not only expanded the functional significance of gender roles but also provided an especially appropriate social environment in which young males and females could learn what their
gender roles were and how to meet the associated requirements.

One other possible function of family structure which has
received a great deal of attention by certain theorists (Zuckerman,
1932; Chance and Mead, 1953) is the opportunity it provides for
regular and frequent heterosexual intercourse. A cornerstone in
such theories is the fact that unlike females of other primate
species most women do not undergo regular behavioral cycles involving periods of sexual responsiveness and attractiveness which alternate with phases of complete indifference or even opposition to
sexual congress.

These hypotheses can be identified generically by their acronym
ERV. They are theories of the "Ever Ready Vagina." Their basic
postulates are: (1) Since she lacks a clear-cut behavioral estrous
cycle, the human female is "constantly receptive" sexually. (2) The
human male is constantly potent and needful of sexual relief through
copulation. (3) Evolution of the nuclear family occurred because
the female's unvarying receptivity gratified the male's perpetual
sexual demands.

The ERV theory is logically and biologically untenable for a
number of reasons. (1) No human female is "constantly receptive."

(Any male who entertains this illusion must be a very old man with a short memory or a very young man due for a bitter disappointment.) (2) Human males do not live in a constant state of sexual readiness or need. (3) Even if women were receptive at all times, this would represent only one of two essential features of behavioral estrus as seen in other primates.

Estrus in nonhuman mammals is marked not only by willingness to receive the male but also by changes in behavior and sensory qualities which make the female sexually exciting and attractive to the male. (Males do not become aroused by nonestrous females and rarely attempt to copulate with them.) The ERV theory's assumption of constant receptivity in our species is twice damned because it fails to explain the implicit assumption that females exhibit constant sexual attractiveness and capacity to arouse males. According to the theory, the constantly receptive female should be "constantly exciting," constantly capable of arousing males sexually. What are the characteristics that make the human female attractive to males even when estrogen levels are low and she is not about to ovulate? (4) Female animals in estrus show by their behavior that coitus at that time has "rewarding" or "reinforcing" consequences. When they are in heat, female rats will exert extra energy, perform learned tasks, or endure electric shock in order to achieve copulation with a male. Anestrous females will not do these things. What is the reward or reinforcement for copulation in the human female? Many explanations have been offered in other contexts, but the entire problem is overlooked by proponents of the ERV theory.

An even more serious weakness of the ERV theory is its almost exclusive concentration upon sex and neglect of sexuality. A second theory which shares the same defect will be mentioned only in passing. It is one which can be titled acronymically the URP, or "Unready Penis" theory. Proposed in the late nineteenth century by a French anatomist, this interpretation of human evolution is based upon the fact that among primates, man is the only anthropoid lacking a penile bone or baculum.

The theory holds that primate species in which males possess an os penis devoted inordinate amounts of time and energy to sexual activity and that by virtue of his deficient genital anatomy, the protohuman male was left with leisure time which he employed to invent tools, develop language, and create culture. Except for its curiosity value, the URP theory has few virtues. Even its major premise is wrong. Instead of being less active sexually than other primates, human males and females probably engage in copulation at a much higher frequency than animals of any other anthropoid species.

One point stressed by the ERV theory is correct and extremely important. I refer to the absence of behavioral estrus in the

human female. The evolution of *Homo sapiens* involved no basic
changes in most aspects of the female's reproductive cycle. Her
monthly rhythm of ovogenesis, ovulation, estrogen and progesterone
secretion, uterine stimulation, and menstrual bleeding follows the
basic primate pattern. There are, however, two conspicuous deviations from this pattern and both are behavioral. (1) The preovulatory rise in estrogen levels is not accompanied by a reliable
and pronounced increase in the women's desire for and responsiveness
to sexual stimulation; and 2) the same change in estrogen secretion
does not suddenly render her sexually more attractive to males.

Nonhuman primate females actively solicit and engage in copulation when estrogen from the maturing ovarian follicle is present
in high concentrations and ovulation is imminent. Males do not
show marked erotic arousal in the presence of anestrous females;
but when the female has been exposed to endogenous or exogenous
estrogen, she evokes strong sexual excitement in the male which is
manifest in his prolonged inspection of her vagina, prompt erection,
vigorous mounting, intromission, and eventual ejaculation.

This complex synchronization of hormonal states in the female
with behavioral tendencies in both males and females automatically
results in copulations occurring predominantly at the time the female is fertile. The adaptive value of such synchrony is so obvious that its disappearance during human evolution is perplexing,
and it is apparent that compensatory modifications had to occur
which would ensure that a sufficient number of matings would take
place when females were about to ovulate and thus could be impregnated. One obvious solution would involve an increase in the overall frequency of copulation across successive stages of the female's
cycle, but this change would in turn necessitate alteration of both
female and male sexual behavior.

It would be necessary 1) that men find women sexually attractive and copulation with them satisfying even when little or no
estrogen was being secreted by the ovaries and 2) that at these
same times women experience the desire for sexual stimulation and
capacity for gratification in response to it.

This essay is not the place for a detailed consideration of
human sexual motivation, but several points relevant to evolutionary
changes may be mentioned briefly. Unlike female monkeys and apes,
women about to ovulate do not exude sexually exicting odors or
develop changes in size or coloration of the perianal skin. Women
do, however, possess permanently dimorphic secondary sex characters
which are lacking in other primates and are considered erotically
exciting in all human cultures. The breast of the human female is
unique among primates both in terms of its anatomical location and
its prominence in nulliparous and nonpregnant individuals. Furthermore, cross-cultural comparisons identify it as a source of sexual

arousal for males and a locus of erotic sensation for females
(Ford and Beach, 1952).

Another species difference relates to certain characteristics
of the human vagina. An important factor contributing to the frequency of intercourse is the amount of pleasure and satisfaction
derived during the act of copulation, and this in turn depends in
part on the quality and intensity of genital sensations. If coitus
is physically difficult or painful, its reinforcement value is decreased or eliminated. In the experimental laboratory, male monkeys
occasionally attempt copulation with ovariectomized females; but
their performance under such conditions is desultory. And when it
occurs,intromission does not result in ejaculation. When the female is injected with estrogen, the male's sexual interest quickens,
his mating performance becomes much more vigorous, and insertion
leads to ejaculation.

Part of the change in the male's attitude is due to changes
in the female's odor and general demeanor, but nonejaculatory copulation may also be influenced by the lack of vaginal secretion in
the absence of estrogenic stimulation. The human vagina is, of
course, stimulated by estrogen; but it also has the capacity for
self-lubrication by means of a nonsecretory process which Masters
and Johnson (1966) have termed "sweating." The fact that vaginal
sweating occurs under conditions of sexual excitement and is not
dependent upon ovarian hormones may help to explain why human males
and females find intercourse rewarding at all stages of the feminine cycle.

Although women do not automatically reveal chemically or by
innately organized patterns of movement their state of sexual
readiness, the human capacity for symbolic behavior provides a
more than adequate replacement for such signals. By the use of
language and ritualized facial and body gestures (some of which
may represent cultural universals; cf. Eibl-Eibesfeldt, 1970),
women in all societies are capable of conveying to male members of
the same societies their sexual desire and availability.

Even though female mammals are sexually receptive during
physiological estrus, they may not respond immediately and positively to a male's investigations or attempts to mate. In all species,
the male is provided for such eventualities with a repertoire of
ancillary behavior which tends to stimulate a temporarily unresponsive or resistant estrous female and arouse her sexually to the
point of permitting or even soliciting coitus. The same "courtship"
or precopulatory stimulation will not elicit receptive behavior in
the female if she has not been exposed to ovarian hormones.

Although human females definitely are not continuously sexually "receptive," they are continuously "copulable"; and their

sexual arousability does not depend on ovarian hormones. This re-
laxation of endocrine control contributes to the occurrence of
coitus at any stage of the menstrual cycle.

As far as the basis for sexual satisfaction in human males
and females is concerned, this again is a subject too complex and
extensive to be covered in one essay. It is, however, interesting
to note that sources of reward for the copulating female have not
been dealt with by evolutionary theories in general.

It appears to have been simply taken for granted that the
primary reinforcement for sexual behavior in male mammals of all
species is the experience of orgasm which is assumed to accompany
ejaculation. Such an assumption needs testing when applied to
other species and is inadequate to account *in toto* for the human
male's sexual activity, but the linkage of positively reinforcing
orgasm to the ejaculatory reflex necessary for emission of sperm
has obvious adaptive connotations. Since the male's ejaculation
is necessary for insemination of the female, association of the
act of copulation with sensations which tend to reinforce that
act and increase the probability of its recurrence would indirectly
contribute to perpetuation of the species.

How are we to account for the female's participation in the
procreative act? Since survival of our species can hardly have
rested upon rape or prostitution, any evolutionary theory needs
to consider the problem of sexual reinforcement in females. It is
not necessary that women enjoy intercourse in order to become
pregnant; but if pregnancy is to occur often enough to ensure con-
tinuity of the species, it is important that copulation be rewarding for both sexes. No simple explanation could possibly account
for feminine participation in coital relations, but there is one
source of reinforcement that can be directly tied in with changes
occurring during human evolution. This element is the phenomenon,
the female orgasm.

Thirty-five years of research on sexual behavior in animals
have led me to the tentative conclusion that female orgasm may be
a biological rarity, a "human invention," so to speak. Of course,
orgasm is by definition a subjectively recognized emotional response
and therefore never accessible to direct observation. We cannot
know whether a female chinchilla or chimpanzee experiences orgasm,
but neither can we tell whether a woman does so except on the basis
of her report of her own feelings. The same statement applies to
the human male, but we nevertheless tend to assume that observed
ejaculation is a reliable indicator of orgasm in other individuals
and in males of other species.

The observable responses of women to gential stimulation also
include specific reactions which can be correlated with the report

of orgasmic sensations, and there is substantiating evidence in the form of electromyographic recordings from the vagina and uterus which reveal a close temporal relationship between spasmodic muscular contractions and the women's subjective experience of orgasm (Masters and Johnson, 1966).

From the comparative or cross-species point of view, a significant fact is that as far as the male is concerned, ejaculation and its behavioral sequelae, including temporary impotence and sexual disinterest, can be identified in the mating behavior of many animal species; whereas behavior indicating the occurrence of sexual climax in copulating females is extremely rare. One or two species may be excepted, and individual members of still other species may not conform to the general rule; but the weight of available evidence favors the theory that female orgasm is a characteristic essentially restricted to our own species.

If the capacity for female orgasm is an evolutionary product, it could be functionally related to several anatomical and behavioral characteristics which separate human beings from other primates. The first is correlated with bipedalism and the associated necessity for upright posture, two characteristics which mark man as unique among the primates. Structural changes occurring in connection with the evolution of bipedal progression and permanent vertical posture included a forward tilting of the human pelvis; and this change, in turn, resulted in repositioning the female genitalia so that the vaginal aperture was moved to a more anterior or ventral position. One result of this change was increased accessibility of the vagina to penile insertion during ventro-ventro copulation.

The species-typical mating position for all terrestrial mammals including monkeys and apes involves copulation *a posteriori;* whereas the coital pattern predominant in all human societies is some variant of the face-to-face position (Ford and Beach, 1952). This may be the so-called "missionary position," in which the man mounts the woman from above; or the relationship may be reversed; or intercourse may take place with both partners seated or lying on their sides. Many other positions are known and employed, but ventro-ventro coitus is the most frequent type in all cultures.

The relevance of this fact to the occurrence of female orgasm is that all types of face-to-face intercourse involve apposition or juxtaposition of the pubic symphyses of the two sexes, and continuous or rhythmic pubic contact tends to result in secondary or indirect stimulation of the clitoris. In many cultures assumption of the superior position by the woman is recognized as one method of maximizing her sexual pleasure and satisfaction, and this is true partly because she can adjust her position and movement to control the degree and frequency of clitoral excitation.

Despite learned disputations regarding the psychological significance of vaginal versus clitoral orgasm, research has shown that the vaginal and uterine contractions which provide the proprioceptive sensations underlying orgasm are identical regardless of the source of peripheral stimulation (Masters and Johnson, 1966). It is equally well established that for nearly all women the most reliable source of stimulation leading to orgasm is mechanical agitation of the clitoris.

The evolutionary argument therefore runs as follows. Human copulation is most frequently carried out *a anteriori* because of anterior placement of the vaginal orifice, which has been indirectly affected by man's upright posture. Face-to-face intercourse results in clitoral stimulation. Stimulation of the clitoris increases the probability of feminine orgasm. Since capacity for orgasm is independent of the woman's current hormonal status, the possibility of its occurrence throughout her cycle tends to reinforce and increase the frequency with which she desires and accepts intercourse.

The ERV theory holds that evolution of the human family was related to the disappearance of behavioral estrus in the female. It postulates that absence of fluctuations between receptivity and nonreceptivity permitted sexual relations at all stages of the female's cycle and thus contributed to formation and maintenance of interpersonal bonds of an enduring nature.

An alternative interpretation could be that the loss of behavioral estrus and the consequent disappearance of synchronization between mating and feminine fertility necessitated an increase in copulatory frequency so that intercourse would occur by chance at the time when it could result in fertilization. Development of the family may thus be viewed as an adaptive change tending to encourage frequency and regularity of heterosexual coitus. It is a reasonable assumption that existence of the family with the associated intensification and prolongation of interpersonal bonds and dependencies would promote more frequent intercourse than would be likely to occur in the absence of family structure.

This assumption is probable for several reasons. For one thing, familiarity between family members would allow simplification and increased efficiency of the signalling systems used by both parties to indicate sexual desire and readiness. For another, the family setting would reduce disruptive interference.

In primate societies which lack heterosexual pair bonding, sexual contact between adult males and females is often prevented or interrupted by harassment of the copulating couple by other individuals. Such activities are common in the Anubis baboon as described by DeVore (1965). A striking contrast is seen in Kummer's account of Hamadryas baboons whose society involves family groups consisting of one adult male, several females, and their young. In this species, males copulate only with their own females; and other individuals do not interfere (Kummer, 1971).

In similar fashion, the social setting of the human family tends to eliminate interference with intercourse from social competition, jealousy, and other negative influences. Finally, family structure would be likely to result in couples' copulating more often simply because of longer periods of proximity.

Since copulation tends to result in mutual physical gratification and since simple learning promotes association of positive values with perceived sources of reward, intrafamilial copulation provides one source of reinforcement of emotional bonding between males and females. In other words, sexual behavior reinforces family structure, and family structure reinforces sexual behavior.

CONCLUDING REMARKS

This essay has been concerned with the ontogenetic development of sexual attributes in the individual and the phylogenetic development of sexuality in the course of human evolution.

From the instant of conception, the single-celled zygote incorporates a genetic program which in the normal course of ontogeny will result in the selective differentiation of either male or female characteristics ranging from anatomical features of the reproductive system to a repertoire of sex-related behavioral tendencies. Maleness and femaleness connote two patterns of interrelated, biologically determined sex differences which incline but do not compel the individual's psychological development in the general directions of masculinity or femininity.

Sexuality is not synonymous with sex but is a construct composed of two elements. One of these is gender role and the other is gender identity. Separable gender roles for males and females are defined and perpetuated by every human society and form the basis for a society's implicit definition of masculine and feminine. Gender identity is one dimension of the self-concept; specifically it refers to the individual's concept of his or her own masculinity or femininity.

Gender identity develops gradually by successive stages which bear formal resemblances to those involved in the ontogeny of maleness and femaleness. Ontogenesis of masculinity and femininity depends upon learning and rests heavily upon language. The child must learn the essentials of masculine and feminine gender roles as defined by his society and must also learn to govern his own behavior in conformance with his assigned gender role. The degree of

agreement between the understood gender role and the individual's
perception of his own behavior and motivations vis-a-vis that role
determines the strength and constancy of his gender identity.

Ninety-nine percent of the time covered by the evolution of
Homo sapiens preceded the domestication of plants and the invention of agriculture. In other words, for all but the last one
percent of his existence, man has been a seminomadic, group-living
hunter-and-gatherer; and these features of his history strongly influenced the evolutionary development of human sexuality. An increasing reliance upon animal food involved division of labor along
sexual lines since the male's superior size, strength, and endurance
equipped him for hunting. Hunting, killing, and butchering large
animals necessitated planning, cooperation, and leadership.

The economic role of women in preagricultural society included
gathering plant foods. Hunting males and gathering females brought
their spoils to the group living place where food sharing was an
absolute necessity.

Specialization of sex roles was also an adaptive necessity in
the area of reproduction. Men had to impregnate women, and women
had to bear and nourish children. Evolution of some type of nuclear
family almost certainly occurred before or soon after the emergence
of *Homo sapiens*. The values were both reproductive and economic.

One special factor which may have influenced evolution of the
family was the disappearance of sharply delimited periods of behavioral estrus in the human female. One result was the extension of
copulatory activity far beyond the female's fertile period and the
consequent necessity for an overall increase in the total frequency
of heterosexual intercourse. It seems possible that family structure
contributed to a heightened level of sexual interaction and that
the increase in sexual relationships exerted a feedback effect
strengthening selective heterosexual pair bonding.

The general conclusion suggested but by no means established
is that as human social life became more and more complex, increasingly dichotomous specialization of social roles for males and females became more and more adaptive in maintaining and perpetuating
the species and the human way of life.

The evolutionary changes envisaged in this interpretation
occurred very gradually and over periods of hundreds of thousands
of years. Their development was governed in part by the forces of
natural selection, and their survival in the behavioral repertoire
of the species depended upon their adaptive value to the species
rather than to the individual. The evolution of separate gender
roles also depended, of necessity, upon intergeneration transmission of culturally defined traits, values, and ways of behaving.

Because behavior has feedback effects upon the gene pool of the reproducing population, sex differences in suitability for a specific gender role were favored by selection.

To state the argument in a single sentence, sexuality is a product of human evolution; and the evolution of *Homo sapiens* demanded as one essential component the progressive development of sexuality.

REFERENCES

August, C.P., Grumbach, M.M., and Kaplan, S.L., 1972, Hormonal changes in puberty: III. Correlation of plasma testosterone, LH, FSH, testicular size, and bone age with male pubertal development. J. Clin. Endocrin. and Metab. 34:319-326.

Beach, F.A., 1971, Hormonal factors controlling the differentiation, development and display of copulatory behavior in the ramstergig and related species, in *Biopsychology of Development* (L. Aronson and E. Tobach, eds.), pp.249-296, Academic Press, New York.

Chance, M.R.A., and Mead, A.P., 1953, Social behaviour and primate evolution, in *Evolution; Symposia of the Society for Experimental Biology*, No. VII (R. Brown and J. F. Danielli, eds.), pp. 395-439, Cambridge University Press, Cambridge.

Crandall, V.J., and Rabson, A., 1960, Children's repetition choices in an intellectual achievement situation following success and failure, J. Genet. Psychol. 97:161-168.

Davenport, W., 1965, Sexual patterns and their regulation in a society of the Southwest Pacific, in *Sex and Behavior* (F.A. Beach, ed.), pp 164-207, John Wiley & Sons, New York.

DeVore, I., 1965, Male dominance and mating behavior in baboons, in *Sex and Behavior* (F.A. Beach, ed.), pp. 266-289, John Wiley & Sons, New York.

Eibl-Eibesfeldt, I., 1970, *Ethology. The Biology of Behavior*, Holt, Rinehart and Winston, New York.

Field, P.M., and Raisman, G., 1973, Structural and functional investigations of a sexually dimorphic part of the rat preoptic area, in *Recent Studies of Hypothalamic Function* (K.L. Lederis and K.E. Cooper, eds.), S. Karger, Basel.

Ford, C.S., and Beach, F.A., 1952, *Patterns of Sexual Behavior*, Harper and Row and Paul B. Hoeber, Inc., New York.

Goldschmidt, W., 1959, *Man's Way*, Holt, Rinehart and Winston, New York.

Gorski, R.A., 1971, Gonadal hormones and the perinatal development of neuroendocrine function, in *Frontiers in Neuroendocrinology, 1971*, (L. Martini and W.F. Ganong, eds.), vol. 2, pp. 237-290, Oxford University Press, New York.

Hammond, W.A., 1895, *Cerebral Hyperaemia: The Result of Mental Strain or Emotional Disturbance. The So-called Nervous Prostration of Neurasthenia*, Berntano, Washington, D.C.

Kagan, J., 1970, Personality development, in *Personality Dynamics* (I.E. Janis, ed.), Harcourt, Brace and World, New York.

Katcher, A., 1955, The discrimination of sex differences by young children, J. Genet. Psychol. 87:131-143.

Kephart, W.M., 1973, Evaluation of romantic love, Medical Aspects of Human Sexuality 7:92-108.

Kohlberg, L.A., 1966, A cognitive-developmental analysis of children's sex-role concepts and attitudes, in *The Development of Sex Differences* (E.E. Maccoby, ed.), pp. 82-173, Stanford University Press, Stanford, Cal.

Kummer, H., 1971, *Primate Societies*, Aldine Atherton, Chicago.

Masters, W.H., and Johnson, V.E., 1966, *Human Sexual Response*, Little, Brown and Co., Boston.

Mussen, P.H., Conger, J.J., and Kagan, J., 1969, *Child Development and Personality*, Harper and Row, New York.

Ramsey, G.V., 1943, The sexual development of boys, Amer. J. Psychol. 56:217-234.

Sahlins, M.D., 1959, The social life of monkeys, apes and primitive man, in *The Evolution of Man's Capacity for Culture* (J.N. Spuhler, ed.), pp. 55-73, Wayne State University Press, Detroit.

Smith, M.E., 1933, The influence of age, sex and situation on the frequency and form and function of questions asked by preschool children, Child Development 2:201.

Stone, C.P., and Barker, R.G., 1939, The attitudes and interests of premenarchial and postmenarchial girls, J. Genet. Psychol. 5427-71.

Tanner, J.M., 1962, *Growth at Adolescence*, Blackwell Scientific Publications, Oxford, England.

Washburn, S.L., and Lancaster, C.S., 1968, The evolution of hunting, in *Perspectives on Human Evolution* (S.L. Washburn and P.C. Jay, eds.), vol 1. pp. 213-229, Holt, Rinehart and Winston, New York.

Zuckerman, S., 1932, *The Social Life of Monkeys and Apes*, Harcourt, Brace and Company, New York.

CONTRIBUTORS

N. T. Adler
Department of Psychology
University of Pennsylvania
Philadelphia, Pennsylvania 19104

Frank A. Beach
Department of Psychology
University of California, Berkeley
Berkeley, California 94720

F. H. Bronson
Department of Zoology
The University of Texas at Austin
Austin, Texas 78712

Harley L. Browning
Department of Sociology and
Population Research Center.
The University of Texas at Austin
200 East 26½ Street
Austin, Texas 78705

Larry L. Bumpass
Center for Demography & Ecology
The University of Wisconsin
1180 Observatory Drive
Madison, Wisconsin 53706

Lynwood G. Clemens
Department of Zoology
Michigan State University
East Lansing, Michigan 48823

Claude Desjardins
Department of Zoology
The University of Texas at Austin
Austin, Texas 78712

Gisela Epple
University of Pennsylvania
Monell Chemical Senses Center
3500 Market Street
Philadelphia, Pennsylvania 19104

G. Gray Eaton
Oregon Regional Primate
  Research Center
505 N.W. 185th Avenue
Beaverton, Oregon 97005

David A. Goldfoot
Wisconsin Regional Primate
  Research Center
1223 Capitol Court
Madison, Wisconsin 53706

Robert A. Goy
Department of Psychology
University of Wisconsin and
Wisconsin Regional Primate
  Research Center
1223 Capitol Court
Madison, Wisconsin 53706

Barry R. Komisaruk
Rutgers, The State University of
  New Jersey
Newark College of Arts & Sciences
Institute of Animal Behavior
101 Warren Street
Newark, New Jersey 07102

Lee-Ming Kow
The Rockefeller University
New York, New York 10021

Charles W. Malsbury
The Rockefeller University
New York, New York 10021

S. M. McCann
Department of Physiology
Southwestern Medical School
The University of Texas
  Health Science Center at Dallas
5323 Harry Hines Boulevard
Dallas, Texas 75235

Howard Moltz
Department of Psychology
The University of Chicago
Chicago, Illinois 60637

William Montagna, Co-Editor
Oregon Regional Primate
  Research Center
505 N. W. 185th Avenue
Beaverton, Oregon 97005

Robert L. Moss
Department of Physiology
Southwestern Medical School
The University of Texas
  Health Science Center at Dallas
5323 Harry Hines Boulevard
Dallas, Texas 75235

Donald W. Pfaff
The Rockefeller University
New York, New York 10021

Charles H. Phoenix
Oregon Regional Primate
  Research Center
505 N. W. 185th Avenue
Beaverton, Oregon 97005

John A. Resko
Oregon Regional Primate
  Research Center
505 N. W. 185th Avenue
Beaverton, Oregon 97005

William A. Sadler, Co-Editor
Center for Population Research
National Institute of Child
  Health & Human Development, NIH
Bethesda, Maryland 20014

Kim Wallen
Wisconsin Regional Primate
  Research Center
1223 Capitol Court
Madison, Wisconsin 53706

SUBJECT INDEX[1]

A

Abortion	319
ACTH/Adrenocorticotrophic hormone	55
Acupuncture	97
Adolescent/Adolescence	367
Adrenalectomy	55, 77
Adrenal glands	97
Adrenergic blocking drugs	1
Aggression/Aggressive behavior	131, 223, 287
American Indians/Native Americans	299
Androgen(s)	23, 211
Anglos/Anglo majority	299
Anterior hypothalamic area	1
Antiandrogen (Sch 13521)	23
Antiestrogen	23, 97
Antihormone	97
Appalachian whites	299
Arcuate region	1, 55
Aromatization of testosterone	23
Ateles	131
Atropine	1

B

Baby boom	299
Basal tuberal region	1
Behavior (see Mating/Sexual/Social behavior entries)	
Birth control	299, 319
Birth expectations	299, 319
Bisexuality (behavioral, neural)	179
Bisexual overlap	367
Blacks/Black fertility	299, 319
Body odors (see Pheromones)	131
Brain implants	1, 55, 97, 179
Brain lesions	1, 55, 97, 179
Brain stem and limbic system pathways	1, 179

C

Callimico goeldii	131
Callithrix jacchus	131
Castration	249
Catecholamines (drugs altering synthesis)	1
Catholics/Catholicism (Irish, Mexican, Mexican-American, white)	299, 319

[1]Entries are indexed by article only; articles are identified by the numbers of their respective first pages.

N

O